Cold Aqueous Planetary Geochemistry with FREZCHEM

Advances in Astrobiology and Biogeophysics
springer.com

This series aims to report new developments in research and teaching in the interdisciplinary fields of astrobiology and biogeophysics. This encompasses all aspects of research into the origins of life from the creation of matter to the emergence of complex life forms and the study of both structure and evolution of planetary ecosystems under a given set of astro- and geophysical parameters. The methods considered can be of theoretical, computational, experimental and observational nature. Preference will be given to proposals where the manuscript puts particular emphasis on the overall readability in view of the broad spectrum of scientific backgrounds involved in astrobiology and biogeophysics.
The type of material considered for publication includes:

- Topical monographs
- Lectures on a new field, or presenting a new angle on a classical field
- Suitably edited research reports
- Compilations of selected papers from meetings that are devoted to specific topics

The timeliness of a manuscript is more important than its form which may be unfinished or tentative. Publication in this new series is thus intended as a service to the international scientific community in that the publisher, Springer, offers global promotion and distribution of documents which otherwise have a restricted readership. Once published and copyrighted, they can be documented in the scientific literature.

Series Editors:

Dr. André Brack
Centre de Biophysique Moléculaire
CNRS, Rue Charles Sadron
45071 Orléans, Cedex 2, France
Brack@cnrs-orleans.fr

Dr. Gerda Horneck
DLR, FF-ME, Radiation Biology
Linder Höhe
51147 Köln, Germany
Gerda.Horneck@dlr.de

Prof. Dr. Michel Mayor
Observatoire de Genève
1290 Sauverny, Switzerland
Michel.Mayor@obs.unige.ch

Dr. Christopher P. McKay
NASA Ames Research Center
Moffet Field, CA 94035, USA

Prof. Dr. H. Stan-Lotter
Institut für Genetik
und Allgemeine Biologie
Universität Salzburg
Hellbrunnerstr. 34
5020 Salzburg, Austria

Giles M. Marion · Jeffrey S. Kargel

Cold Aqueous Planetary Geochemistry with FREZCHEM

From Modeling to the Search for Life at the Limits

With 58 Figures and 33 Tables

Springer

Giles M. Marion
Earth and Ecosystem Sciences
Desert Research Institute
2215 Raggio Parkway
Reno, NV 89512, USA
gmarion@dri.edu

Jeffrey S. Kargel
Departement of Hydrology
& Water Resources
University of Arizona
Tucson, AZ 85721, USA
kargel@hwr.arizona.edu

Cover illustration: The cover art shows an eroding ancient Martian shoreline with sedimentary rock outcrops and ice remnants of a former lake, including rock deposits similar to those now being explored by the Mars Opportunity rover. The ancient lake is shown in the cutaway cross section at bottom right as it once may have been, with an ice-covered liquid brine lake and bubbles associated with gas hydrates and/or life.

ISBN 978-3-540-75678-1   e-ISBN 978-3-540-75679-8

DOI 10.1007/978-3-540-75679-8

Springer Series in
Advances in Astrobiology and Biogeophysics ISSN 1610-8957

Library of Congress Control Number: 2007939487

© 2008 Springer-Verlag Berlin Heidelberg

This work is subject to copyright. All rights are reserved, whether the whole or part of the material is concerned, specifically the rights of translation, reprinting, reuse of illustrations, recitation, broadcasting, reproduction on microfilm or in any other way, and storage in data banks. Duplication of this publication or parts thereof is permitted only under the provisions of the German Copyright Law of September 9, 1965, in its current version, and permission for use must always be obtained from Springer. Violations are liable for prosecution under the German Copyright Law.

The use of general descriptive names, registered names, trademarks, etc. in this publication does not imply, even in the absence of a specific statement, that such names are exempt from the relevant protective laws and regulations and therefore free for general use.

Typesetting and production: LE-TEX Jelonek, Schmidt & Vöckler GbR, Leipzig, Germany
Cover design: WMXDesign GmbH, Heidelberg

Printed on acid-free paper

9 8 7 6 5 4 3 2 1

springer.com

G.M.M. dedicates this book to his parents,
Maurice L. Marion and Lilianne Nadon Marion,
and to his wife, Dawn C. Hammond.

J.S.K. dedicates this book to Bé, Van,
Christopher, Dianna, and Isaiah, and his parents,
and to those who hold the world in their hearts.

Preface

This book focuses primarily on the FREZCHEM model, which was explicitly developed to quantify aqueous electrolyte properties at subzero ($< 0\,°C$) temperatures. The foundations of this model are based on chemical thermodynamic principles. Professionals and students that will find this book especially useful include geochemists interested in cold aqueous processes, geochemical modelers, cold planetary scientists, astrobiologists, physicochemists, and chemical engineers.

The original version of the model dealt with chloride and sulfate chemistries and was written when the senior author worked at the Cold Regions Research and Engineering Laboratory (U.S. Army Corps of Engineers) in Hanover, New Hampshire. Subsequent versions with new chemistries were largely funded by NASA for applications to cold Solar System bodies such as Mars and Europa.

Over the years, a number of individuals have contributed to developing the model, helped correct FORTRAN coding problems by using the model and providing feedback, collaborated on associated projects and papers, and reviewed associated papers. Those that have contributed, directly and indirectly, to model parameterization include Donald G. Archer, Peter Brimblecombe, Kenneth S. Carslaw, Simon L. Clegg, Anthony J. Gow, Steven A. Grant, David L. Hogenboom, Jeffrey S. Kargel, Boris S. Krumgalz, Mikhail V. Mironenko, Nancy Møller, Christophe Monnin, John W. Morse, Ronald J. Spencer, Nicolaus J. Tosca, and John H. Weare. Users of the model that have provided feedback include Rich Anouar, Charles Barnhart, Craig Brown, David Catling, Peter Croot, Megan Elwood Madden, Qi Fu, Robert Gärtner, Mats Granskog, Birgit Hagedorn, Christopher Hall, Wouter Heijlen, Steve Jepson, Lars Kaleschke, Edwin Kite, Claudine Lee, Pawel M. Lesniak, Anna Markiuw, Nathalie Maurer, Christopher P. McKay, Jill Mikucki, Lisa Miller, Mikhail V. Mironenko, Samuel Morin, Todd Nichols, Christopher Omelon, Stathys Papadimitriou, Jean-Marie Perrier, Venasan Pillay, Volker Rath, Bernhardt Saini-Eidukat, William Seyfried, Douglas Sheppard, Mark Skidmore, Ronald S. Sletten, Alexander Sokolov, Steven Vance, Yongliang Xiong, Min Zhang, and Mikhail Zolotov. Collaborators on associated projects and papers include Jay Arnone, David C. Catling, Hajo Eicken, Ronald E. Farren, Rainer Feistel, Christian H. Fritsen, Eric Gaidos, Steven A. Grant,

Scott D. Jakubowski, Jeffrey S. Kargel, Andrew J. Komrowski, W. Berry Lyons, Mikhail V. Mironenko, John W. Morse, Meredith C. Payne, Bryan Stevenson, Paul Verburg, Kathy Welch, and Aaron Zent. Reviewers of associated papers include R.S. Arvidson, Steven A. Grant, Anthony J. Gow, Jerry P. Greenberg, Jeffrey S. Kargel, Erich Königsberger, Roger Kreidberg, Boris S. Krumgalz, Virgil J. Lunardini, W. Berry Lyons, Debra Meese, Nancy Møller, Christophe Monnin, John W. Morse, Ronald J. Spencer, Mikhail Zolotov, and innumerable anonymous reviewers. We thank all of the above individuals, who have helped make the FREZCHEM model more sound, less buggy, and broader in scope.

We also thank the following publishers for permission to reproduce published graphs, tables, and text: American Journal of Science, ASM Press, Cambridge University Press, Cold Regions Research and Engineering Laboratory, Elsevier Limited, Mary Ann Liebert Inc., NASA/JPL/Caltech, and Springer Science and Business Media.

And finally, we extend a special thanks to Annette Risley and Lisa Wable who have helped with the preparation of this book and many other manuscripts over the past 8 years. We also thank Lisa Wable for developing the book cover art.

Reno, Flagstaff
October 2007

Giles M. Marion
Jeffrey S. Kargel

Contents

1	**Introduction**		1
2	**Aqueous Chemistry**		3
	2.1	Basic Principles	3
		2.1.1 Definitions	3
		2.1.2 The First and Second Laws of Thermodynamics	4
		2.1.3 Activities, Chemical Potentials and Equilibrium Constants	7
	2.2	Pitzer Approach	10
		2.2.1 The Osmotic Coefficient and Activity Coefficients	10
		2.2.2 Temperature and Pressure Dependencies	15
3	**The FREZCHEM Model**		19
	3.1	Historical Development	19
	3.2	Basic Structure	21
		3.2.1 Chemical Equilibrium	21
		3.2.2 Mass Balances	22
		3.2.3 Reaction Pathways	22
	3.3	Chemistries and Their Temperature and Pressure Dependence	24
		3.3.1 Water Ice/Liquid Water/Water Vapor Equilibria	24
		3.3.2 Salt Equilibria	29
		3.3.3 Gas/Solution Phase Equilibria	37
		3.3.4 Gas Hydrate Equilibria	42
	3.4	Mathematical Algorithms	49
		3.4.1 The Sequential Approach	49
		3.4.2 Gibbs Energy Minimization	50
		3.4.3 Other Mathematical Techniques	52
	3.5	Validation	56
	3.6	Limitations	67
		3.6.1 Pitzer-Equation Parameterization Limitations	68
		3.6.2 Modeling Limitations	75
4	**Limits for Life**		79
	4.1	Temperature	84
	4.2	Salinity	86

4.3 Acidity ... 88
4.4 Desiccation ... 89
4.5 Radiation ... 89
4.6 Pressure ... 90
4.7 Time ... 97

5 Biogeochemical Applications to Solar System Bodies ... 101
5.1 Earth ... 102
5.1.1 Seawater Freezing ... 102
5.1.2 Aqueous Saline Environments ... 110
5.1.3 Snowball Earth-Hothouse Earth ... 113
5.1.4 Why Are Clouds not Green? ... 120
5.1.5 Other Earth Applications ... 123
5.2 Mars ... 125
5.2.1 Surficial Aqueous Geochemical Evolution ... 125
5.2.2 Early Mars Oceans ... 135
5.2.3 A Cold, Intermittently Wet Mars ... 135
5.2.4 Hydrate Deposits and Thermal Stratification ... 139
5.3 Europa ... 141
5.3.1 Ocean Compositions ... 142
5.3.2 Ice Compositions ... 148
5.4 Application Limitations ... 150

6 The Search for and Future of Life in the Universe ... 155
6.1 A Search Strategy for Life in the Universe ... 155
6.2 Entropic Death? ... 158
6.3 Solar System Life ... 162
6.4 To the Stars or Bust? ... 169

A FREZCHEM Program Guide ... 175
A.1 Model Input Limitations ... 175
A.2 Model Inputs ... 176
A.3 Model Outputs ... 177
A.3.1 Seawater Freezing ... 177
A.3.2 Strong Acid ... 178
A.3.3 Gas Hydrates ... 178
A.3.4 Pressure Application ... 179

B Parameter Tables ... 193

References ... 223

Index ... 247

1 Introduction

"A theory is the more impressive the greater the simplicity of its premises, the more different kinds of things it relates, and the more extended its area of applicability. Therefore the deep impression that classical thermodynamics made upon me. It is the only physical theory of universal content concerning which I am convinced that, within the framework of the applicability of its basic concepts, it will never be overthrown." (Albert Einstein, in Schilpp 1949)[1]. It is on the rock of classical thermodynamics that the chemical models of this book rest.

In Chap. 2, the book begins with a discussion of classical thermodynamics introducing the first and second laws of thermodynamics and the basic properties of systems such as internal energy (U), enthalpy (H), entropy (S), and Gibbs energy (G), From these quantities we derive chemical potentials, activities, activity coefficients, and equilibrium constants. These are the basic principles upon which our equilibrium chemical thermodynamic model rests. The presentation of these principles closely follows Pitzer (1995). Since one of our primary tasks is to present the Pitzer approach for modeling concentrated electrolyte solutions, his *Thermodynamics* textbook is ideally organized in laying out the basic principles upon which these models rest. Also, different authors sometimes define terms such as the Debye–Hückel slope somewhat differently (see examples in Ananthaswarmy and Atkinson, 1984), which can lead to confusion. In this work, we carefully followed Pitzer's definitions (Pitzer 1991, 1995). Next, we examine the Pitzer approach for defining chemical thermodynamic properties of highly concentrated solutions. The model developed in this work was originally designed to deal with freezing processes that can quickly concentrate solutes. Therefore, the Pitzer approach was ideally suited for this purpose.

Implementation of the Pitzer approach is through the FREZCHEM (<u>FREEZ</u>ING <u>CHEMI</u>STRY) model, which is at the core of this work. This model was originally designed to simulate salt chemistries and freezing processes at low temperatures (-54 to $25\,°C$) and 1 atm pressure. Over the years, this model has been broadened to include more chemistries (from 16

---

[1] Translation modified. In the original translation by Schilpp (1949), the word "is" was removed from two places in the first sentence to improve the parallelism and flow.

to 58 solid phases), a broader temperature range for some chemistries (to $-113\,°C$), and incorporation of a pressure dependence (1 to 1000 bars) into the model. Implementation, parameterization, validation, and limitations of the FREZCHEM model are extensively discussed in Chap. 3.

The current mantra of astrobiology is "follow the water." Where there is water, there may be life. The FREZCHEM model can determine the presence or absence of water down to the eutectic temperature, below which only solid phases are thermodynamically stable. Salinity, the desiccation potential, and acidity are other potentially life-limiting factors that are calculated by FREZCHEM. In Chap. 4, we discuss potential life-limiting factors such as temperature, salinity, acidity, desiccation, radiation, pressure, and time.

FREZCHEM has been used to explore cold biogeochemical processes on Earth (Marion 1997; Marion and Grant 1997; Marion and Farren 1999; Marion et al. 1999; Elberling 2001; Marion and Jakubowski 2004; Marion et al. 2005; Lyons et al. 2005; Marion et al. 2006), Mars (Morse and Marion 1999; Gaidos and Marion 2003; Marion et al. 2003a; Marion and Schulze-Makuch 2007), and Europa (Kargel et al. 2000; Marion 2001, 2002; Zolotov and Shock 2001; Marion et al. 2003b; Marion et al. 2005; Marion and Schulze-Makuch 2007). Examples of these biogeochemical applications and the consequences for life are presented in Chap. 5 along with new applications to snowball Earth/ hothouse Earth and tropospheric clouds.

In Chap. 6, we first discuss a search strategy for life in the Universe and then speculate about future life on Earth, in our Solar System, and in the Universe.

And finally, appendices provide a user's guide for the FREZCHEM model and tables of model parameters. Version 9.2 of this model includes the precipitation-dissolution of chloride, nitrate, sulfate, and bicarbonate-carbonate salts of calcium, magnesium, sodium, potassium, and ferrous iron. This version also contains strong acid chemistries (hydrochloric, nitric, and sulfuric), gas hydrate chemistries (carbon dioxide and methane), and temperature/pressure dependencies. Electronic copies of the FORTRAN code are available from the senior author (giles.marion@dri.edu).

2 Aqueous Chemistry

2.1 Basic Principles

2.1.1 Definitions

In thermodynamics, a "system" is that part of the Universe that is selected for purposes of study; the remainder is called the "surroundings" or "environment" (Fig. 2.1). The choice of what constitutes a system is always arbitrary; but the choice can greatly simplify application of thermodynamic principles.

Systems can be grouped into three types: isolated, closed, and open. In an isolated system, neither energy nor matter can cross the system boundary; in a closed system, energy but not matter can cross the system boundary; and in an open system, both energy and matter can cross the boundary. Examples of all three of these systems will be encountered in this book. The FREZCHEM model, which is an integral part of this book, is, in general, closed with respect to matter; but there are exceptions dealing with gaseous components (Sect. 3.3).

Systems are further separated into homogenous or heterogenous systems. A homogenous system is one where the macroscopic properties are the same

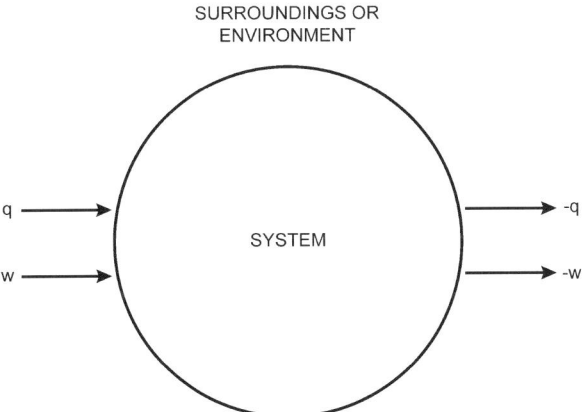

Fig. 2.1. The flow of heat (q) and work (w) between a system and the surroundings or environment

in all parts (e.g., pure water or a dilute salt solution). A heterogenous system consists of two or more distinct homogenous regions called phases among which macroscopic properties are different (e.g., ice and water). The FREZCHEM model is designed to handle heterogenous systems consisting of a gas phase, a liquid phase, and multiple solid phases.

The state of a system is defined by its properties. Extensive properties are proportional to the size of the system. Examples include volume, mass, internal energy, Gibbs energy, enthalpy, and entropy. Intensive properties, on the other hand, are independent of the size of the system. Examples include density (mass/volume), concentration (mass/volume), specific volume (volume/mass), temperature, and pressure.

"Equilibrium" is defined as the state of absolute rest from which the system has no tendency to depart; such "stable" systems are based on "true" thermodynamic equilibrium and are the subject of this book. This is to be distinguished from "unstable" states where processes may be imperceptibly slow; such systems are sometimes called "inert," "unreactive," or "unstable" (Pitzer, 1995) and are not the subject of this book. Models based on equilibrium thermodynamics (e.g., FREZCHEM) predict stable states. In the real world, unstable states may persist indefinitely.

The FREZCHEM model is based on molalities [moles/kg(water)] as the unit for solute concentrations. Other units that could have been used include mole fractions ($n_i / \sum n_i$, where n_i is the number of moles of constituent i), and molarity (moles/liter). Molalities are independent of temperature and pressure, which is an advantage compared to molarities. Typically mole fractions are only used for extremely high solute concentrations (e.g., acid aerosol models; Clegg and Brimblecombe, 1995). All ensuing derivations include only concentrations expressed as molalities.

2.1.2 The First and Second Laws of Thermodynamics

The first law of thermodynamics is the conservation of energy. In any reaction, the following equality always holds:

$$\Delta U_{\text{universe}} = \Delta U_{\text{system}} + \Delta U_{\text{environment}} = 0.0, \qquad (2.1)$$

where U is the internal energy. If the system gains (or loses) energy, there must be an equal loss (or gain) of energy in the environment. The energy that flows into the system can be represented as:

$$\Delta U_{\text{system}} = q + w, \qquad (2.2)$$

where q is the flow to the system of energy as heat due to a thermal gradient and w is the work done on the system (Fig. 2.1). The equivalency of heat and mechanical work is one of the seminal findings of the early thermodynamists (e.g., Rumford, Mayer, Joule) that led to the first law of thermodynamics

(Pitzer, 1995). In what follows, subscripts to distinguish system, environment, and the Universe are only used where necessary; the absence of a subscript will imply the system (i.e., $\Delta U = \Delta U_{\text{system}}$).

While the first law appears intuitive and easily stated (conservation of energy), this is not the case for the second law, which deals with the dispersal (or degradation) of energy during irreversible processes. One way of stating the second law is energy spontaneously tends to flow only from being concentrated in one place to becoming diffused and spread out (F.L. Lambert, *http://www.entropysimple.com*). In terms of probabilities, the second law can be stated as follows: "Every system that is left to itself will, on the average, change toward a condition of maximum probability." (Pitzer, 1995). Probabilities and energy dispersal are related to an extensive property of the system called entropy. Stated in terms of entropy, we have

$$\Delta S = \frac{q}{T} + \Delta S_{\text{irr}}, \qquad (2.3)$$

where ΔS is the entropy change of the system, q is heat that flows to the system, T is the absolute temperature, and ΔS_{irr} is due to irreversible processes within the system. In the limit of a reversible process, $\Delta S_{\text{irr}} = 0.0$, and the entropy change is simply q/T. But in the real world where all spontaneous processes are irreversible, $\Delta S_{\text{irr}} > 0.0$, and there is degradation of energy due to the production of entropy. We can also write

$$\Delta S_{\text{universe}} = \Delta S_{\text{system}} + \Delta S_{\text{environment}}, \qquad (2.4)$$

which is analogous to Eq. 2.1 for internal energy. Substituting Eq. 2.3 into Eq. 2.4 and $-q/T$ for $\Delta S_{\text{environment}}$ yields

$$\Delta S_{\text{universe}} = \frac{q}{T} + \Delta S_{\text{irr}} - \frac{q}{T} = \Delta S_{\text{irr}}. \qquad (2.5)$$

In contrast to the conservation of internal energy (Eq. 2.1, the first law of thermodynamics), the entropy of the Universe always increases (Eq. 2.5), which is an alternative definition of the second law of thermodynamics. Inherent in the concept of entropy is a preferred direction for spontaneous change ($\Delta S_{\text{irr}} > 0$). For example, at 1 bar pressure, ice melts at $10\,^\circ\text{C}$, water freezes at $-10\,^\circ\text{C}$, and not vice versa. A spontaneous process leads from a state of lower probability to a state of higher probability, and equilibrium is the state of maximum probability (Pitzer, 1995).

Consider a simple system where the only work is a volume expansion against an external pressure (Fig. 2.2). In this case, for either a reversible or irreversible process, it can be shown that

$$dU = U_B - U_A = dq + dw = T\,dS - P\,dV, \qquad (2.6)$$

where $T\,dS = dq + T\,dS_{\text{irr}}$ and $-P\,dV = dw - T\,dS_{\text{irr}}$ (Pitzer, 1995). The potential useful work ($-\int P\,dV$) is degraded by the quantity $\int T\,dS_{\text{irr}}$. The

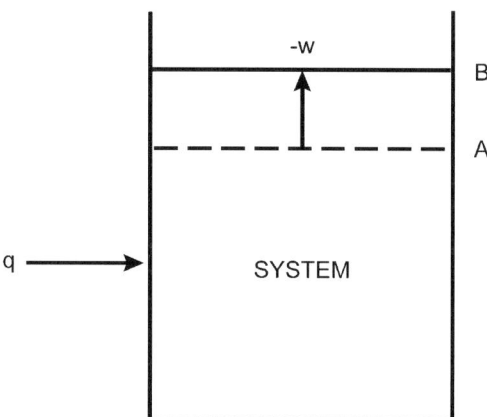

Fig. 2.2. A hypothetical case where heat (q) flows into a system causing work ($-w$) on the environment. In this case, the actual work ($-w$) differs from the potential work ($-P\Delta V$) by the amount $T\Delta S_{\mathrm{irr}}$ that accounts for irreversible processes

total energy of all systems remains the same (Eq. 2.1, first law), but a fraction of potential useful work has been degraded due to irreversible processes (Eq. 2.5, second law).

Gibbs (1948) in his seminal work on thermodynamics defined a series of thermodynamic functions:

$$H \equiv U + PV, \tag{2.7}$$

$$A \equiv U - TS, \tag{2.8}$$

$$G \equiv H - TS = U - TS + PV, \tag{2.9}$$

where H is enthalpy (heat content), A is Helmholtz energy, and G is Gibbs energy. The Gibbs energy function is particularly useful in thermodynamics. Taking the differential of G (Eq. 2.9) yields

$$\mathrm{d}G = \mathrm{d}U - T\,\mathrm{d}S - S\,\mathrm{d}T + P\,\mathrm{d}V + V\,\mathrm{d}P. \tag{2.10}$$

Then, combining Eqs. 2.6 and 2.10 yields

$$\mathrm{d}G = -S\,\mathrm{d}T + V\,\mathrm{d}P. \tag{2.11}$$

At constant T and P (easy to experimentally control), the condition for equilibrium is $\Delta G = 0$. The Gibbs energy function (Eq. 2.11) is a generally more useful function than the internal energy function (Eq. 2.6) that requires constant S and V at equilibrium ($\Delta U = 0$).

2.1.3 Activities, Chemical Potentials and Equilibrium Constants

The activity of a constituent $i(a_i)$ is given by

$$a_i = \frac{f_i}{f_i^0} \approx \frac{p_i}{p_i^0} \qquad (2.12)$$

or

$$a_i = m_i \gamma_i, \qquad (2.13)$$

where f_i is the fugacity, f_i^0 is the fugacity in the standard state, the p_i values are the partial pressures, m_i is the molality, and γ_i is the activity coefficient of the ith constituent. For all constituents, the activity (a_i) is equal to unity in the standard state; why this is the case will be made clearer shortly. For a gas, the standard state is where the fugacity (f_i) is equal to unity. For a solvent such as water or a solid phase such as a mineral, the standard state is the pure substance. For example, the activity of water at any state is ca. p_i/p_i^0, the ratio of the partial pressure at a given state relative to the partial pressure of pure water at the same temperature and 1 bar pressure. For solid phases (with the exception of mixed gas hydrates), the assumption is made in the FREZCHEM model that these solids are all pure phases with activities equal to unity. For solutes, the standard state occurs at infinite dilution, where $a_i/m_i = 1$ as $m_i \to 0.0$ ($\gamma_i \to 1.0$).

Introducing a compositional dependence into Eq. 2.11 leads to

$$dG = -S\,dT + V\,dP + \sum \mu_i\,dn_i, \qquad (2.14)$$

where n_i is the number of moles and μ_i is the chemical potential of the ith constituent, which is defined as

$$\left(\frac{\delta G}{\delta n_i}\right)_{T,P,n_{j \neq i}} = \mu_i,. \qquad (2.15)$$

The relation between chemical potential and activity is given by

$$\mu_i = \mu_i^0 + RT \ln(a_i) = \mu_i^0 + RT \ln(\gamma_i m_i). \qquad (2.16)$$

Note that, as pointed out above, when $\mu_i = \mu_i^0$ (the constituent is in its standard state), $a_i = 1$. In Sect. 3.4, we will present two approaches for mathematically modeling equilibrium chemistry based on an explicit recognition of (1) chemical potentials or (2) activities. While the mathematics of these two approaches are fundamentally different, they are physiochemically equivalent (Eq. 2.16).

Next we need to define the equilibrium constant (K) and its temperature and pressure dependence. For a chemical reaction such as

$$lL + mM + \ldots = qQ + rR + \ldots \qquad (2.17)$$

let ΔG_r be the molar Gibbs energy change for a reaction when the substances are in any given state, then

$$\Delta G_r = (q\mu_Q + r\mu_R + \ldots) - (l\mu_L + m\mu_M + \ldots). \tag{2.18}$$

Then let ΔG_r^0 be the change when each substance is in its standard state

$$\Delta G_r^0 = \left(q\mu_Q^0 + r\mu_R^0 + \ldots\right) - \left(l\mu_L^0 + m\mu_M^0 + \ldots\right). \tag{2.19}$$

If a_L, a_M represent activities in nonstandard states, then

$$l\left(\mu_L - \mu_L^0\right) = RT \ln\left(a_L^l\right). \tag{2.20}$$

Summing all the terms in reaction (2.17) leads to

$$\Delta G_r - \Delta G_r^0 = RT \ln\left(\frac{a_Q^q a_R^r \ldots}{a_L^l a_M^m \ldots}\right). \tag{2.21}$$

Bearing in mind that at constant T and P, $\Delta G_r = 0$ at equilibrium, then

$$\Delta G_r^0 = -RT \ln\left(\frac{a_Q^q a_R^r \ldots}{a_L^l a_M^m \ldots}\right). \tag{2.22}$$

Since ΔG_r^0 is a constant at a given temperature, then so is the activity product in Eq. 2.22, which leads to

$$K = \left(\frac{a_Q^q a_R^r \ldots}{a_L^l a_M^m \ldots}\right) \tag{2.23}$$

and

$$\Delta G_r^0 = -RT \ln K, \tag{2.24}$$

where K is the equilibrium constant for a reaction at a specific temperature.

There are alternative methods for estimating the temperature dependence of the equilibrium constant. One can use Pitzer's theoretical approach to estimate activity coefficients, which will be discussed in the following section, for the constituents of a reaction (Eqs. 2.13 and 2.16) coupled with experimental measurements of the molalities to estimate activities and the equilibrium constant directly at various temperatures. Alternatively, one can estimate the temperature dependence with the theoretical equation

$$\frac{\mathrm{d}\ln(K)}{\mathrm{d}T} = \frac{\mathrm{d}\Delta H_r^0}{RT^2}, \tag{2.25}$$

where ΔH_r^0 is the standard enthalpy change for a reaction. If the standard heat capacity at constant pressure (C_P^0) is known as a function of temperature, then the standard enthalpy can be calculated by integrating

$$\Delta C_P^0 = \left(\frac{\mathrm{d}\Delta H_r^0}{\mathrm{d}T}\right)_P \tag{2.26}$$

(Pitzer, 1995). This approach necessitates knowing the temperature dependence of the standard enthalpies or heat capacities for all reactants and products for a given reaction. Often this is not the case. Then a simpler, but less accurate, approach is to assume that ΔH_r^0 is a constant over the temperature range of interest [see p. 112 in Pitzer (1995) for an example]. In that case, Eq. 2.25 simplifies to

$$\ln\left(\frac{K^{T_2}}{K^{T_1}}\right) = \frac{\Delta H_r^0}{R}\left(\frac{1}{T_1} - \frac{1}{T_2}\right). \tag{2.27}$$

This equation is sometimes known as the van't Hoff equation. Standard enthalpies at 25 °C are given in many compilations (e.g., Garrels and Christ, 1965; Nordstrom and Munoz, 1994; Drever, 1997) that allow calculation of the temperature dependence of the equilibrium constant near 25 °C.

As with temperature (see above), one can use Pitzer's theoretical approach to estimate activity coefficients at various pressures for the constituents of a reaction (Eqs. 2.13 and 2.16) coupled with experimental measurements of the molalities to estimate activities and the equilibrium constant directly. Alternatively, the equilibrium constant as a function of pressure can be calculated by

$$\left(\frac{\mathrm{d}\ln(K)}{\mathrm{d}P}\right) = -\frac{\Delta \bar{V}_r^0}{RT} \tag{2.28}$$

or

$$\ln\left(\frac{K^P}{K^{P^0}}\right) = \left(\frac{-\Delta \bar{V}_r^0 (P-P^0)}{RT} + \frac{\Delta \bar{K}_r^0 (P-P^0)^2}{2RT}\right), \tag{2.29}$$

where

$$\Delta \bar{V}_r^0 = \Sigma \bar{V}_i^0 + n\bar{V}_{H_2O}^0 - V_{MX(cr)}^0, \tag{2.30}$$

$$\Delta \bar{K}_r^0 = \Sigma \bar{K}_i^0 + n\bar{K}_{H_2O}^0 - K_{MX(cr)}^0, \tag{2.31}$$

and K^P is the equilibrium constant at pressure P (bars), K^{P0} is the equilibrium constant at standard pressure P^0 (1.01325 bars), R is the gas constant (83.1451 cm$^3 \cdot$ bar mol$^{-1} \cdot$ deg^{-1}, Pitzer, 1995), T is temperature (K), $\bar{V}_i^0$ is the partial molar volume at infinite dilution of the ith constituent, and $\bar{K}_i^0$ is the molar compressibility at infinite dilution of the ith constituent (Krumgalz et al. 1999). There are numerous compilations of $\bar{V}_i^0$ and $\bar{K}_i^0$ at infinite dilution (e.g., Millero 1983; Krumgalz et al. 1996, 2000; Appendix B), which facilitates calculation of the pressure dependence of equilibrium constants.

In addition to knowing the T–P dependence of equilibrium constants (Eqs. 2.25 and 2.28), we must also know the T–P dependence of solute activity coefficients and the osmotic coefficient of the solution. A theoretical model, such as Pitzer's approach, is necessary for this purpose because activity coefficients and the osmotic coefficient must be defined at finite concentrations and not simply for the infinitely dilute state, which suffices for equilibrium constants (Eqs. 2.25 and 2.28).

2.2 Pitzer Approach

2.2.1 The Osmotic Coefficient and Activity Coefficients

The chemical equilibrium model, FREZCHEM, requires calculation of solute activity coefficients (γ) and the osmotic coefficient (ϕ) in concentrated solutions (Chap. 3). In this work, the Pitzer approach is used to calculate these quantities.

The excess Gibbs energy (G^{ex}) is defined as the difference between the actual Gibbs energy (G) and that of an ideal solution (G^{id})

$$G^{\text{ex}}(T, P, n_i) = G(T, P, n_i) - G^{\text{id}}(T, P, n_i). \tag{2.32}$$

The three terms in these equations reading from left to right are related to γ_i, a_i, and m_i of Eq. 2.13, respectively. The activity coefficient and the osmotic coefficient measure the degree to which solute concentrations and the activity of water (a_w) depart from ideal solutions, respectively. For ideal solutions, $a_i = m_i$ and $\gamma_i = 1.0$ (Eq. 2.13) or $G^{\text{ex}} = 0$ (Eq. 2.32). Similarly, $a_w = 1.0$ for an ideal solution. In the real world, solutions are rarely ideal, except in the infinitely dilute case; we therefore need a model for calculating γ_i and $\phi[= f(a_w)]$. An early model based on statistical mechanics was developed by Debye and Hückel (1923). Their equations are

$$\ln(\gamma_i) = -3A_\phi z_i^2 I^{1/2} \tag{2.33}$$

and

$$1 - \phi = \frac{2A_\phi I^{3/2}}{\sum m_i}, \tag{2.34}$$

where z_i is the ionic charge, A_ϕ is the product of several fundamental constants and is given by

$$A_\phi = \frac{1}{3} \left(\frac{2\pi N_A \, d_w}{1000} \right)^{1/2} \left(\frac{e^2}{\varepsilon k T} \right)^{3/2}, \tag{2.35}$$

where N_A is the Avogadro number, d_w is the solvent density (g cm^{-3}), e is the electronic charge, ε is the dielectric constant, k is the Boltzmann constant, and T is the absolute temperature. At 25 °C, $A_\phi = 0.3915$ (Pitzer, 1995). The ionic strength (I) is defined by

$$I = \frac{1}{2} \sum m_i z_i^2 . \qquad (2.36)$$

The osmotic coefficient (ϕ), in turn, is related to the a_w through the relation

$$a_w = \exp\left(\frac{-\phi \sum m_i}{55.50844}\right) , \qquad (2.37)$$

where the constant in the denominator is equal to $1000/M_w$ (1000/18.01528). Eqs. 2.33 and 2.34 account for long-range electrostatic interactions in electrolyte solutions and provide reasonable estimates of solute/solvent properties in very dilute solutions. There have been many attempts to extend these equations to higher concentrations. We will, however, skip intermediate concentration equations that are valid up to $I = 0.1 - 0.5$ m (see Nordstrom and Munoz, 1994; Drever, 1997 for examples) and go directly to the Pitzer approach for high concentrations ($I > 0.5$ m).

The excess Gibbs free energy (G^{ex}) was assumed by Pitzer to relate to composition through the relation

$$\frac{G^{\text{ex}}}{RT} = w_w f(I) + \frac{1}{w_w} \sum\sum \lambda_{ij}(I) n_i n_j + \frac{1}{w_w^2} \sum\sum\sum \mu_{ijk} n_i n_j n_k , \qquad (2.38)$$

where w_w is the amount of water (kg) and n_i is the moles of species i. Note that $n_i/w_w = m_i$ (molality). Here $f(I)$ is a function of ionic strength and accounts for long-range electrostatic forces and corresponds to the right-hand terms in the Debye–Hückel expressions (Eqs. 2.33 and 2.34). The λ_{ij} terms account for short-range forces between species i and j. The μ_{ijk} terms account for short-range triple particle interactions. These interaction terms can account for ion–ion, ion–neutral, and neutral–neutral species interactions and similar triple particle interactions. It is these specific species interaction terms, which are critical at high concentrations, that distinguish the Pitzer approach from the Debye–Hückel approach (Eqs. 2.33 and 2.34).

When Pitzer developed these equations, the ultimate form for describing the interaction terms was based on both theoretical models and experimental data. On the other hand, the number of terms to include in the equations is left to the user's discretion. For example, are neutral–neutral species interaction terms needed? In some applications, yes; in other applications, no. See Harvie et al. (1984), He and Morse (1993), and Pitzer (1995) for examples where different terms were selected. In what follows, we will specify the exact form of the Pitzer equations used in the FREZCHEM model. For a discussion of the connection between these equations (2.39 to 2.42) and Eq. 2.38, see Pitzer (1991, 1995).

The osmotic coefficient is given by

$$(\phi - 1) = -\left(\sum m_i\right)^{-1} \left(\frac{\delta(G^{\text{ex}}/RT)}{\delta w_w}\right)_{T,P,n_i}$$

$$= \frac{2}{\sum m_i} \left\{ \frac{-A_\phi I^{3/2}}{1 + bI^{1/2}} + \sum\sum m_c m_a \left(B_{ca}^\phi + ZC_{ca}\right) \right.$$

$$+ \sum\sum m_c m_{c'} \left(\Phi_{cc'}^\phi + \sum m_a \Psi_{cc'a}\right)$$

$$+ \sum\sum m_a m_{a'} \left(\Phi_{aa'}^\phi + \sum m_c \Psi_{caa'}\right)$$

$$+ \sum\sum m_n m_c \lambda_{nc} + \sum\sum m_n m_a \lambda_{na}$$

$$\left. + \sum\sum\sum m_n m_c m_a \zeta_{n,c,a} \right\}. \tag{2.39}$$

The activity coefficient for a cation "M" is given by

$$\ln(\gamma_M) = \left(\frac{\delta(G^{\text{ex}}/RT)}{\delta n_M}\right)_{T,P,w_w,n_i \neq M}$$

$$= z_M^2 F + \sum m_a (2B_{Ma} + ZC_{Ma}) + \sum m_c \left(2\Phi_{Mc} + \sum m_a \Psi_{Mca}\right)$$

$$+ \sum\sum m_a m_{a'} \Psi_{Maa'} + z_M \sum\sum m_c m_a C_{ca} + 2\sum m_n \lambda_{nM}$$

$$+ \sum\sum m_n m_a \zeta_{nMa}. \tag{2.40}$$

Similarly the activity coefficient for an anion "X" is given by

$$\ln(\gamma_X) = \left(\frac{\delta(G^{\text{ex}}/RT)}{\delta n_X}\right)_{T,P,w_w,n_i \neq X}$$

$$= z_X^2 F + \sum m_c (2B_{cX} + ZC_{cX}) + \sum m_a \left(2\Phi_{Xa} + \sum m_c \Psi_{cXa}\right)$$

$$+ \sum\sum m_c m_{c'} \Psi_{cc'X} + |z_X| \sum\sum m_c m_a C_{ca} + 2\sum m_n \lambda_{nX}$$

$$+ \sum\sum m_n m_c \zeta_{ncX}. \tag{2.41}$$

And, finally, the activity coefficient for a neutral species "N" is given by

$$\ln(\gamma_N) = \left(\frac{\delta(G^{\text{ex}}/RT)}{\delta n_N}\right)_{T,P,w_w,n_i \neq N}$$

$$= \sum m_c (2\lambda_{Nc}) + \sum m_a (2\lambda_{Na}) + \sum\sum m_c m_a \zeta_{Nca}. \tag{2.42}$$

The new factors in these equations not previously defined include the subscripts n, c, and a that refer respectively to neutral, cation, and anion species;

an accent over a subscript such as c' implies a cation other than c (and similarly for anions). b is a constant equal to $1.2 \, \text{kg}^{1/2} \, \text{mol}^{-1/2}$, and z is the ion charge. F is defined as

$$F = f^\gamma + \sum\sum m_c m_a B'_{ca} + \sum\sum m_c m_{c'} \Phi'_{cc'} + \sum\sum m_a m_{a'} \Phi'_{aa'}, \quad (2.43)$$

where

$$f^\gamma = -A_\phi \left[\frac{I^{1/2}}{1 + bI^{1/2}} + \frac{2}{b} \ln(1 + bI^{1/2}) \right]. \quad (2.44)$$

Z is defined as

$$Z = \sum m_i |z_i|. \quad (2.45)$$

The remaining terms in these equations are interaction parameters. The "B" terms account for cation–anion interactions and are defined by

$$B^\phi_{MX} = B^{(0)}_{MX} + B^{(1)}_{MX} \exp\left(-\alpha_1 I^{1/2}\right) + B^{(2)}_{MX} \exp\left(-\alpha_2 I^{1/2}\right), \quad (2.46)$$

where $\alpha_1 = 2.0$ and $\alpha_2 = 0 \, \text{kg}^{1/2} \text{mol}^{-1/2}$ for all binary systems, except for 2:2 electrolytes, where $\alpha_1 = 1.4$ and $\alpha_2 = 12 \, \text{kg}^{1/2} \, \text{mol}^{-1/2}$;

$$B_{MX} = B^{(0)}_{MX} + B^{(1)}_{MX} g\left(\alpha_1 I^{1/2}\right) + B^{(2)}_{MX} g\left(\alpha_2 I^{1/2}\right), \quad (2.47)$$

where

$$g(x) = \frac{2\left[1 - (1+x)\exp(-x)\right]}{x^2}, \quad (2.48)$$

and

$$B'_{MX} = \frac{B^{(1)}_{MX} g'\left(\alpha_1 I^{1/2}\right)}{I} + \frac{B^{(2)}_{MX} g'\left(\alpha_2 I^{1/2}\right)}{I}, \quad (2.49)$$

where

$$g'(x) = \frac{-2\left[1 - \left(1 + x + \frac{x^2}{2}\right)e^{-x}\right]}{x^2}. \quad (2.50)$$

Data compilations often report the cation–anion C_{MX} term as C^ϕ_{MX}; these quantities are related as follows:

$$C_{MX} = \frac{C^\phi_{MX}}{2|z_M z_X|^{1/2}}. \quad (2.51)$$

The Φ terms in the above equations account for cation–cation or anion–anion interactions. The Φ terms are defined by

$$\Phi_{ij} = \theta_{ij} + {}^E\theta_{ij}(I), \quad (2.52)$$

$$\Phi'_{ij} = {}^E\theta'_{ij}(I) = \frac{\mathrm{d}({}^E\theta_{ij}(I))}{\mathrm{d}I}, \quad (2.53)$$

$$\Phi^\phi_{ij} = \theta_{ij} + {}^E\theta_{ij}(I) + I{}^E\theta'_{ij}(I), \quad (2.54)$$

where
$$^E\theta_{ij}(I) = \frac{z_i z_j}{4I}\left[J(x_{ij}) - \frac{1}{2}J(x_{ii}) - \frac{1}{2}J(x_{jj})\right], \quad (2.55)$$

$$^E\theta'_{ij}(I) = -\frac{^E\theta_{ij}(I)}{I} + \frac{z_i z_j}{8I^2}\left[x_{ij}J'(x_{ij}) - \frac{1}{2}x_{ii}J'(x_{ii}) - \frac{1}{2}x_{jj}J'(x_{jj})\right], \quad (2.56)$$

J and J' are given by

$$J(x) = \frac{x}{4} - 1 + \frac{1}{x}\int_0^\infty \left[1 - \exp\left(-\frac{x}{y}e^{-y}\right)\right] y^2 \mathrm{d}y, \quad (2.57)$$

$$J'(x) = \frac{1}{4} - \frac{1}{x^2}\int_0^\infty \left[1 - \left(1 + \frac{x}{y}e^{-y}\right) x \exp\left(-\frac{x}{y}e^{-y}\right)\right] y^2 \mathrm{d}y, \quad (2.58)$$

and

$$x_{ij} = 6 z_i z_j A_\phi I^{1/2}. \quad (2.59)$$

The integrals in Eqs. 2.57 and 2.58 are solved numerically using two Chebyshev polynomial approximations.

Where $x \leq 1$:
$$z = 4x^{\frac{1}{5}} - 2, \quad (2.60)$$

$$\frac{\mathrm{d}z}{\mathrm{d}x} = \frac{4}{5}x^{-\frac{4}{5}}, \quad (2.61)$$

$$b_k = z b_{k+1} - b_{k+2} + a_k^I, \quad (2.62)$$

$k = 0, 20$

$$d_k = b_{k+1} + z d_{k+1} - d_{k+2}. \quad (2.63)$$

Where $x \geq 1$:
$$z = \frac{40}{9}x^{-\frac{1}{10}} - \frac{22}{9}, \quad (2.64)$$

$$\frac{\mathrm{d}z}{\mathrm{d}x} = -\frac{40}{90}x^{-\frac{11}{10}}, \quad (2.65)$$

$$b_k = z b_{k+1} - b_{k+2} + a_k^{II}, \quad (2.66)$$

$k = 0, 20$

$$d_k = b_{k+1} + z d_{k+1} - d_{k+2}. \quad (2.67)$$

Then:
$$J(x) = \frac{1}{4}x - 1 + \frac{1}{2}(b_0 - b_2), \quad (2.68)$$

$$J'(x) = \frac{1}{4} + \frac{1}{2}\frac{\mathrm{d}z}{\mathrm{d}x}(d_0 - d_2). \quad (2.69)$$

The coefficients a_k^I and a_k^{II} are given in Table B.3. By definition, $b_{21} = b_{22} = d_{21} = d_{22} = 0.0$. By using Eq. 2.62 or 2.66, the numbers b can be generated in decreasing sequence; a similar approach using Eq. 2.63 or 2.67 generates the numbers d. See Pitzer (1991) for further details. The expressions for $^E\theta_{ij}$ and $^E\theta'_{ij}$ are only defined for asymmetrical cases where the ions have the same sign but a different charge ($z_i \neq z_j$). For ions of the same sign and charge, $^E\theta_{ij}$ and $^E\theta'_{ij}(\phi'_{ij})$ are equal to 0.0 and $\Phi_{ij} = \theta_{ij} = \Phi^\phi_{ij}$ (Eqs. 2.52 to 2.54).

The Ψ_{ijk} parameters in Eqs. 2.39 to 2.42 account for triple ion interactions between two cations and an anion or two anions and a cation. The neutral species parameters λ_{nc} (or λ_{na}) and ζ_{nca} are needed in cases where neutral species are present in significant concentrations [e.g., CO_2(aq), CH_4(aq), and O_2(aq)].

Calculating the osmotic coefficient and activity coefficients of an aqueous solution using the Pitzer approach requires knowing the cation–anion parameters, $B^{(0)}_{ca}$, $B^{(1)}_{ca}$, $B^{(2)}_{ca}$, and C_{ca}; the cation–cation (or anion–anion) parameter, $\theta_{cc'}$ (or $\theta_{aa'}$); and the triple particle parameter, $\Psi_{cc'a}$ (or $\Psi_{caa'}$) for all the important constituents of a solution. If neutral solutes are present at significant concentrations, then the neutral–cation (or neutral–anion) parameter, λ_{nc} (or λ_{na}), and the triple particle parameter, ζ_{nca}, are also needed. Fortunately, there have been many studies using the Pitzer approach in the past 30 years. As a consequence, many of the most important parameters and their temperature dependence have been determined (see, for example, Harvie et al. 1984; Pitzer 1991, 1995; Appendix B).

While the mathematics of calculating the osmotic coefficient and activity coefficients are complicated (Eqs. 2.39 to 2.69), the great virtue of the Pitzer approach is that it allows one to calculate these quantities at high solute concentrations ($I > 5$ m) (Pitzer 1991, 1995; Marion and Farren 1999; Marion 2001, 2002; Marion et al. 2003a,b; Marion et al. 2005, 2006). This is particularly important in characterizing the freezing process, which can concentrate solutes rapidly once ice begins to form.

2.2.2 Temperature and Pressure Dependencies

In this work the temperature dependencies of Pitzer parameters and equilibrium constants were largely determined from isothermal data sets that were then fit to the following equation:

$$P(T)_i = a_{1i} + a_{2i}T + a_{3i}T^2 + a_{4i}T^3 + \frac{a_{5i}}{T} + a_{6i}\ln(T) + \frac{a_{7i}}{T^2} + a_{8i}T^4, \quad (2.70)$$

where $P(T)_i$ is the ith Pitzer parameter or $\ln(K)$. This equation form is not completely arbitrary but is related to the standard enthalpies. For example, if we assume in Eq. 2.25 that ΔH^0_r can be described by a power series in T, then

$$\Delta H^0_r = \Delta H^0_0 + A'T + B'T^2 + C'T^3 + \ldots, \quad (2.71)$$

where ΔH_0^0 is the standard enthalpy at a reference temperature (e.g., 25 °C). Then, substituting Eq. 2.71 into Eq. 2.25 and integrating yields

$$\ln\left(\frac{K^T}{K^{T_0}}\right) = -\frac{\Delta H_0^0}{R}\left(\frac{1}{T} - \frac{1}{T_0}\right) + \frac{A'}{R}\ln\left(\frac{T}{T_0}\right)$$
$$+ \frac{B'}{R}(T - T_0) + \frac{C'}{R}(T^2 - T_0^2) + \ldots, \quad (2.72)$$

which leads to equations of the general form

$$\ln(K_T) = \frac{A}{T} + B + C\ln(T) + DT + ET^2 + \ldots. \quad (2.73)$$

So there is an underlying basis related to the standard enthalpy of reaction (or heat capacities, Eq. 2.26) for the equation form used in this work to characterize the temperature dependence of equilibrium constants and Pitzer parameters (cf. Eqs. 2.70 and 2.73).

The effect of pressure on chemical equilibria is related to the volumetric properties of solutions. For example from Eq. 2.14

$$\left(\frac{dG}{dP}\right)_{T,n_i} = V. \quad (2.74)$$

The excess Gibbs energy (Eq. 2.32) is here associated with volumetric properties

$$V = V^{id} + V^{ex} = \Sigma n_i \bar{V}_i^0 + V^{ex} = 1000 v_w + \Sigma m_i \bar{V}_i^0 + V^{ex} \quad (2.75)$$

for the case where a molal solution contains 1 kg of water and v_w is the specific volume of water (cm³/g), $\bar{V}_i^0$ is the standard molar volume of the ith constituent at infinite dilution, and V^{ex} is the excess volume of mixing. In terms of the Pitzer approach, an expression for V^{ex} [$= (\delta G^{ex}/\delta P)_{T,ni}$] is given by

$$\frac{V^{ex}}{RT} = f^v + 2\Sigma\Sigma m_c m_a \left[B_{ca}^v + (\Sigma m_c z_c) C_{ca}^v\right] \quad (2.76)$$

(Monnin 1989), which is analogous to Eq. 2.38. Equation 2.76 does not, however, include ternary interaction parameters (θ and Ψ) because the database for calculating these terms is inadequate at present (Monnin 1989; Pitzer 1991, 1995; Krumgalz et al. 1999). In Eq. 2.76, the Debye–Hückel term is given by

$$f^v = \frac{A_v}{RT}\frac{1}{b}\ln\left(1 + bI^{1/2}\right), \quad (2.77)$$

with

$$A_v = -4RT\left(\frac{\delta A_\phi}{\delta P}\right)_T = 2A_\phi RT\left[3\left(\frac{d\ln(\varepsilon)}{dP}\right)_T + \left(\frac{d\ln(V_w)}{dP}\right)_T\right], \quad (2.78)$$

where V_w is the volume of pure water. Values of A_v as a function of T and P are compiled in Ananthaswarmy and Atkinson (1984). The partial molar volume for an ion is given by

$$\bar{V}_i = \left(\frac{dV}{dn_i}\right)_{T,P,n_{j\neq i}}. \tag{2.79}$$

The partial molar volumes of a cation (M) and an anion (X) are given by

$$\bar{V}_M = \bar{V}_M^0 + z_M^2 f + 2RT \sum m_a \left[B_{Ma}^v + \left(\sum m_c z_c\right) C_{Ma}^v\right]$$
$$+ RT \sum\sum m_c m_a \left[z_M^2 (B_{ca}^v)' + z_M C_{ca}^v\right], \tag{2.80}$$

$$\bar{V}_X = \bar{V}_X^0 + z_X^2 f + 2RT \sum m_c \left[B_{cX}^v + \left(\sum m_c z_c\right) C_{cX}^v\right]$$
$$+ RT \sum\sum m_c m_a \left[z_X^2 (B_{ca}^v)' + |z_X| C_{ca}^v\right], \tag{2.81}$$

where

$$f = \frac{A_v}{4} \left[\frac{I^{1/2}}{1+bI^{1/2}} + \frac{2}{b} \ln\left(1 + bI^{1/2}\right)\right], \tag{2.82}$$

$$B_{ca}^v = B_{ca}^{(0)v} + B_{ca}^{(1)v} g\left(\alpha_1 I^{1/2}\right) + B_{ca}^{(2)v} g\left(\alpha_2 I^{1/2}\right), \tag{2.83}$$

$$B_{ca}^{(i)v} = \frac{\delta B_{ca}^{(i)}}{\delta P} \quad i = 0, 1, 2, \tag{2.84}$$

$$C_{ca}^v = \frac{\delta C_{ca}}{\delta P}, \tag{2.85}$$

and

$$(B_{ca}^v)' = \frac{\delta B_{ca}^v}{\delta I} = B_{ca}^{(1)v} \frac{\delta g(\alpha_1 I^{1/2})}{\delta I} + B_{ca}^{(2)v} \frac{\delta g(\alpha_2 I^{1/2})}{\delta I}. \tag{2.86}$$

The pressure dependence of the activity coefficient (γ) is then estimated by

$$\ln\left(\frac{\gamma_i^P}{\gamma_i^{P^0}}\right) = \frac{(\bar{V}_i - \bar{V}_i^0)(P - P^0)}{RT}, \tag{2.87}$$

where $(\bar{V}_i - \bar{V}_i^0)$ is calculated for cations and anions using Eqs. 2.80 and 2.81.

To our knowledge, no one has ever worked out the mathematics for directly estimating the pressure dependence of the osmotic coefficient (or a_w) using the Pitzer approach. However, Monnin (1990) developed an alternative model based on the Pitzer approach that allows calculation of the pressure dependence for the activity of water (a_w). The density of an aqueous solution (ρ) can be calculated with the equation

$$\rho = \frac{1000 + \sum m_i M_i}{\frac{1000}{\rho^0} + \sum m_i \bar{V}_i^0 + V^{\text{ex}}}, \tag{2.88}$$

where m_i is the molal concentration, M_i is the molecular weight, ρ^0 is the density of pure water at a given temperature and pressure, $\bar{V}_i^0$ is the partial molal volume at infinite dilution, and V^{ex} is the excess volume of mixing given by Eq. 2.76. In Sect. 3.3, we will present a model for calculating ρ^0, the density of pure water, as a function of temperature and pressure. Given the density of a solution, the partial molal volume of water can be calculated by

$$\bar{V}_w = \frac{M_w}{1000}\left(\frac{1000 + \sum m_i M_i}{\rho} - \sum m_i \bar{V}_i\right). \tag{2.89}$$

Then the pressure dependence of the activity of water can be calculated by

$$\ln\left(\frac{a_w^P}{a_w^{P0}}\right) = \left(\frac{(\bar{V}_w - \bar{V}_w^0)(P - P^0)}{RT}\right). \tag{2.90}$$

Note that the equations for estimating the pressure dependencies of γ_i and a_w (Eqs. 2.87 and 2.90) depend on the Pitzer equations (Eqs. 2.76, 2.80, and 2.81); but this is not the case for the pressure dependence of the equilibrium constants (Eq. 2.29); the latter equation is based entirely on partial molar volumes at infinite dilution, which are independent of concentration. Also, compared to the pressure-dependent equation for the equilibrium constant (Eq. 2.29), the pressure equations for activity coefficients (Eq. 2.87) and the activity of water (Eq. 2.90) do not contain compressibilities (K) because the database for these terms and the associated Pitzer parameters are lacking at present (Krumgalz et al. 1999). The consequences of truncating Eqs. 2.80 and 2.81 for ternary terms and Eqs. 2.87 and 2.90 for compressibilities will be discussed in Sect. 3.6 under limitations.

3 The FREZCHEM Model

3.1 Historical Development

The FREZCHEM model was developed from earlier geochemical models that used the Pitzer approach. This legacy is important because there are relationships built into FREZCHEM inherited from these earlier models that influence calculations. For example, workers have used several different equations to describe the Debye–Hückel constant, A_ϕ (Eq. 2.35) (e.g., Ananthaswamy and Atkinson 1984; Archer and Wang 1990; Spencer et al. 1990; Pitzer 1991). To replace A_ϕ in the FREZCHEM model with the most accurate current equation [probably Archer and Wang (1990)] would literally require man-years of work because this would necessitate a reevaluation of all the constituent Pitzer parameters and solubility products. What is most fundamental in applying the Pitzer approach is that model parameters and equations must be internally compatible. It is not possible to take an A_ϕ from one paper, Pitzer parameters from a second paper, and solubility products from a third paper and expect for them to accurately define chemical equilibria. In the section on limitations (3.6), we will discuss the magnitude of error inherent in using different equations for A_ϕ as well as other critical equations in FREZCHEM.

To date, nine versions of the FREZCHEM model have been identified. Version 1 of the FREZCHEM model (Marion and Grant 1994) was literally a working version of the Spencer–Møller–Weare (SMW) model (Spencer et al. 1990) that was explicitly developed for dealing with equilibria at subzero temperatures. In turn, the SMW model was based, in part, on predecessor models, especially Harvie et al. (1984) and Møller (1988). Versions 1 (Marion and Grant, 1994) and 2 (Mironenko et al. 1997) are two separate working versions of the original model that includes ice and 15 minerals, which largely consist of chloride and sulfate salts (#s 31–46, Table 3.1); these two versions differ in the mathematical algorithm used (Sect. 3.4). Versions 3 and 4 added six salts (five sulfates and calcite, #s 47–52, Table 3.1) (Marion and Farren 1999), version 5 focused primarily on adding bicarbonate/carbonate chemistries (#s 53–65, Table 3.1) to the existing chloride/ sulfate model (Marion 2001), version 6 added strong acids and nitrate species to the model (#s 66–80, Table 3.1) (Marion 2002), and version 7 added six new iron minerals (#s 81–86, Table 3.1) (Marion et al. 2003a). Version 8 added a pressure

Table 3.1. A listing of chemical species in the FREZCHEM model (version 9.2)

A. Solution and atmospheric species

# Species	# Species	# Species	# Species
1 Na^+(aq)	11 Cl^-(aq)	21 CO_2(aq)	151 O_2(g)
2 K^+(aq)	12 SO_4^{2-}(aq)	22 $FeCO_3^\circ$(aq)	152 O_2(aq)
3 Ca^{2+}(aq)	13 OH^-(aq)	23 HCl(g)	153 H_2(g)
4 Mg^{2+}(aq)	14 HCO_3^-(aq)	24 $CaCO_3^\circ$(aq)	154 CH_4(g)
5 H^+(aq)	15 CO_3^{2-}(aq)	25 $MgCO_3^\circ$(aq)	155 CH_4(aq)
6 $MgOH^+$(aq)	16 HSO_4^-(aq)	26 HNO_3(g)	
7 Fe^{2+}(aq)	17 NO_3^-(aq)	27 H_2SO_4(g)	
8 $FeOH^+$(aq)	18	28 H_2O(g)	
9	19	29 CO_2(g)	
10	20	30 H_2O(l)	

B. Solid-phase species

# Species	# Species	# Species	# Species
31 H_2O(cr,I)	46 $MgSO_4 \cdot K_2SO_4 \cdot 6H_2O$(cr)	61 $CaMg(CO_3)_2$(cr)	76 $K_3H(SO_4)_2$(cr)
32 $NaCl \cdot 2H_2O$(cr)	47 $Na_2SO_4 \cdot MgSO_4 \cdot 4H_2O$(cr)	62 $Na_2CO_3 \cdot 7H_2O$(cr)	77 $K_5H_3(SO_4)_4$(cr)
33 NaCl(cr)	48 $CaSO_4 \cdot 2H_2O$(cr)	63 $KHCO_3$(cr)	78 $K_8H_6(SO_4)_7 \cdot H_2O$(cr)
34 KCl(cr)	49 $CaSO_4$(cr)	64 $CaCO_3$(cr,aragonite)	79 $KHSO_4$(cr)
35 $CaCl_2 \cdot 6H_2O$(cr)	50 $MgSO_4 \cdot 12H_2O$(cr)	65 $CaCO_3$(cr,vaterite)	80 $MgSO_4 \cdot H_2O$(cr)
36 $MgCl_2 \cdot 6H_2O$(cr)	51 $Na_2SO_4 \cdot 3K_2SO_4$(cr)	66 $HNO_3 \cdot 3H_2O$(cr)	81 $FeSO_4 \cdot 7H_2O$(cr)
37 $MgCl_2 \cdot 8H_2O$(cr)	52 $CaCO_3$(cr,calcite)	67 KNO_3(cr)	82 $FeSO_4 \cdot H_2O$(cr)
38 $MgCl_2 \cdot 12H_2O$(cr)	53 $MgCO_3$(cr)	68 $NaNO_3$(cr)	83 $FeCl_2 \cdot 6H_2O$(cr)
39 $KMgCl_3 \cdot 6H_2O$(cr)	54 $MgCO_3 \cdot 3H_2O$(cr)	69 $HCl \cdot 3H_2O$(cr)	84 $FeCl_2 \cdot 4H_2O$(cr)
40 $CaCl_2 \cdot 2MgCl_2 \cdot 12H_2O$(cr)	55 $MgCO_3 \cdot 5H_2O$(cr)	70 $H_2SO_4 \cdot 6.5H_2O$(cr)	85 $FeCO_3$(cr)
41 $Na_2SO_4 \cdot 10H_2O$(cr)	56 $CaCO_3 \cdot 6H_2O$(cr)	71 $H_2SO_4 \cdot 4H_2O$(cr)	86 $Fe(OH)_3$(cr)
42 Na_2SO_4(cr)	57 $NaHCO_3$(cr)	72 $HCl \cdot 6H_2O$(cr)	87 $CO_2 \cdot 6H_2O$(cr)
43 $MgSO_4 \cdot 6H_2O$(cr)	58 $Na_2CO_3 \cdot 10H_2O$(cr)	73 $NaNO_3 \cdot Na_2SO_4 \cdot 2H_2O$(cr)	88 $CH_4 \cdot 6H_2O$(cr)
44 $MgSO_4 \cdot 7H_2O$(cr)	59 $NaHCO_3 \cdot Na_2CO_3 \cdot 2H_2O$(cr)	74 $Na_3H(SO_4)_2$(cr)	
45 K_2SO_4(cr)	60 $3MgCO_3 \cdot Mg(OH)_2 \cdot 3H_2O$(cr)	75 $NaHSO_4 \cdot H_2O$(cr)	

dependence to the model but no new chemical species (Marion et al. 2005); for the first time, this version of the model allows one to estimate solution density. Version 9 incorporates two gas hydrates and associated chemistries into the model (#s 87–88, Table 3.1) (Marion et al. 2006).

Over the years, some parameters have had to be changed as the model expanded. For example, the ice-water line has been changed three times because the lower temperature limit expanded (Sect. 3.3.1). Also we expanded the maximum concentration range for Na_2SO_4 chemistry from 2 to 4 m (Sect. 3.3.2). Parameterization of the model is discussed in general terms in this chapter. Actual parameters for individual reactions of version 9 are given in Appendix B.

3.2 Basic Structure

3.2.1 Chemical Equilibrium

The FREZCHEM model is a chemical equilibrium model. For a reaction such as gypsum dissolution

$$CaSO_4 \cdot 2H_2O \Leftrightarrow Ca^{2+} + SO_4^{2-} + 2H_2O \tag{3.1}$$

the calculated ion activity product (IAP) is given by

$$\text{IAP} = \frac{(a_{Ca^{2+}})\left(a_{SO_4^{2-}}\right)(a_w)^2}{a_{CaSO_4 \cdot 2H_2O}}, \tag{3.2}$$

where a_i is the activity of the ith constituent. Bearing in mind that the activity of pure solid phases is equal to unity (Chap. 2), and replacing a_i for ions with $(\gamma_i)(m_i)$ leads to

$$\text{IAP} = (\gamma_{Ca^{2+}})(m_{Ca^{2+}})\left(\gamma_{SO_4^{2-}}\right)\left(m_{SO_4^{2-}}\right)(a_w)^2, \tag{3.3}$$

where γ_i is the activity coefficient calculated with the Pitzer equations (Eqs. 2.40 and 2.41), m_i are the molal concentrations specified as initial input or calculated in the model, and a_w is the activity of water calculated from Eq. 2.37 and the Pitzer equation for the osmotic coefficient (Eq. 2.39).

For a specific mineral such as gypsum:

$$\text{if IAP} > K_{sp}, \text{ the system is supersaturated}, \tag{3.4}$$
$$\text{if IAP} = K_{sp}, \text{ the system is in equilibrium}, \tag{3.5}$$
$$\text{if IAP} < K_{sp}, \text{ the system is undersaturated}, \tag{3.6}$$

where K_{sp} is the mineral solubility product (or an equilibrium constant for a different type of reaction). If the solution is supersaturated, the model is structured to bring the solution into equilibrium by reducing the molal concentrations of constituents such as Ca^{2+} and SO_4^{2-} (Eq. 3.3) until IAP = K_{sp}. One can bypass this equilibration step if one wants to only consider solution-phase chemistry (see input instructions in Appendix A). For example, this bypass is useful for estimating a mineral solubility product from experimental data and model estimates of activity coefficients and the activity of water; this bypass is also useful in cases where the solution is known to be supersaturated with respect to minerals such as calcite in seawater.

3.2.2 Mass Balances

The basic system of the FREZCHEM model consists of an aqueous phase and solid phases that together are "closed" with respect to matter. In the above gypsum case, for every mole of Ca^{2+} and SO_4^{2-} that is removed from solution, a mole of $CaSO_4 \cdot 2H_2O$ accumulates. Only with respect to gaseous components is the system "open" allowing transfer of matter into or out of the system. Mass balances are not generally maintained for equilibria between the gaseous and aqueous phases. You can think of these gases as infinite buffers for the reactions. Gas hydrates are an exception that will be discussed below. The model is structured to handle as input these gaseous constituents: $O_2(g)$, $CO_2(g)$, $CH_4(g)$, $HCl(g)$, $HNO_3(g)$, and $H_2SO_4(g)$. The corresponding constituent activity (and concentration) in the aqueous phase is fixed by a Henry's law constant (K_H), e.g.,

$$K_H = \frac{a_{CO_2(aq)}}{P_{CO_2(g)}}, \qquad (3.7)$$

where $a_{CO_2(aq)}$ is the activity of CO_2 in the aqueous phase and $P_{CO_2(g)}$ is the partial pressure of CO_2 in the gaseous phase. In the special case of gas hydrate chemistries (CO_2 and CH_4) where the partial pressures may be hundreds of bars, P_{CO_2} is replaced with f_{CO_2}, the fugacity of CO_2; see the description of gas hydrate chemistry later in Sect. 3.3.4. In the case of acids, the concentrations in the aqueous phase can be fixed by specifying aqueous-phase concentrations or by specifying gaseous-phase partial pressures; in the latter case, the aqueous phase is fixed by the Henry's law constant (Eq. 3.7).

3.2.3 Reaction Pathways

Several reaction pathways are built into the FREZCHEM model including (1) temperature change, (2) evaporation, (3) pressure change, (4) equilibrium or fractional crystallization and, for gas hydrates, (5) open or closed carbon systems, and (6) pure or mixed gas hydrates. Under the "temperature change" option, the user can specify the upper and lower temperature range and a decremental temperature interval (ΔT) at which equilibrium at a fixed pressure is calculated (e.g., 298.15 to 253.15 K with $\Delta T = 5$ would result in equilibrium calculations at 298.15, 293.15, ... and 253.15 K). At subzero temperatures, this allows freezing of the solution to the system eutectic, below which only solid phases are thermodynamically stable. The "evaporation" option allows removal of water to near-dryness and subsequent concentration of solutes at a fixed temperature and pressure. A limitation of the FREZCHEM model is that it always requires some residual water for calculation purposes, so completely dry systems such as eutectics and evaporation can only be approximated (see examples in Appendix A).

Under the "pressure change" option, the user can specify a lower and upper pressure range and an incremental pressure interval (ΔP) at which equilibrium is calculated at a fixed temperature (e.g., 1 to 1001 bars with $\Delta P = 100$ bars would result in equilibrium calculations at 1, 101, ... and 1001 bars). Under "equilibrium crystallization", solid phases that have precipitated are allowed to dissolve and repreciptate as different solids when environmental drivers such as temperature, pressure, or evaporation change. Under "fractional crystallization", a precipitated solid phase is not allowed to subsequently dissolve and repreciptate when environmental drivers change; this option allows for hypothetical removal (layering) of precipitates in basins when environmental drivers change.

For the special case of gas hydrates, one can specify either an "open" or "closed" carbon system. In the open option, system carbon is fixed by the partial pressure of $CO_2(g)$ and/or $CH_4(g)$. In the closed option, system carbon is fixed by the initial carbon content in the gas (arbitrarily assumed to be 0.1 L) and aqueous (1.0 kg water) phases; this option was primarily developed to examine small quantities of carbon encased in ice cores. Given the initial concentrations of carbon in the gaseous and aqueous phases and their respective volumes, one can calculate the total carbon of the system. As temperature, pressure, or composition changes, the total carbon is partitioned between the gaseous and aqueous phases according to the Henry's law constant up to the point where gas hydrate precipitates. When gas hydrate saturation is reached, all the carbon (CO_2 or CH_4) in the system precipitates as gas hydrate in one step. The amount of carbon in these systems is small; once gas hydrates begin to precipitate, maintaining equilibria among the gas, aqueous, and solid phases is not warranted. If you want to consider a mixed CO_2-CH_4 gas hydrate, you need to specify the mole fraction of each gas. Ideally $x_{CO_2(g)} + x_{CH_4(g)} = 1.00$; otherwise, if $x_{CO_2(g)} + x_{CH_4(g)} < 1.00$, then there exists an unidentified "dummy" gas that may indirectly influence the calculations.

3.3 Chemistries and Their Temperature and Pressure Dependence

3.3.1 Water Ice/Liquid Water/Water Vapor Equilibria

Equilibria among water ice, liquid water, and water vapor are critical for model development because these relations are fundamental to any cold aqueous model, and they can be used as a base for model parameterization. For example, given a freezing point depression (fpd) measurement for a specific solution, one can calculate directly the activity of liquid water (or osmotic coefficient) that can then be used as data to parameterize the model (Clegg and Brimblecombe 1995). These phase relations also allow one to estimate in a model the properties of one phase (e.g., gas) based on the calculated properties of another phase (e.g., aqueous), or to control one phase (e.g., aqueous) based on the known properties of another phase (e.g., gas).

Incorporation of strong acids into the FREZCHEM model extended the model lower temperature range from 219 K (the eutectic of seawater) to 195 K [the eutectic of HCl·6H$_2$O(cr)]. As fpd data were used to define water activities at these low temperatures, it became apparent that the original equation defining the equilibrium of ice and liquid water at temperatures below $\approx$ 228 K (Spencer et al. 1990) did not fit the fpd data well. This necessitated a reevaluation of the relationship.

For liquid water (w) and ice (i) at equilibrium,

$$\mu_i = \mu_i^0 + RT\ln(a_i) = \mu_w = \mu_w^0 + RT\ln(a_w), \qquad (3.8)$$

where μ_i and μ_w are the chemical potentials of ice and liquid water, μ_i^0 and μ_w^0 are the standard chemical potentials, and a_i and a_w are the activities of water in the form of ice and liquid water. Equation 3.8 can be rearranged to yield

$$\frac{\mu_i^0 - \mu_w^0}{RT} = \ln\left(\frac{a_w}{a_i}\right) = \frac{-\Delta_{\text{fus}}G^0}{RT} = \ln(K_{\text{ice}}), \qquad (3.9)$$

where $\Delta_{\text{fus}}G^0$ is the standard Gibbs energy of fusion ($\mu_w^0 - \mu_i^0$) and K_{ice} is the equilibrium constant for liquid water in equilibrium with ice. At equilibrium with respect to pure ice, $a_i = 1.0$. The following therefore applies:

$$a_w = \exp\left(\frac{-\Delta_{\text{fus}}G^0}{RT}\right) = K_{\text{ice}}. \qquad (3.10)$$

The activity of liquid water in equilibrium with pure ice is equal to the equilibrium constant (K_{ice}) at a given temperature. In turn, the activity of liquid water in equilibrium with pure ice can be estimated from

$$a_w = \frac{f_i}{f_w} = K_{\text{ice}} \approx \frac{P_i}{P_w}, \qquad (3.11)$$

3.3 Chemistries and Their Temperature and Pressure Dependence

where f_i and f_w are the fugacities of water vapor above pure ice and pure liquid water, respectively, and P_i and P_w are the corresponding partial pressures. In this derivation, the water vapor was assumed to behave as an ideal gas, which allows substitution of partial pressures for fugacities. To estimate K_ice, the partial pressures of water above pure ice and pure liquid water must therefore be specified.

The equation originally used in the FREZCHEM model to define K_ice, the Spencer et al. (1990) equation, begins to significantly depart from the Goff–Gratch (List 1951; McDonald 1965) and the Clegg–Brimblecombe (Clegg and Brimblecombe 1995) equations at temperatures below 228 K (Fig. 3.1). Similarly, the Goff–Gratch equation significantly departs from the Clegg–Brimblecome equation at temperatures below 208 K. The Goff–Gratch equations directly estimate P_i and P_w (the vapor pressures above pure water ice and pure supercooled liquid water, respectively) to temperatures as low as −100°C (List 1951; McDonald 1965; Clegg and Brimblecome 1995), which allows a direct calculation of K_ice (Eq. 3.11). The Clegg–Brimblecombe equation is based on P_i (180 to 273 K) and P_w (213 to 273 K) estimated from the Goff–Gratch equations supplemented with P_w estimates from Clegg and Brimblecome (1995, p. 47) for temperatures ranging from 180 to 220 K; the latter estimates were calculated from an extrapolated equation based on vapor pressures, enthalpies, and heat capacities. The coldest experimental fpd datum used by Clegg and Brimblecombe (1995) in developing their sulfuric acid model was 6.0188 m H_2SO_4 at 201.94 K, which has an experimentally estimated a_w equal to 0.518. The estimated a_w for the Goff–Gratch and Clegg–Brimblecome equations at this temperature are 0.526 and 0.521, respectively

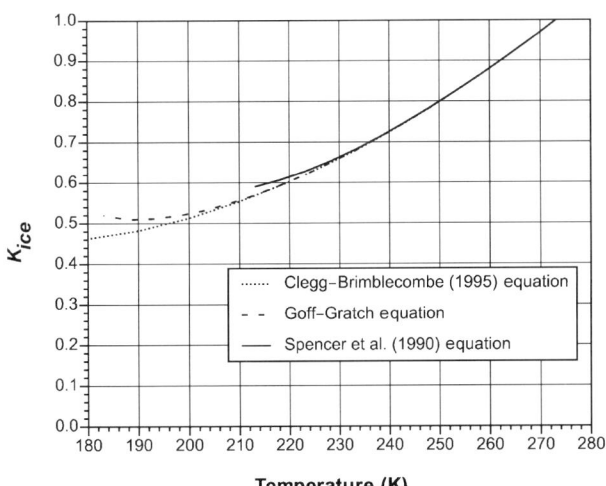

Fig. 3.1. Three equations for the equilibrium of water ice and liquid water for temperatures ranging from 180 to 273 K. Reprinted from Marion (2002) with permission

(Fig. 3.1). Similarly at 203.93 K (5.9528 m H_2SO_4), the experimentally estimated a_w is 0.526 (Clegg and Brimblecombe 1995), while the estimated a_w for the Goff–Gratch and Clegg–Brimblecombe equations at this temperature are 0.531 and 0.528, respectively. Since the Clegg–Brimblecombe equation for K_{ice} (Fig. 3.1) is most consistent with the experimental fpd data at temperatures below 223 K, this model was used in all subsequent analyses to define the activity of water in equilibrium with pure ice at specific temperatures:

$$K_{ice} = 1.906354 - 1.880285 \times 10^{-2} T + 6.603001 \times 10^{-5} T^2 \\ - 3.419967 \times 10^{-8} T^3. \tag{3.12}$$

An equilibrium constant for liquid water–water vapor ($K = a_w/P_w$) was estimated, where a_w is the activity of water in the liquid phase and P_w is the gas partial pressure above liquid water. For temperatures ≥ 273.15 K, K was estimated using the Goff–Gratch equation for P_w (McDonald 1965) assuming a pure liquid water system; in this case, $a_w = 1.0$; therefore, $K = 1.0/P_w$. For temperatures from 180 to 273.15 K, K was estimated using the published values of P_w above pure liquid water from Clegg and Brimblecome (1995). The resulting water equilibrium equation is valid for temperatures ranging from 180 to 298 K:

$$\ln(K) = 9.053594 \times 10^1 - 7.215505 \times 10^{-1} T + 2.112659 \times 10^{-3} T^2 \\ - 2.254724 \times 10^{-6} T^3. \tag{3.13}$$

Introducing pressure into the FREZCHEM model requires quantifying volumetric properties of water. In a recent paper (Marion et al. 2005), we examined two models for quantifying the volumetric properties of liquid water at subzero temperatures: one model is based on the measured properties of supercooled water, and the other model is based on the properties of liquid water in equilibrium with respect to temperature and pressure. There are two datasets at subzero temperatures in Fig. 3.2 for the properties of liquid water at 1.01 bars (gauge pressure = 0.0 bars). The data for supercooled water are from Speedy (1987). Note that as this relation approaches a singularity at 227 K, the molar volume of liquid water becomes equal to water ice (Fig. 3.2). The "data points" for equilibrium water at subzero temperatures are calculated based on an analysis of the pressure–temperature melting curve of pure ice. Given the melting temperature of pure ice at a given pressure (Wagner et al. 1994), the molar volume of ice at 1.01 bars (Eq. 3.18), the compressibility of ice (Eq. 3.19), and the compressibility of water (Eq. 3.17), it is possible to estimate the molar volume of water at 1.01 bar pressure indirectly using the equilibrium relation in Eq. 2.29. While the difference between the two water molar volume datasets may seem insignificant at temperatures > 250 K (Fig. 3.2), we demonstrated that these differences can lead to significant differences in calculated solution properties (Marion et al. 2005). Evidence based on the pressure/temperature melting of ice and salt solution densities favor

3.3 Chemistries and Their Temperature and Pressure Dependence

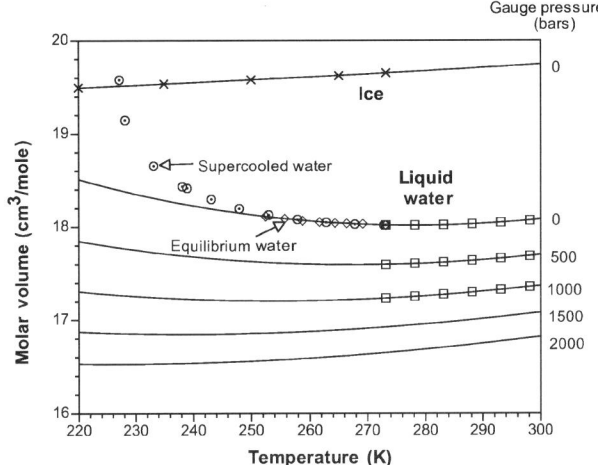

Fig. 3.2. The molar volume of ice and liquid water as a function of temperature and pressure. *Symbols* are experimental data; *lines* are model estimates. Reprinted from Marion et al. (2005) with permission

the equilibrium water model, which depends on extrapolations, for characterizing the volumetric properties of liquid water in electrolyte solutions at subzero temperatures, rather than the supercooled water model. For a fuller discussion of the arguments, see Marion et al. (2005).

In the FREZCHEM model, the density of pure water at $T \geq 273\,\mathrm{K}$ and $P = 1.01$ bars is calculated with the Kell (1975) equation, which is given by

$$1000\rho_w^0 = (999.83952 + 16.94518t - 7.98704 \times 10^{-3}t^2$$
$$-4.617046 \times 10^{-5}t^3 + 1.05563 \times 10^{-7}t^4$$
$$-2.805425 \times 10^{-10}t^5)/(1.0 + 1.687985 \times 10^{-2}t), \quad (3.14)$$

where t is the temperature (°C). From this relation the molar volume of pure water at 1.01 bars is calculated with

$$\bar{V}_l^0 = \frac{18.01528}{\rho_w^0}. \quad (3.15)$$

An equation fitted to the equilibrium water dataset (Fig. 3.2) at 1.01 bar and subzero temperatures is given by

$$\bar{V}_l^0 = 18.0182 - 1.407964 \times 10^{-3}(T - 273.15) + 1.461418 \times 10^{-4}(T - 273.15)^2. \quad (3.16)$$

The compressibility of water ($\bar{K}_l^0$) is based on the experimental database of Ter Minassian et al. (1981). In the latter study, water compressibilities are reported from 233 to 393 K and from 1 to 5001 bars (total pressure). We used a subset of this database that covered the temperature range from 273 to 303 K and the pressure range from 1 to 2001 bars to derive an equation given by

$$\bar{K}_l^0 = 8.62420 \times 10^{-3} - 5.06616 \times 10^{-5}T - 3.78615 \times 10^{-6}P$$
$$+ 2.27679 \times 10^{-8}PT + 1.18481 \times 10^{-10}P^2 + 8.20762 \times 10^{-8}T^2$$
$$- 3.63261 \times 10^{-11}PT^2 - 2.87099 \times 10^{-13}TP^2. \quad (3.17)$$

Equation 3.17 is based on the equilibrium properties of water at $T \geq 273\,\text{K}$; use of this equation at subzero temperatures necessitates an extrapolation. We did not use the full range of the Ter Minassian et al. (1981) database because at temperatures $< 273\,\text{K}$, their database relies on the properties of supercooled water. The molar volume for water ice at 1.01 bars is given by

$$\bar{V}_{I,\text{cr}}^0 = 19.30447 - 7.988471 \times 10^{-4}T + 7.563261 \times 10^{-6}T^2 \quad (3.18)$$

(Fig. 3.2). This equation is derived from crystal lattice parameters for ice Ih (Petrenko and Whitworth 1999) and is valid over a temperature range of 160 to 273 K. The ice isothermal compressibility shows a significant temperature dependence (Marion and Jakubowski 2004):

$$\bar{K}_{I,\text{cr}}^0(\text{cm}^3/(\text{mol bar})) = 2.790102 \times 10^{-2} - 2.235440 \times 10^{-4}T$$
$$+ 4.497731 \times 10^{-7}T^2. \quad (3.19)$$

In this work, the molar volumes of pure liquid water and pure water ice as functions of pressure are calculated by

$$\bar{V}_T^P = \bar{V}_T^0 - \bar{K}_T \Delta P, \quad (3.20)$$

where $\bar{V}_T^P$ is the molar volume at pressure P and temperature T, $\bar{V}_T^0$ is the molar volume at 1.01 bar pressure and temperature T, $\bar{K}_T$ is the compressibility at temperature T, and ΔP is the change in pressure. An underlying assumption in the application of Eq. 3.20 is that the compressibilities of liquid water and water ice can be approximated by a constant that is independent of pressure. Clearly there is a strong pressure dependence for the isothermal compressibility of water (Eq. 3.17). In this work, we calculated an average water compressibility at a given temperature by dividing the pressure range of interest (e.g., 1000 bars) into 20 equally spaced intervals (i.e., 50 bars), estimated the compressibility at the midpoint of each interval, and used the average of the 20 values in Eq. 3.20. The isothermal compressibility of ice (Eq. 3.19) is assumed to represent an average over the pressure range of interest. The change in compressibility of ice with pressure is not large (Richards and Speyers 1914; Nagornov and Chizhov 1990).

Estimates of the molar volumes of pure water at 0, 500, 1000, 1500, and 2000 bars of gauge pressure are given in Fig. 3.2. At subzero temperatures these estimates are based on the equilibrium water model. For comparative purposes, we also included the experimental measurements of Chen et al. (1977) at 0, 500, and 1000 bars between 273 and 298 K. At 298 K, the

3.3 Chemistries and Their Temperature and Pressure Dependence

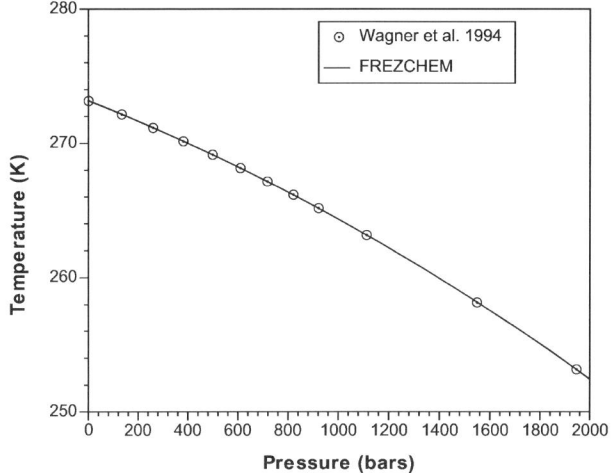

Fig. 3.3. Freezing point depression of a pure ice/water system as a function of pressure. Reprinted from Marion et al. (2003b) with permission

estimates of the molar volumes at 500 bars are 17.689 and 17.688 cm³ mol⁻¹, and at 1000 bars, the molar volumes are 17.358 and 17.356 cm³ mol⁻¹ for the Chen et al. (1977) data and our model, respectively. The equilibrium model predicts molar volumes for water of 17.214 and 16.564 cm³ mol⁻¹ at 1000 bars ($T = 264.35$ K) and 2000 bars ($T = 252.45$ K) along the pressure melting curve of ice/water (Wagner et al. 1994) (Fig. 3.3). These molar volumes compare favorably with a model developed by Nagornov and Chizhov (1990) that predicts 17.219 and 16.565 cm³ mol⁻¹ at 1000 and 2000 bars of pressure. The pressure melting curve for ice/water (Wagner et al. 1994) was used in deriving the molar volume of water at subzero temperatures (Eq. 3.16) and the compressibility of ice (Eq. 3.19), so naturally our model fits this database very well (Fig. 3.3).

3.3.2 Salt Equilibria

Table 3.2 summarizes the chemistries that lead to the precipitation of solid phases. The actual parameters for these equilibria are in Appendix B. The range of temperatures used in model parameterizations or validation and the maximum molal concentrations are both fundamentally important in properly applying the model. Extrapolations outside these temperature/compositional ranges require careful scrutiny of model calculations.

In most cases, the maximum concentration represents the composition of the saturated solid at 25 °C. For example, the concentration of NaCl(cr) at 25 °C is 6.11 m (Fig. 3.4). An example where this is not the case is Na$_2$SO$_4$, where the maximum concentration of 4.0 m is twice the concentration at equilibria with Na$_2$SO$_4$·10H$_2$O(cr) at 25 °C. In this particular case, we extended

Table 3.2. The temperature range, maximum concentrations (or pressure), solid phases, and primary references for chemical systems in the FREZCHEM model

Chemistry	Temperature range (°C)	Maximum concentration (m)	Solid phases	References
H_2O	−93 to 25	–	H_2O (ice)	Marion 2002
NaCl	−45 to 25	6.2	NaCl (halite), $NaCl \cdot 2H_2O$ (hydrohalite)	Spencer et al. 1990
KCl	−23 to 25	4.8	KCl (sylvite)	Spencer et al. 1990
$MgCl_2$	−40 to 25	5.8	$MgCl_2 \cdot 6H_2O$ (bischofite), $MgCl_2 \cdot 8H_2O$, $MgCl_2 \cdot 12H_2O$	Spencer et al. 1990
$CaCl_2$	−51 to 25	7.4	$CaCl_2 \cdot 6H_2O$ (antarcticite)	Spencer et al. 1990
$FeCl_2$	−37 to 25	5.1	$FeCl_2 \cdot 4H_2O$, $FeCl_2 \cdot 6H_2O$	Marion et al. 2003a; Pitzer 1991
mixed Cl salts	−21 to 25 22 to 25	0.7 (KCl) 5.8 ($MgCl_2$) 2.9 ($MgCl_2$) 5.2 ($CaCl_2$)	$KMgCl_3 \cdot 6H_2O$ (carnallite) $CaCl_2 \cdot 2MgCl_2 \cdot 12H_2O$ (tachyhydrite)	Spencer et al. 1990
Na_2SO_4	−23 to 25	4.0	Na_2SO_4 (thenardite), $Na_2SO_4 \cdot 10H_2O$ (mirabilite)	Marion and Farren 1999; Marion 2002
K_2SO_4	−11 to 25	0.7	K_2SO_4 (arcanite)	Marion and Farren 1999
$MgSO_4$	−20 to 25	3.1	$MgSO_4 \cdot H_2O$ (kieserite) $MgSO_4 \cdot 6H_2O$ (hexahydrite) $MgSO_4 \cdot 7H_2O$ (epsomite) $MgSO_4 \cdot 12H_2O$	Marion 2002; Marion and Farren 1999
$CaSO_4$	−21 to 25	–	$CaSO_4$ (anhydrite), $CaSO_4 \cdot 2H_2O$ (gypsum)	Marion and Farren 1999

3.3 Chemistries and Their Temperature and Pressure Dependence   31

Table 3.2. (continued)

Chemistry	Temperature range (°C)	Maximum concentration (m)	Solid phases	References
$FeSO_4$	−2 to 25	1.9	$FeSO_4 \cdot H_2O$ (szomolnokite), $FeSO_4 \cdot 7H_2O$ (melanterite)	Marion et al. 2003a; Reardon and Beckie 1987
mixed SO_4 salts	3 to 25	0.75 (K_2SO_4), 2.5 (Na_2SO_4)	$Na_2SO_4 \cdot 3K_2SO_4$ (aphthitalite)	Marion and Farren 1999
	21 to 25	2.0 (Na_2SO_4), 2.7 ($MgSO_4$)	$Na_2SO_4 \cdot MgSO_4 \cdot 4H_2O$ (bloedite)	
	−5 to 25	0.84 (K_2SO_4), 3.2 ($MgSO_4$)	$MgSO_4 \cdot K_2SO_4 \cdot 6H_2O$ (picromerite)	
$NaNO_3$	−18 to 25	10.8	$NaNO_3$ (soda niter)	Marion 2002
KNO_3	−3 to 25	3.7	KNO_3 (niter)	Marion, 2002
$Mg(NO_3)_2$	≈ 0 to 25[a]	0.1	–	Pitzer 1991
$Ca(NO_3)_2$	≈ 0 to 25[a]	0.1	–	Pitzer 1991
mixed NO_3-SO_4 salt	15 to 25	1.6 (Na_2SO_4), 10.3 ($NaNO_3$)	$NaNO_3 \cdot Na_2SO_4 \cdot 2H_2O$ (darapskite)	Marion, 2002
$NaHCO_3$, $Na_2(CO_3)$	−22 to 25	1.2 ($NaHCO_3$)	$NaHCO_3$ (nahcolite)	Marion 2001; He and Morse 1993
	21 to 25	2.6 (Na_2CO_3)	$Na_2CO_3 \cdot 7H_2O$	
	−22 to 25	2.75 (Na_2CO_3)	$Na_2CO_3 \cdot 10H_2O$ (natron)	
	21 to 25	0.6 ($NaHCO_3$), 2.75 (Na_2CO_3)	$NaHCO_3 \cdot Na_2CO_3 \cdot 2H_2O$ (trona)	
$KHCO_3$	−6 to 25	3.6	$KHCO_3$ (kalicinite)	Marion 2001; He and Morse 1993
$MgCO_3$	−2 to 25	0.4	$MgCO_3$ (magnesite), $MgCO_3 \cdot 3H_2O$ (nesquehonite), $MgCO_3 \cdot 5H_2O$ (lansfordite), $MgCO_3 \cdot Mg(OH)_2 \cdot 3H_2O$ (hydromagnesite)	Marion 2001; He and Morse 1993

Table 3.2. (continued)

Chemistry	Temperature range (°C)	Maximum concentration (m)	Solid phases	References
$CaCO_3$	−7 to 25	–	$CaCO_3$ (calcite) $CaCO_3$ (aragonite) $CaCO_3$ (vaterite) $CaCO_3 \cdot 6H_2O$ (ikaite)	Marion 2001; He and Morse 1993; Plummer and Busenberg 1982
$CaMg(CO_3)_2$	≈ 0 to 25[a]	–	$CaMg(CO_3)_2$ (dolomite)	Marion 2001
$FeCO_3$	≈ 0 to 25[a]	–	$FeCO_3$ (siderite)	Marion et al. 2003a; Ptacek 1992
$Fe(OH)_3$	≈ 0 to 25[a]	–	$Fe(OH)_3$ (ferrihydrite)	Marion et al. 2003a
HCl	−93 to 25	12.0	$HCl \cdot 3H_2O$, $HCl \cdot 6H_2O$	Marion 2002; AIM model[b]
HNO_3	−65 to 25	9.0	$HNO_3 \cdot 3H_2O$	Marion 2002; AIM model[b]
H_2SO_4	−93 to 25	8.0	$H_2SO_4 \cdot 4H_2O$, $H_2SO_4 \cdot 6.5H_2O$	Marion 2002; AIM model[b]
$NaHSO_4$	−30 to 25	5.1 (Na_2SO_4), 8.0 (H_2SO_4)	$Na_3H(SO_4)_2$ $NaHSO_4 \cdot H_2O$	Marion 2002; Pitzer 1991
$KHSO_4$	0 to 25	2.3 (K_2SO_4), 7.5 (H_2SO_4)	$K_3H(SO_4)_2$, $K_5H_3(SO_4)_4$, $K_8H_6(SO_4)_7 \cdot H_2O$ (misenite) $KHSO_4$ (mercallite)	Marion 2002; Pitzer 1991
$CO_2 \cdot 6H_2O$	−91 to 10	45 bars (pressure)	$CO_2 \cdot 6H_2O$ (carbon dioxide hydrate)	Marion et al. 2006; Sloan 1998
$CH_4 \cdot 6H_2O$	−83 to 25	443 bars (pressure)	$CH_4 \cdot 6H_2O$ (methane hydrate)	Marion et al. 2006; Sloan 1998

[a] Based on a measured value and its temperature derivative at 25 °C
[b] Derived from the AIM mole fraction model
(http://www.aim.env.uea.ac.uk/aim/aim.htm)

3.3 Chemistries and Their Temperature and Pressure Dependence

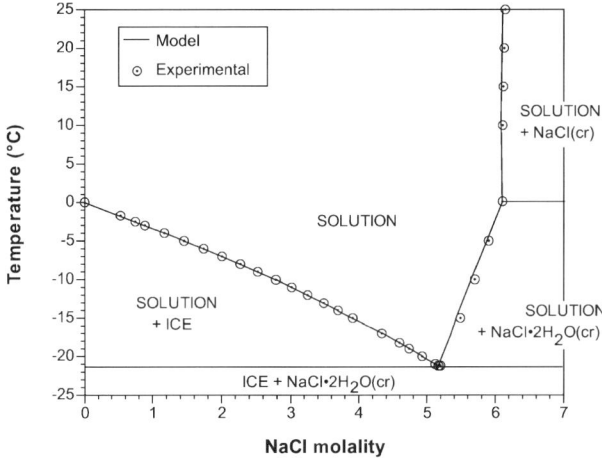

Fig. 3.4. Equilibria for NaCl solutions between 25 °C and the eutectic

Na_2SO_4 parameterization to higher concentrations relying on the properties of supersaturated solutions (Marion 2002). For relatively insoluble solid phases such as $CaSO_4$ and $CaCO_3$, we did not include maximum concentrations in Table 3.2 because parameterization of these insoluble compounds are not limited by their concentrations per se.

The only compounds in Table 3.2 that do not form solid phases within our model are $Mg(NO_3)_2$ and $Ca(NO_3)_2$. On Earth, common nitrate salts such an $NaNO_3$ and KNO_3 typically form in arid, alkaline environments. Under these environmental conditions, Mg and Ca concentrations are low because of the insolubility of their respective carbonate minerals. $Mg(NO_3)_2$ and $Ca(NO_3)_2$ Pitzer-equation parameters were added to the model to account for trace concentrations of Mg and Ca in such nitrate environments. It would be a serious misuse of the model to calculate solution properties in systems where $Mg(NO_3)_2$ and $Ca(NO_3)_2$ are present at high concentrations.

In general, chloride and sulfate salts are parameterized to lower temperatures than their single salt eutectics. For example, NaCl is parameterized to −45 °C (Table 3.2), and the eutectic of $NaCl \cdot 2H_2O$ is −21.3 °C (Fig. 3.4). $MgSO_4$ is parameterized to −20 °C (Table 3.2), and the eutectic of $MgSO_4 \cdot 12H_2O$ is −3.6 °C (Marion and Farren 1999). These extensions below the pure salt eutectics are due to parameterizations based on mixed salt solutions, which have lower eutectics than their component pure salts. These lower temperature ranges are quite useful in exploring chemical equilibria at subzero temperatures. An example of a salt whose lower temperature is the eutectic is $FeSO_4$ at −2 °C. Extension of $FeSO_4$ chemistry to temperatures below −2 °C needs to be done with caution.

In most cases, the lower temperature limit (Table 3.2) is based on data used to parameterize or validate the model. In a few cases, the temperature range is given as "≈ 0 to 25 °C." There cases are based on a measured value for the equilibrium constant at 25 °C and extrapolation using a temperature function. For example, the van't Hoff equation (Drever 1997)

$$\ln(K_{T_2}) = \ln(K_{T_1}) + \frac{\Delta H_r^0}{R}\left[\frac{1}{T_1} - \frac{1}{T_2}\right] \quad (3.21)$$

was used in several of these cases, where ΔH_r^0 is the change in the standard enthalpy of reaction and R is the gas constant. The accuracy of these extrapolations to lower temperatures is difficult to judge. But Plummer et al. (1988) claim that the temperature range for their PHRQPITZ model is generally 0 to 60 °C, if ΔH_r^0 is known. And Pitzer (1995) has pointed out that ΔH_r^0 is often nearly constant over a limited temperature range.

Figure 3.4 exemplifies the type of chemical equilibria that are at the core of the FREZCHEM model. Both equilibria with ice and minerals are considered. Model calculations for NaCl solutions place the peritectic, where NaCl(cr) and NaCl·2H$_2$O(cr) are in equilibrium, at 0.1 °C and 6.096 m, in excellent agreement with experimental measurements of 0.1 °C and 6.096 m (Linke 1965). Similarly, model calculations of the eutectic are −21.3 °C and 5.175 m, in excellent agreement with experimental measurements of −21.2 °C and 5.168 m (Hall et al. 1988). A different type of chemical system is exemplified by H$_2$SO$_4$ (Fig. 3.5), where the only solid-phase chemical equilibria are at subzero temperatures. In this case, the model predicts a eutectic at −62.0 °C and 5.68 m, which is in excellent agreement with experimental measurements of −62.0 °C and 5.68 (Linke 1965).

Introducing pressure into the FREZCHEM model necessitates quantifying volumetric properties of ions in solution and solids in order to calculate the pressure dependence of K (Eq. 2.29), γ (Eq. 2.87), and a_w (Eq. 2.90). Figures 3.6 and 3.7 depict the molar volumes and compressibilities of ions

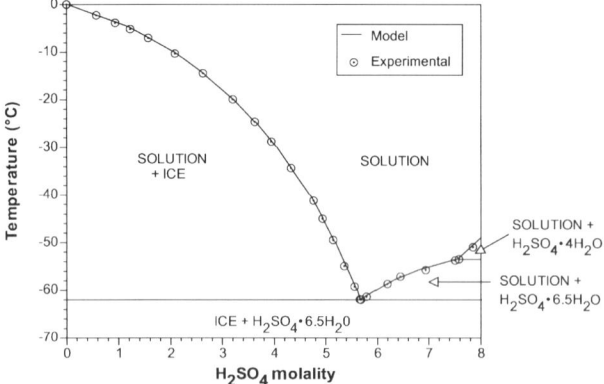

Fig. 3.5. Equilibria for H$_2$SO$_4$ solutions at subzero temperatures. Reprinted from Marion et al. (2002) with permission

3.3 Chemistries and Their Temperature and Pressure Dependence 35

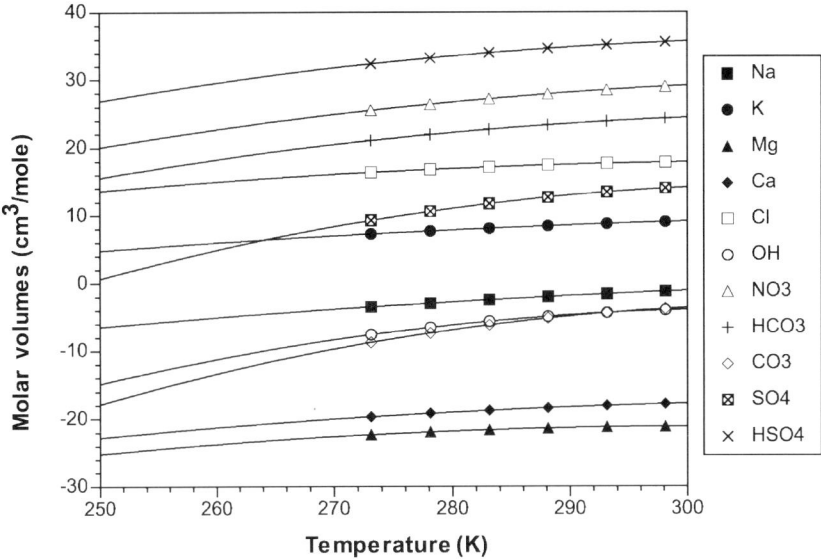

Fig. 3.6. Molar volumes of ions at infinite dilution as a function of temperature. *Symbols* are from Millero (1983). *Lines* are model estimates

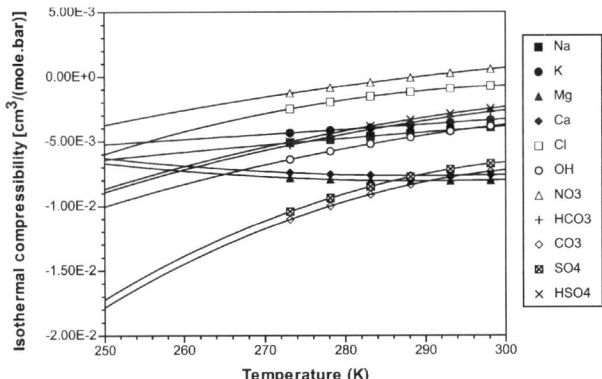

Fig. 3.7. Isothermal compressibility of ions at infinite dilution as a function of temperature. *Symbols* are from Millero (1983). *Lines* are model estimates

as functions of temperatures (Millero 1983). Equation parameters are given in Appendix B. Use of these relations at subzero temperatures is based on extrapolations. These equations were used to characterize NaCl and $CaCl_2$ solution densities to temperatures as low as 237 K (Marion et al. 2005).

In applications of the FREZCHEM model to pressure, the assumption is made that solid phases (Table 3.1), other than ice, are incompressible; therefore, in applying Eq. 2.29, only a constant molar volume for solids is used

(Appendix B, Table B.9). Justification for this assumption will be discussed in Sect. 3.6. Of the 58 solid phases in the model (Table 3.1), we were able to estimate the molar volumes for 54; only the molar volumes of four acid salts [$K_8H_6(SO_4)_7 \cdot H_2O$, $K_5H_3(SO_4)_4$, $K_3H(SO_4)_2$, and $Na_3H(SO_4)_2$] are missing in our database.

Equations 2.87 (activity coefficient), 2.88 (density), and 2.90 (activity of water) are all indirectly dependent on the temperature and pressure dependence of $B^{(0)v}$, $B^{(1)v}$, $B^{(2)v}$, and C^v (Eqs. 2.76, 2.80, and 2.81). Table B.10 (Appendix B) lists the temperature dependence of these volumetric Pitzer parameters. The pressure dependence of these parameters were evaluated with the density equation (Eq. 2.88). All three terms in the denominator of Eq. 2.88 are temperature and pressure dependent. The density of pure water (ρ^0) as a function of temperature and pressure is evaluated with Eqs. 3.14–3.16 and 3.20. Similarly, the molar volume of ions as a function of temperature and pressure is calculated by

$$\bar{V}_i^P = \bar{V}_i^0 - \bar{K}_i^0(\Delta P), \tag{3.22}$$

where $\bar{V}_i^P$ is the molar volume at T and P, $\bar{V}_i^0$ is the molar volume at T and $P = 1.01$ bars, and $\bar{K}_i^0$ is the compressibility at T and $P = 1.01$ bars. This equation assumes that $\bar{K}_i^0$ is independent of pressure. The final term in Eq. 2.88 is V^{ex}, which is a function of $B^{(0)v}$, $B^{(1)v}$, $B^{(2)v}$, and C^v. To simplify the calculations, we assume that only $B^{(1)v}$ and C^v are pressure dependent:

$$B_P^{(1)v} = B_0^{(1)v} - \bar{K}_0^{B^{(1)v}}(\Delta P) \tag{3.23}$$

$$C_P^v = C_0^v - \bar{K}_0^{C^v}(\Delta P). \tag{3.24}$$

The pressure coefficients are listed in Table B.11 (Appendix B).

The density database for the six chemistries of Table B.11 is limited with respect to compositions, temperature, and pressure. Furthermore, not all datasets are compatible. For example, the Chen et al. (1980) database for $MgSO_4$ covers a temperature range of 0 to 50 °C but only has a maximum concentration of 0.25 m. Extrapolating Pitzer-equation parameters based on this low concentration database to the higher concentrations of the Hogenboom et al. (1995) dataset (1.50 to 2.34 m) leads to ludicrous values for the Hogenboom density data. Fortunately, the converse is not true. That is, Pitzer paramerizations based on high concentrations extrapolate well to low concentrations. Figure 3.8 depicts our model fit to the Hogenboom dataset at 30 °C. In general, the fit is reasonable except at low pressures at 2.343 m. Three density data points from the Hogenboom dataset range from 1.244 to 1.246 g cm^{-3} at 1 bar pressure. Our model fit predicts 1.239 g cm^{-3}. An independent experimental estimate derived from the dataset of Lo Surdo et al. (1982) gives 1.241 g cm^{-3}. More important than this minor discrepancy is the fact that the Pitzer parameters derived from the Hogenboom dataset

3.3 Chemistries and Their Temperature and Pressure Dependence

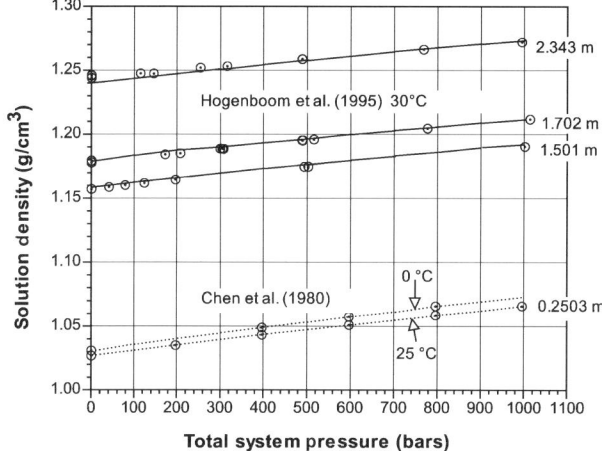

Fig. 3.8. The density of $MgSO_4$ solutions as a function of pressure. *Symbols*: experimental data; *lines*: model estimates

at 30 °C do extrapolate well to lower concentrations and temperatures; the fit to the Chen et al. (1980) dataset is excellent (Fig. 3.8).

3.3.3 Gas/Solution Phase Equilibria

In addition to equilibria for solid phases (Table 3.2) in the FREZCHEM model, there are also critical equilibria that control interactions among gas and aqueous phases (Table 3.3).

Seven of these reactions are Henry's law (K_H) reactions (first seven entries in Table 3.3) that control equilibria between the gas and aqueous phases, which for a constituent "c" is given by

$$K_H = \frac{a_{c,(aq)}}{f_{c(g)}} = \frac{(\gamma_c)(m_c)}{(\phi_c)(P_c)}, \qquad (3.25)$$

where a is the activity in the aqueous phase, f is the fugacity in the gaseous phase, γ is the activity coefficient of the aqueous species, m is the molal concentration, ϕ is the fugacity coefficient, and P is the gas partial pressure. For CO_2 and CH_4 in the FREZCHEM model, fugacity coefficients (ϕ) are explicitly calculated because these constituents can be present at high pressures where gas hydrates form. See the following section for a discussion of gas hydrate chemistries. For the remaining five Henry's law equations (H_2O, O_2, HCl, HNO_3, H_2SO_4), ϕ is assumed equal to unity and Eq. 3.25 simplifies to

$$K_H = \frac{(\gamma_c)(m_c)}{(P_c.)} \qquad (3.26)$$

Table 3.3. Temperature range and primary references of gas and solution phase equilibria in FREZCHEM model[a]

Reaction	Temperature range (°C)	References
$H_2O(g) \leftrightarrow H_2O(l)$	−93 to 25	Marion 2002
$CO_2(g) \leftrightarrow CO_2(aq)$	0 to 100	Plummer and Busenberg 1982; Marion 2001; He and Morse 1993
$O_2(g) \leftrightarrow O_2(aq)$	0 to 100	Clegg and Brimblecombe 1990b; Marion 2001
$CH_4(g) \leftrightarrow CH_4(aq)$	0 to 250	Duan et al. 1992a; Marion et al. 2006
$HCl(g) \leftrightarrow HCl(aq)$	−90 to 25	Marion 2002; Carslaw et al. 1995
$HNO_3(g) \leftrightarrow HNO_3(aq)$	−60 to 120	Marion 2002; Clegg and Brimblecombe 1990a
$H_2SO_4(g) \leftrightarrow H_2SO_4(aq)$	−60 to 25	Marion 2002; AIM[b]
$Mg^{2+}+H_2O \leftrightarrow MgOH^+ + H^+$	≈0 to 25	Plummer et al. 1988; Marion 2001
$Fe^{2+}+H_2O \leftrightarrow FeOH^+ + H^+$	≈0 to 25[c]	Marion et al. 2003a
$Ca^{2+}+CO_3^{2-} \leftrightarrow CaCO_3^0$	0 to 25	Plummer et al. 1988
$Mg^{2+}+CO_3^{2-} \leftrightarrow MgCO_3^0$	0 to 25	Plummer et al. 1988
$FeCO_3^0 \leftrightarrow Fe^{2+}+CO_3^{2-}$	≈0 to 25[c]	Marion et al. 2003a
$H_2O \leftrightarrow H^+ + OH^-$	0 to 50	Marion 2001
$CO_2 + H_2O \leftrightarrow H^+ + HCO_3^-$	0 to 100	Plummer and Busenberg 1982; Marion 2001
$HCO_3^- \leftrightarrow H^+ + CO_3^{2-}$	0 to 100	Plummer and Busenberg 1982; Marion 2001
$HSO_4^{-1} \leftrightarrow H^+ + SO_4^{2-}$	−93 to 55	Marion 2002; Clegg et al. 1994
$2H_2O(l) \leftrightarrow 2H_2(g) + O_2(g)$	≈0 to 25[c]	Marion et al. 2003a

[a] Lack of a phase designation in the reactions implies an aqueous phase (aq) species
[b] AIM model (http://www.aim.env.uea.ac.uk/aim/aim.htm)
[c] Based on a measured value and its temperature derivative at 25 °C

3.3 Chemistries and Their Temperature and Pressure Dependence

This assumption limits application of the latter chemistries to low pressures. Activity coefficients for aqueous-phase gases (CO_2, O_2, and CH_4) are calculated using the Pitzer equation for neutral species (Eq. 2.42). Activity coefficients for aqueous acids are calculated using the Pitzer equations for ions (Eqs. 2.40 and 2.41). For the case of HCl, the Henry's law constant is given by

$$K_H = \frac{(\gamma_H)(m_H)(\gamma_{Cl})(m_{Cl})}{(P_{HCl(g)})}. \qquad (3.27)$$

For the case of H_2O gas/solution equilibrium, the a_w (numerator of Eq. 3.25) is calculated directly in the model (Eq. 2.37).

There are five reactions that deal with ion associations (numbers 8 to 12 in Table 3.3). There are, of course, many more such associations in concentrated electrolyte solutions. But the Pitzer approach allows one to either explicitly identify an ion association (Table 3.3) or to implicitly include the interaction effect in the interaction coefficients (B, C, Ψ, Φ; Eqs. 2.40 and 2.41) (Pitzer, 1991, 1995); the latter is how most associations are dealt with in our model. In most cases, these ion associations are minor components of the aqueous phase.

There are four reactions that deal explicitly with the H^+ ion (Table 3.3): one is the dissociation constant for water (K_w), two are the first and second dissociation constants of carbonic acid (K_1 and K_2), and the fourth deals with the dissociation of the bisulfate (HSO_4^-) ion ($K_{\text{bisulfate}}$).

Incorporating carbonate chemistry into the FREZCHEM model necessitates an explicit recognition of pH. In the pH range from 4 to 12, the following charge balance exists for the cations and anions in solution:

$$[Na^+] + [K^+] + [MgOH^+] + [FeOH^+] + 2[Mg^{2+}] + 2[Ca^{2+}] + 2[Fe^{2+}]$$
$$= [HCO_3^-] + 2[CO_3^{2-}] + 2[SO_4^{2-}] + [Cl^-] + [NO_3^-] + [OH^-], \qquad (3.28)$$

where brackets refer to concentrations. This equation ignores the insignificant contribution of $[H^+]$ to charge balance in this pH range. For a system in equilibrium with solid-phase $CaCO_3$, Eq. 3.28 can be rewritten as

$$\frac{2(K_{sp})(\gamma_{H^+})^2[H^+]^2}{(\gamma_{Ca^{2+}})K_H K_1 K_2 P_{CO_2} a_w} + \Delta_i = \frac{K_H K_1 P_{CO_2} a_w}{(\gamma_{H^+})[H^+]\left(\gamma_{HCO_3^-}\right)}$$
$$+ \frac{2K_H K_1 K_2 P_{CO_2} a_w}{(\gamma_{H^+})^2[H^+]^2\left(\gamma_{CO_3^{2-}}\right)} + \frac{K_w a_w}{(\gamma_{OH^-})(\gamma_{H^+})[H^+]}, \qquad (3.29)$$

where

$$\Delta_i = [Na^+] + [K^+] + [MgOH^+] + [FeOH^+] + 2[Mg^{2+}] + 2[Fe^{2+}]$$
$$- 2[SO_4^{2-}] - [Cl^-] - [NO_3^-]. \qquad (3.30)$$

K_{sp} is the $CaCO_3$ solubility product, K_H is Henry's law constant for CO_2, K_1 and K_2 are respectively the first and second dissociation constants for carbonic acid, K_w is the dissociation constant for water, [H$^+$] is the hydrogen ion concentration, P_{CO2} is the partial pressure of CO$_2$, a_w is the activity of water, and the γ values are activity coefficients. Δ_i is the difference in non-calcium-carbonate species, which is determined by the existing FREZCHEM model. Given the P_{CO2}, activity coefficients and a_w from the Pitzer equations, equilibrium constants, and Δ_i, Eq. 3.29 can be solved by successive approximations for the [H$^+$], which is then used to control CaCO$_3$ solubility. For a system that is undersaturated with respect to CaCO$_3$, the Ca^{2+} concentration can be included in the Δ_i term (Eq. 3.30). Then a modified Eq. 3.29 is solved for [H$^+$] that is used to estimate HCO$_3^-$, CO$_3^{2-}$, and OH$^-$ concentrations needed to balance the charges in the solution phase. Similarly, for a Na, K, or Mg carbonate mineral, the Na, K, or Mg carbonate solubility product can be incorporated into Eq. 3.29 and other ions (e.g., Ca) incorporated into Eq. 3.30.

In strong acid cases (pH < 4), the FREZCHEM model is structured to input acidity directly by specifying the hydrogen (H) and acid anion (Cl, NO$_3$, SO$_4$) concentrations or indirectly by specifying the atmospheric concentrations of acids (P_{HCl}, P_{HNO3}, $P_{H_2SO_4}$), which are then equilibrated with the solution 67g67 phase. For a strongly acidic system (pH < 4), electrostatic charge balance requires that

$$[H^+] + [Na^+] + [K^+] + [MgOH^+] + [FeOH^+] + 2[Mg^{2+}]$$
$$+ 2[Ca^{2+}] + 2[Fe^{2+}] = [Cl^-] + [NO_3^-] + [HSO_4^-] + 2[SO_4^{2-}]. \quad (3.31)$$

This equation can be rewritten as

$$[H^+] + \Delta_i = \frac{K_{HCl} P_{HCl}}{\gamma_H \gamma_{Cl} [H^+]} + \frac{K_{HNO_3} P_{HNO_3}}{\gamma_H \gamma_{NO_3} [H^+]} + \frac{K_{H_2SO_4} P_{H_2SO_4}}{K_{HSO_4} \gamma_{HSO_4} \gamma_H [H^+]}$$
$$+ \frac{2 K_{H_2SO_4} P_{H_2SO_4}}{(\gamma_H)^2 \gamma_{SO_4} [H^+]^2}, \quad (3.32)$$

where

$$\Delta_i = [Na^+] + [K^+] + [MgOH^+] + [FeOH^+] + 2[Mg^{2+}] + 2[Ca^{2+}] + 2[Fe^{2+}], \quad (3.33)$$

which are "nonacid" cations estimated by the existing FREZCHEM model, and P_{HA} are acid gas partial pressures. The model calculates the activity coefficients (γ) and the equilibrium constants (K), and the P_{HCl}, P_{HNO_3}, and $P_{H_2SO_4}$ are gas partial pressures specified as input to the model. This leaves only the [H$^+$] concentration as an unknown that is then solved for by successive approximations. By this means, the model can use gas-phase properties to constrain the solution phase. On the other hand, if acid gas concentrations are not specified as input in strong acid cases, then the input of H and acid anions are used to estimate as output acid gas concentrations using the Henry's law constant (Eq. 3.27).

3.3 Chemistries and Their Temperature and Pressure Dependence

Only a few of the reactions summarized in Table 3.3 are actually based on data at subzero temperatures. In most cases, the lower temperature for data is 0 °C. This could potentially be a serious limitation for the FREZCHEM model. For example, quantifying carbonate chemistry requires specification of K_{H,CO_2}, K_1, K_2, and K_w; all of these reactions are only quantified for temperatures ≥ 0 °C (Table 3.3). Figure 3.9 demonstrates how six of the most important relationships of Table 3.3 extrapolate to subzero temperatures. We were able, based on these extrapolations, to quantify the solubility product of nahcolite ($NaHCO_3$) and natron ($Na_2CO_3 \cdot 10H_2O$) to temperatures as low as -22 °C (251 K) (Marion 2001). Even for highly soluble bicarbonate and carbonate minerals such as nahcolite and natron, their solubilities decrease rapidly with temperature (Marion 2001). For example, for a hypothetical saline, alkaline brine that initially was 4.5 m alkalinity at 25 °C, the final alkalinity at the eutectic at -23.6 °C was 0.3 m (Marion 2001). At least for carbonate systems it is not necessary to extrapolate much beyond about -25 °C to quantify this chemistry, which we believe can reasonably be done using existing equation extrapolations (Fig. 3.9).

The relationship used to quantify the bisulfate equilibria is given by

$$K_{\text{bisulfate}} = \frac{(a_{H^+})\left(a_{SO_4^{2-}}\right)}{\left(a_{HSO_4^-}\right)} \tag{3.34}$$

and is largely based on experimental data at temperatures ≥ 0 °C (Clegg et al. 1994; Clegg and Brimblecombe 1995). In the latter papers dealing with sulfuric acid chemistry, the assumption was made that their equation for the bisulfate relationship was valid across the entire temperature range of their work from 180 to 328 K and as such was used to develop other model param-

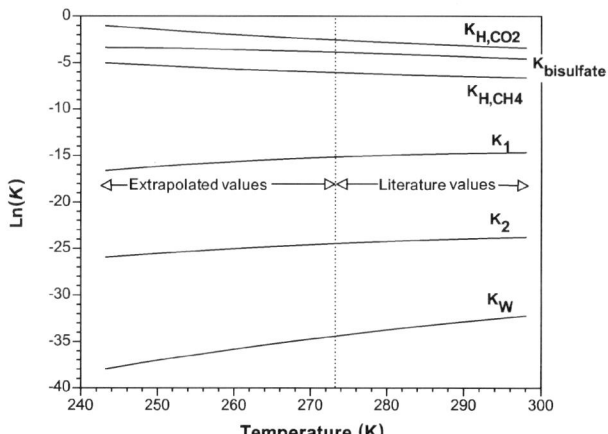

Fig. 3.9. Extrapolation of several key chemistries to temperatures < 273 K (0 °C)

eters. Because the molal sulfuric acid parameterization of the FREZCHEM model (Marion 2002) is based on the Clegg and Brimblecombe (1995) mole fraction model, we retained the assumption that our equation is also valid between 180 and 298 K.

The only important relationship from Table 3.3 that requires extrapolation at subzero temperatures that is not shown in Fig. 3.9 is K_{H,O_2}, which is very similar in magnitude to K_{H,CH_4} (Fig. 3.9). For the seven most important reactions in Table 3.3 that require extrapolation (Fig. 3.9 and $K_{H,O2}$), we simply extrapolated equations developed for temperatures $\geq 0\,°C$ into the subzero temperature range (Fig. 3.9). This anchors the extrapolation onto a solid foundation of work at temperatures $\geq 0\,°C$. None of these equations are linear, but all only show a modest change in slope with temperature. Nevertheless, the further the equations are extrapolated, the greater the uncertainty in their accuracy. Despite these uncertainties, we feel that extrapolations to temperatures as low as $-25\,°C$ are reasonable. And in the case of the bisulfate equilibrium that was used to parameterize sulfuric acid solutions, this relationship used with our sulfuric acid Pitzer-equation parameters is valid to 180 K.

In pressure applications of the FREZCHEM model, the molar volumes of the neutral species $CO_2(aq)$, $O_2(aq)$, and $CH_4(aq)$ are constants independent of temperature and pressure (Appendix B). This is similar to how solids are handled in the FREZCHEM model (see previous section). Of the gas-phase gases, only $CO_2(g)$ and $CH_4(g)$ at high pressures in gas hydrate equilibria are assumed to be compressible (see the following section on gas hydrate chemistry). That means that other gas constituents such as H_2O, O_2, HCl, HNO_3, and H_2SO_4 are only validly parameterized for low pressures (a few bars).

3.3.4 Gas Hydrate Equilibria

Two gas hydrates are now part of the FREZCHEM model: $CO_2·6H_2O$ and $CH_4·6H_2O$ (Table 3.2). Gas hydrate equilibrium for $CO_2·6H_2O$ is described by the reaction

$$CO_2·6H_2O(cr) \leftrightarrow CO_2(g) + 6H_2O(l), \quad (3.35)$$

which is terms of the solubility product is given by

$$K_{CO_2·6H_2O} = \frac{\left(f_{CO_2(g)}\right)(a_w)^6}{CO_2·6H_2O(cr)} = \left(\phi_{CO_2(g)}\right)\left(P_{CO_2(g)}\right)(a_w)^6, \quad (3.36)$$

where $CO_2·6H_2O$ on the right-hand side of Eq. 3.36 is assumed to be a pure solid phase with an activity of unity. Gas concentrations were used to define gas hydrate equilibria because gas pressures are the experimental measurements in gas hydrate studies (Sloan, 1998). An additional benefit of using

3.3 Chemistries and Their Temperature and Pressure Dependence

gas pressures directly in defining the solubility product is that it negates the necessity of having to extrapolate Henry's law constants into the subzero temperature range to estimate $CO_2(aq)$ (Fig. 3.9).

To estimate a gas hydrate solubility product requires knowing ϕ_g, P_g, and a_w (Eq. 3.36). The gas partial pressure, P_g, is experimentally measured. The activity of water, a_w, is calculated by the FREZCHEM model (Eq. 2.37), as is the gas fugacity coefficient (ϕ_g) using a model developed by Duan et al. (1992b). The equation used to calculate gas fugacity coefficients is given by

$$\ln \phi(T, P) = Z - 1 - \ln Z + \frac{B}{V_r} + \frac{C}{2V_r^2} + \frac{D}{4V_r^4} + \frac{E}{5V_r^5} + G, \quad (3.37)$$

where

$$Z = \frac{PV}{RT} = \frac{P_r V_r}{T_r}$$

$$= 1 + \frac{B}{V_r} + \frac{C}{V_r^2} + \frac{D}{V_r^4} + \frac{E}{V_r^5} + \frac{F}{V_r^2}\left(B + \frac{\gamma}{V_r^2}\right)\exp\left(\frac{-\gamma}{V_r^2}\right) \quad (3.38)$$

and Z is the compressibility factor that measures the departure of a real gas from an ideal gas ($Z = 1$). In these equations

$$B = a_1 + \frac{a_2}{T_r^2} + \frac{a_3}{T_r^3}, \quad (3.39)$$

$$C = a_4 + \frac{a_5}{T_r^2} + \frac{a_6}{T_r^3}, \quad (3.40)$$

$$D = a_7 + \frac{a_8}{T_r^2} + \frac{a_9}{T_r^3}, \quad (3.41)$$

$$E = a_{10} + \frac{a_{11}}{T_r^2} + \frac{a_{12}}{T_r^3}, \quad (3.42)$$

$$F = \frac{\alpha}{T_r^3}, \quad (3.43)$$

$$G = \frac{F}{2\gamma}\left[B + 1 - \left(B + 1 + \frac{\gamma}{V_r^2}\right)\exp\left(\frac{-\gamma}{V_r^2}\right)\right], \quad (3.44)$$

$$P_r = \frac{P}{P_c}, \quad (3.45)$$

$$T_r = \frac{T}{T_c}, \quad (3.46)$$

$$V_r = \frac{V}{V_c}, \quad (3.47)$$

and

$$V_c = \frac{RT_c}{P_c}. \quad (3.48)$$

Values of $a_1 - a_{12}$, α, B, γ, and critical temperatures (T_c) and pressures (P_c) for $CO_2(g)$ and $CH_4(g)$ are compiled in Appendix B, Table B.12. Given these parameters, it is possible to define all the terms in Eq. 3.38, except for V_r. In the FREZCHEM model, "Brent's" method (Press et al. 1992) is used to find the root of this equation (V_r), which is then used to calculate Z (Eq. 3.38), and finally ϕ (Eq. 3.37).

There is an abundance of experimental gas partial pressures for gas hydrate equilibria across a broad range of temperatures (Fig. 3.10; Sloan 1998). The lower temperature limit in our model database for these systems is 180 K (Fig. 3.10) because this is the lower limit of our model's ability to estimate a_w (Fig. 3.1, Eq. 3.11), which is needed to calculate the solubility product of gas hydrates (Eq. 3.36). In our model, the upper temperature limit for methane hydrate is at 298 K (25 °C), which is the upper temperature limit for FREZCHEM; the upper temperature limit for carbon dioxide hydrate is at 283 K (10 °C), which is the temperature where liquid $CO_2(l)$ becomes the thermodynamically stable phase.

Given gas partial pressures (Fig. 3.10), a model to calculate ϕ (Eq. 3.37), and a model to calculate a_w (Eq. 2.37), the solubility product (Eq. 3.36) can be calculated for gas hydrates (Fig. 3.11). The actual solubility product calculations are made at the experimental gas partial pressures, which vary widely (Fig. 3.10). Equation 2.29 was used to adjust all these pressure-dependent estimates (K^P) to a hypothetical 1.0 atm total pressure (K^{P0}), which is what is presented in Fig. 3.11.

Estimating solubility products for pure gas hydrates as outlined above is fairly straightforward. Application of this model to real-world situations is, however, generally more complicated because gases in natural environments are seldom pure gases. There are a number of models that can be used for gas mixtures (Duan et al. 1992a,b; Pitzer 1995). In this work, we assume that

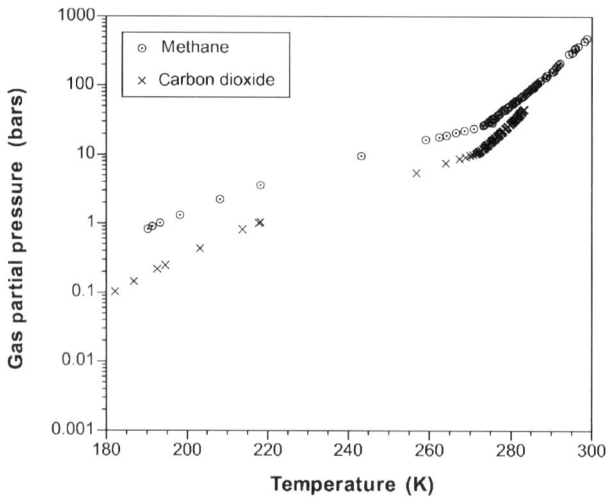

Fig. 3.10. Experimental data for methane and carbon dioxide hydrate equilibria. Reprinted from Marion et al. (2006) with permission

3.3 Chemistries and Their Temperature and Pressure Dependence   45

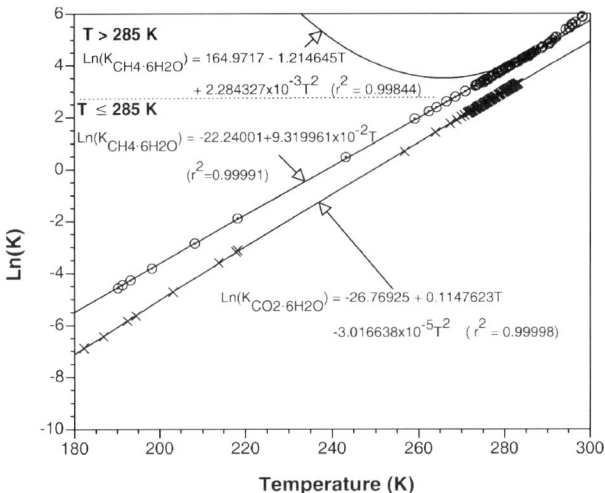

Fig. 3.11. Solubility products of $CH_4 \cdot 6H_2O$ and $CO_2 \cdot 6H_2O$ at a hypothetical 1 atm total pressure. Reprinted from Marion et al. (2006) with permission

Amagat's law of additive volumes holds for all pressures, which means that the ideal-solution law holds for the gaseous mixture, but not necessarily for the pure gases per se, and therefore the fugacity (f) is given by

$$f_i = x_i f_i^0 = x_i \phi_i^0 P_i^0, \qquad (3.49)$$

where x_i is the mole fraction, f_i^0 is the fugacity of the pure gas, and ϕ_i^0 is the fugacity coefficient at the temperature and total pressure (P_i^0) of the solution (Pitzer 1995). See Fig. 14 in Duan et al. (1992b) for partial justification of this model for CH_4-CO_2 gas mixtures.

Another mixture problem is how to deal with mixed solid-phase gas hydrates. Both methane and carbon dioxide form structure I gas hydrates. Thus CH_4-CO_2 gas mixtures will ultimately lead to CH_4-CO_2 gas hydrates. Equilibria for the simple gas hydrates can be represented by

$$K_{CH_4 \cdot 6H_2O} = \frac{(CH_{4(g)})(H_2O)^6}{(CH_4 \cdot 6H_2O)} \qquad (3.50)$$

and

$$K_{CO_2 \cdot 6H_2O} = \frac{(CO_{2(g)})(H_2O)^6}{(CO_2 \cdot 6H_2O)} \qquad (3.51)$$

where parentheses represent activities (or fugacities). For pure gas hydrates, the terms in the denominators of Eqs. 3.50 and 3.51 are equal to unity. This is not the case, however, for mixed gas hydrates. In this work, we assume that gas hydrate mixtures can be represented as solid solutions (Stumm and Morgan 1970). For the reaction

$$CO_2 \cdot 6H_2O + CH_{4(g)} \Leftrightarrow CH_4 \cdot 6H_2O + CO_{2(g)}, \qquad (3.52)$$

the equilibrium constant is given by

$$\frac{(CH_4 \cdot 6H_2O)(CO_{2(g)})}{(CO_2 \cdot 6H_2O)(CH_{4(g)})} = \frac{K_{CO_2 \cdot 6H_2O}}{K_{CH_4 \cdot 6H_2O}}. \tag{3.53}$$

The activity of the solid phases can be represented as

$$(X \cdot 6H_2O) = \gamma_{X \cdot 6H_2O} x_{X \cdot 6H_2O}, \tag{3.54}$$

where X is CH_4 or CO_2, γ is the activity coefficient of the solid phase, and x is the mole fraction in the solid phase. Substituting Eqs. 3.54 into Eq. 3.53 and rearranging terms leads to

$$\frac{[x_{CH_4 \cdot 6H_2O}][\gamma_{CH_4 \cdot 6H_2O}]}{[x_{CO_2 \cdot 6H_2O}][\gamma_{CO_2 \cdot 6H_2O}]} = \frac{K_{CO_2 \cdot 6H_2O}(CH_{4(g)})}{K_{CH_4 \cdot 6H_2O}(CO_{2(g)})}. \tag{3.55}$$

The terms on the right-hand side of this equation are known (Fig. 3.11) or calculated by the model (gas fugacity $= \phi_{(g)} P_{(g)}$) (see previous discussions). If we assume an ideal solid solution ($\gamma_{CH_4 \cdot 6H_2O} = \gamma_{CO_2 \cdot 6H_2O}$), then Eq. 3.55 simplifies to

$$\frac{[x_{CH_4 \cdot 6H_2O}]}{[x_{CO_2 \cdot 6H_2O}]} \approx \frac{K_{CO_2 \cdot 6H_2O}(CH_{4(g)})}{K_{CH_4 \cdot 6H_2O}(CO_{2(g)})}, \tag{3.56}$$

which can be used directly to estimate the fraction of CH_4 and CO_2 in the gas hydrates ($x_{CH_4 \cdot 6H_2O} + x_{CO_2 \cdot 6H_2O} = 1.0$). Then these mole fractions can be substituted into the denominator of Eqs. 3.50 and 3.51 to convert these pure gas equations into mixed gas equations. Or, alternatively, we could estimate $\gamma_{CH_4 \cdot 6H_2O}$ and $\gamma_{CO_2 \cdot 6H_2O}$ in Eq. 3.55 to get an improved model. Both of these approaches (Eqs. 3.55 and 3.56) will be discussed in the following text.

We tested the stability of mixed gas hydrates with experimental data from Adisasmito et al. (1991), Dholabhai and Bishnoi (1994), and Dholabhai et al. (1997) for mixed CH_4-CO_2 gas hydrates. Equation 3.56 gives estimates of mole fractions that are used with Eqs. 3.50 or 3.51 to estimate equilibrium gas pressures for mixed CH_4-CO_2 solutions assuming an ideal solid solution ($\gamma_{CH_4 \cdot 6H_2O} = \gamma_{CO_2 \cdot 6H_2O}$). This model overestimates the experimental measurements by 1 to 43% (Fig. 3.12). On the other hand, Eq. 3.55 in Fig. 3.12 is a fit of our model to the experimental data. In this case, we set $\gamma_{CO_2 \cdot 6H_2O} = 1.00$ and adjusted $\gamma_{CH_4 \cdot 6H_2O}$ to optimize the fit to the experimental data. Figure 3.13 shows estimates of $\gamma_{CH_4 \cdot 6H_2O}$ as a function of $x_{CH_4(g)}$ and temperature. Within each of the eight datasets, the $x_{CH_4(g)}$ values are similar, and the numerical designation in the legend of Fig. 3.13 represents the average value. For the top four datasets [high $x_{CH_4(g)}$], it was possible to derive independent equations (Fig. 3.13, Table 3.4); for the bottom four datasets [low $x_{CH_4(g)}$] the data showed such scatter (Fig. 3.13) that we fit a single equation to these four datasets (Table 3.4).

3.3 Chemistries and Their Temperature and Pressure Dependence   47

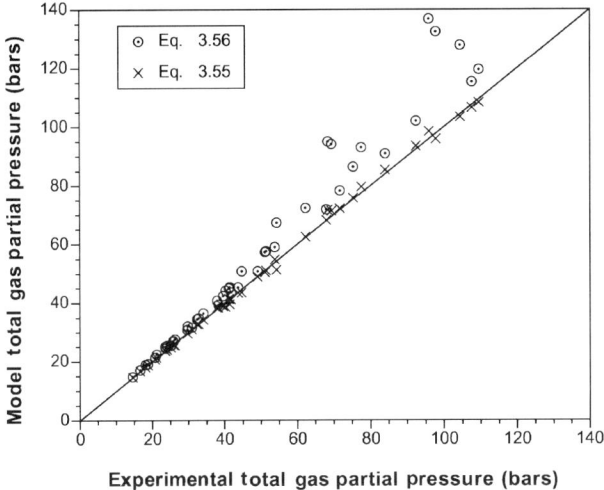

Fig. 3.12. Comparison of two models for estimating the total gas pressure of mixed CH_4-CO_2 gas hydrates. Reprinted from Marion et al. (2006) with permission

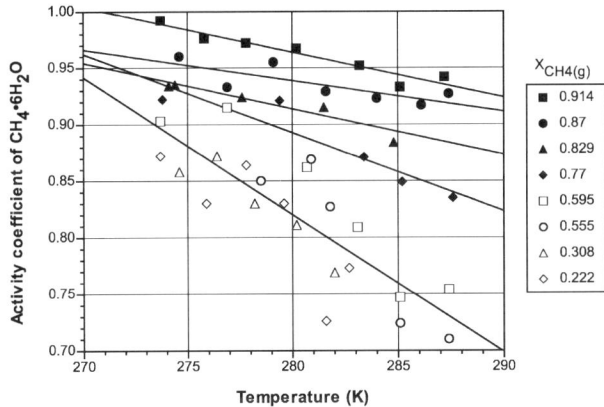

Fig. 3.13. The relationship between the activity coefficient of $CH_4 \cdot 6H_2O$ and the mole fraction of $CH_4(g)$ and temperature. *Symbols* are data; *lines* are model fits (see Table 3.4 for equations). Reprinted from Marion et al. (2006) with permission

Note that as $x_{CH_4(g)}$ approaches 1.00 [pure $CH_4(g)$ system)], $\gamma_{CH_4 \cdot 6H_2O}$ approaches 1.00 as it should theoretically for a pure solid phase. To estimate $\gamma_{CH_4 \cdot 6H_2O}$ at an intermediate $x_{CH_4(g)}$ from the equations in Table 3.4, one can generate estimates above and below the desired $x_{CH_4(g)}$ and linearly interpolate between them. Applying these equations to estimate $\gamma_{CH_4 \cdot 6H_2O}$ (with $\gamma_{CO_2 \cdot 6H_2O} = 1.0$) and using Eq. 3.55 leads to a great improvement in the model estimate of total gas pressures for mixed CH_4-CO_2 hydrates

Table 3.4. Equations relating the activity coefficient of $CH_4 \cdot 6H_2O$ to the mole fraction of CH_4 in the gas phase and temperature (K) (Fig. 3.13). Reprinted from Marion et al. (2006) with permission

$X_{CH_4(g)}$	Equation	r^2
1.000	$\gamma_{CH_4 \cdot 6H_2O} = 1.000$	
0.914	$\gamma_{CH_4 \cdot 6H_2O} = 2.07740 - 3.977471 \times 10^{-3}T$	0.920
0.870	$\gamma_{CH_4 \cdot 6H_2O} = 1.70079 - 2.722018 \times 10^{-3}T$	0.634
0.829	$\gamma_{CH_4 \cdot 6H_2O} = 2.04624 - 4.044674 \times 10^{-3}T$	0.869
0.770	$\gamma_{CH_4 \cdot 6H_2O} = 2.83506 - 6.937219 \times 10^{-3}T$	0.868
≤ 0.595	$\gamma_{CH_4 \cdot 6H_2O} = 4.22891 - 1.2173844 \times 10^{-2}T$	0.688

(Fig. 3.12); only 5 of 48 comparisons are off by more than 3%. Using Eq. 3.56 with $\gamma_{CH_4 \cdot 6H_2O} = \gamma_{CO_2 \cdot 6H_2O}$ leads to errors as large as 43% (Fig. 3.12).

The system of equations in Table 3.4 can be simplified because both the intercepts and slopes of these equations are highly correlated to $x_{CH_4(g)}$. The linear equation

$$\gamma_{CH_4 \cdot 6H_2O} = a + bT, \qquad (3.57)$$

with $a = 1.00 + 7.998(1.00 - x_{CH_4(g)})$ and $b = -0.03021(1.00 - x_{CH_4(g)})$, can be written as

$$\gamma_{CH_4 \cdot 6H_2O} = 1.00 + (7.998 - 0.03021T)(1.00 - x_{CH_4(g)}). \qquad (3.58)$$

We can test these two alternative models (Table 3.4 equations or Eq. 3.58) for estimating $\gamma_{CH_4 \cdot 6H_2O}$ by rewriting Eq. 3.55 as

$$\frac{x_{CH_4 \cdot 6H_2O}}{x_{CO_2 \cdot 6H_2O}} = \frac{K_{CO_2 \cdot 6H_2O}(CH_4(g))(\gamma_{CO_2 \cdot 6H_2O})}{K_{CH_4 \cdot 6H_2O}(CO_2(g))(\gamma_{CH_4 \cdot 6H_2O})} = K_D. \qquad (3.59)$$

The gas hydrate equilibrium constants (K) are known (Fig. 3.11), the gas fugacities are calculated with Eq. 3.49, $\gamma_{CO_2 \cdot 6H_2O} = 1.0$ (assumed), and $\gamma_{CH_4 \cdot 6H_2O}$ is calculated with the equations in Table 3.4 (or Eq. 3.58). Given that $x_{CH_4 \cdot 6H_2O} + x_{CO_2 \cdot 6H_2O} = 1.0$, the mole fractions of gas hydrates can be calculated by

$$x_{CO_2 \cdot 6H_2O} = 1.0/(1.0 + K_D) \qquad (3.60)$$

and

$$x_{CH_4 \cdot 6H_2O} = K_D/(1.0 + K_D). \qquad (3.61)$$

For the case where $x_{CH_4(g)} = 0.770(x_{CO_2(g)} = 0.230)$, $T = 280\,K$, $P = 41.1$ bars (equilibrium total gas pressure for a mixed CO_2-CH_4 gas hydrate), the calculated $x_{CH_4 \cdot 6H_2O} = 0.663$ for both Table 3.4 equations and Eq. 3.58, and $x_{CO_2 \cdot 6H_2O} = 0.337$. For another case where $x_{CH_4(g)} = 0.265$ ($x_{CO_2(g)} = 0.735$), $T = 280\,K$, $P = 31.3$ bars (equilibrium total gas pressure for a mixed CO_2-CH_4 gas hydrate), the calculated $x_{CH_4 \cdot 6H_2O} = 0.179$ (Table 3.4 equations) and 0.180 (Eq. 3.58) and the corresponding $x_{CO_2 \cdot 6H_2O} = 0.821$ and

0.820. Both Table 3.4 equations and the simpler Eq. 3.58 give, within round-off error, identical results when used with Eq. 3.59. In both of the above hypothetical cases, $x_{CH_4 \cdot 6H_2O}/x_{CH_4(g)} < 1.0$ and $x_{CO_2 \cdot 6H_2O}/x_{CO_2(g)} > 1.0$, which means that CO_2 is preferentially incorporated into mixed gas hydrates. This is probably largely due to the greater solubility of CO_2(aq) compared to CH_4(aq).

Note that both the equations in Table 3.4 and Eq. 3.58 will produce $\gamma_{CH_4 \cdot 6H_2O}$ values > 1 at subzero temperatures (extrapolation of equations in Fig. 3.13). As equilibrium gas pressure drops (and temperature drops), the system approachs ideal behavior (see Eq. 3.56 in Fig. 3.12). In programming the FREZCHEM model for these relationships, we set an upper bound of $\gamma_{CH_4 \cdot 6H_2O} = 1.0$ at subzero temperatures. Also all $x_{CH_4(g)}$ values ≤ 0.595 are assumed to fit a single equation (Table 3.4, Fig. 3.13) or are assigned a value of 0.595 in Eq. 3.58.

3.4 Mathematical Algorithms

3.4.1 The Sequential Approach

The original FREZCHEM model (FREZCHEM1) used a sequential approach to solve the nonlinear equations controlling chemical equilibria (Marion and Grant, 1994). Initially, if the solution is supersaturated with respect to ice, then the solution is equilibrated with ice (Fig. 3.14). This is done by increasing solute concentrations until the activity of water (a_w) is in equilibrium with the ice phase at a specified temperature and pressure (Eq. 3.11). The model then estimates the amount of ice that must have formed to concentrate solutes to the equilibrium point. Then the model calls a series of salt subroutines to test for mineral supersaturation and equilibrates if necessary. After mineral equilibration, the model again tests for ice equilibrium and iterates until ice and minerals are all in equilibrium with the solution and gaseous phases (Fig. 3.14). Obviously, if there are two chloride salts precipitating simultaneously, then equilibration of the first salt will throw the second salt out of equilibrium and vice versa. Nevertheless, the sequential iterative algorithm will ultimately converge onto a mathematical solution that satisfies all equilibria simultaneously; for exceptions, see the discussion of limitations where convergence occasionally fails. Adding a new chemistry to this model basically involves adding a new mineral subroutine; at present, there are 35 such mineral subroutines in FREZCHEM9. However, this mathematical algorithm has some convergence problems at high ionic strengths, at junctions where new phases begin to precipitate, and at temperatures below the eutectic where only solid phases can exist.

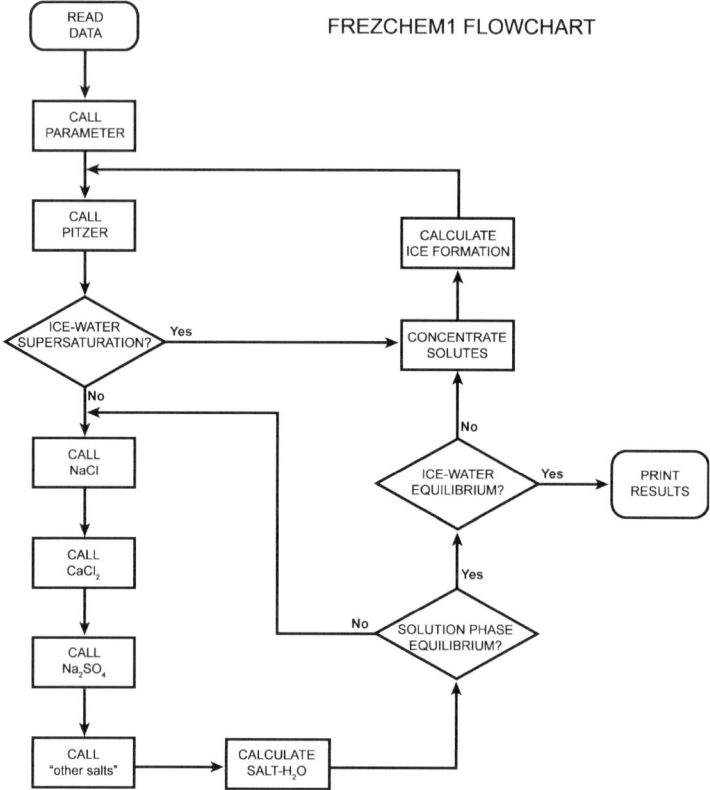

Fig. 3.14. Flowchart of FREZCHEM1 model

3.4.2 Gibbs Energy Minimization

The FREZCHEM2 model (Mironenko et al. 1997) uses the Gibbs energy minimization approach (Harvie et al. 1987) to compute chemical equilibria, which rectified at least some of the original model convergence problems. In this algorithm, local minimum is considered as an equilibrium composition of the system, in which all existing phases are specified before computation (Fig. 3.15). The Gibbs energy function of the system that contains M solids and aqueous solution (water and J species) is as follow:

$$g = \frac{G}{RT} = \sum_{k=1}^{M} \mu_k^0 n_k + \mu_w n_w + \sum_{j=1}^{J} \mu_j n_j, \qquad (3.62)$$

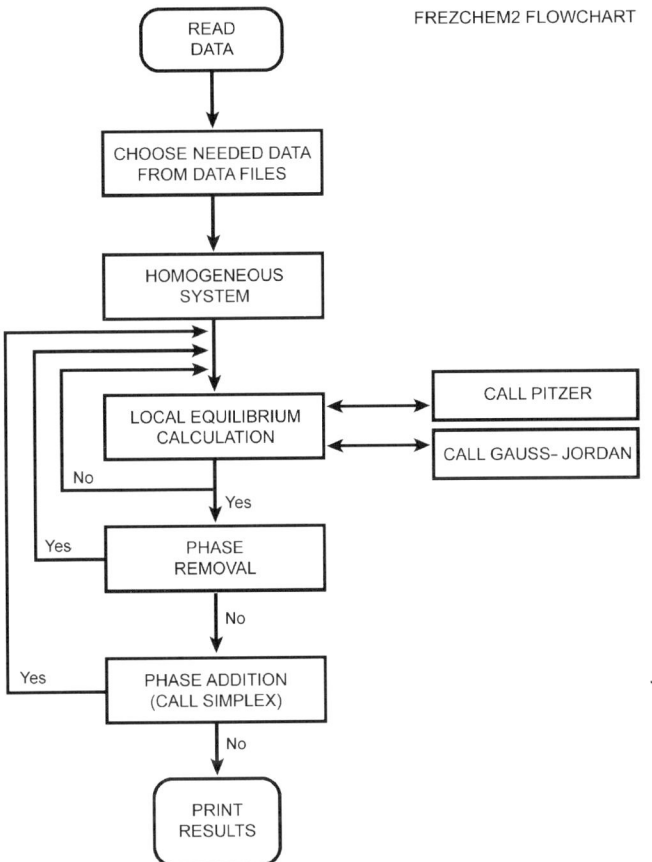

Fig. 3.15. Flowchart of FREZCHEM2 model. Reprinted from Mironenko et al. (1997) with permission

where

G = free energy of the system,
n = the molal quantity of components,
μ_j = the chemical potential of species j,
μ_w = the chemical potential of water,
μ_k^0 = the standard chemical potential of a one-component solid phase k,
R = the universal gas constant, and
T = absolute temperature (K).

This approach finds the minimum in the Gibbs energy of the system by removing and adding phases (Fig. 3.15) subject to mass balance constraints, which can be written as a system of linear equations:

$$\sum_{j=1}^{M+1+J} v_{ij}n_j = b_i, \quad i = 1, P, \tag{3.63}$$

where P is the number of independent chemical components in the system, v_{ij} is the number of moles of independent component i in one mole of component j, and b_i represents the number of moles of independent component i in the system (Mironenko et al. 1997). For an electroneutrality constraint, $b_i = 0$ and $v_{ij} = z_j$, where z_j is the charge of the jth component. The mathematics of this approach are considerably more sophisticated and complex (Harvie et al. 1987; Mironenko et al. 1997) than that used in FREZCHEM1 (see above). However, the rapid development of new chemistries in the model quickly overran the capacity of the FREZCHEM2 model. For example, the FREZCHEM1 and FREZCHEM2 models have 16 solid phases (ice, chloride salts, and sulfate salts). The FREZCHEM9 model has 58 solid phases (the 16 original plus additional sulfates, carbonates, acids, nitrates, acid-salts, a ferric mineral, and two gas hydrates) (Table 3.1). All model versions subsequent to FREZCHEM2 use the sequential approach of FREZCHEM1. This reversion to the sequential approach, which, at least on paper, is not as sound as the Gibbs energy minimization, reflects the mathematical comfort zones of the primary programmers. Currently, FREZCHEM10 is coming online (http://frezchem.dri.edu), which will reinstate the Gibbs energy minimization algorithm.

3.4.3 Other Mathematical Techniques

In this section we will discuss the specific mathematical techniques used to estimate chemical equilibria using the sequential approach, which is the foundation for all versions of the FREZCHEM model, except for versions 2 and 10 (see above). The techniques used to solve (find the roots of) the equilibrium relations can be grouped into three classes: simple one-dimensional (1-D) techniques, Brents method for more complex 1-D cases, and the Newton–Raphson technique that is used for both 1-D and multidimensional cases.

For a mineral such as carnallite ($KMgCl_3 \cdot 6H_2O$), the equilibrium relation can be written as

$$\frac{K_{KMgCl_3 \cdot 6H_2O}}{(\gamma_K)(\gamma_{Mg})(\gamma_{Cl})^3 a_w^6} = \left[K_T^+\right]\left[Mg_T^{2+}\right]\left[Cl_T^-\right]^3, \tag{3.64}$$

where K_T^+, Mg_T^{2+}, and Cl_T^- are the total ion concentrations in the solution plus carnallite phases. If the product on the right-hand side is greater than the left-hand side, the solution is supersaturated with carnallite. In that case,

Eq. 3.64 is rewritten as

$$\frac{K_{\text{KMgCl}_3\cdot 6\text{H}_2\text{O}}}{(\gamma_{\text{K}})(\gamma_{\text{Mg}})(\gamma_{\text{Cl}})^3 a_w^6} = \left[K_T^+ - X\right]\left[\text{Mg}_T^{2+} - X\right]\left[\text{Cl}_T^- - 3X\right]^3, \quad (3.65)$$

where X is the quantity that must be removed from the solution phase to bring this relationship into equilibrium with respect to carnallite. As a starting point, the minimum of the ion concentrations is selected and multiplied by a constant percentage

$$X_{i+1} = X_i + \min(K_T, \text{Mg}_T, \text{Cl}_T) \times 0.01, \quad X_0 = 0 \quad (3.66)$$

in an iterative ("DO") loop until X overcorrects for equilibrium. Then the percentage is changed to a lower value to refine the estimate. The multipliers used in Eq. 3.66 range from 0.01 (1.0%) to 0.00001 (0.001%). When the calculation converges onto an X value that satisfies the equilibrium in Eq. 3.65, the solution concentrations are adjusted (e.g., $K^+(\text{aq}) = K_T - X$) and the content of carnallite is X. Using this technique, a mass balance is preserved for all components. Most of the mineral equilibrium calculations in the model use this mathematical technique.

There are two simple variations on this 1-D theme that are used for ion associations. The ion-pair relation for $\text{CaCO}_3^0(X)$ can be represented as

$$\frac{K_{\text{CaCO}_3^0}}{(\gamma_{\text{Ca}^{2+}})(\gamma_{\text{CO}_e^{2-}})} = \frac{[\text{Ca}_T - X][\text{CO}_{3T} - X]}{X}, \quad (3.67)$$

where $\text{Ca}_T = \text{Ca}^{2+} + \text{CaCO}_3^0$ and $\text{CO}_{3T} = \text{CO}_3^{2-} + \text{CaCO}_3^0$. This relationship is solved for X using the quadratic equation

$$X = \frac{-b \pm \sqrt{b^2 - 4ac}}{2a}. \quad (3.68)$$

This approach is used for CaCO_3^0, MgCO_3^0, FeCO_3^0, and HSO_4^-. In principle, the quadratic equation could also have been used for mineral equilibria involving two solution species [e.g., NaCl(cr)]. However, the latter was not done in order to maintain a consistency in how mineral equilibria are calculated. For three or more separate solution-phase species (e.g., carnallite, Eq. 3.65), the quadratic equation does not work.

For MgOH^+ and FeOH^+, the equilibrium relationship is written as

$$\text{Fe}^{2+} + \text{H}_2\text{O} \leftrightarrow \text{FeOH}^+ + \text{H}^+, \quad (3.69)$$

with

$$K = \frac{(\text{FeOH}^+)(\text{H}^+)}{(\text{Fe}^{2+})(\text{H}_2\text{O})} \quad (3.70)$$

and
$$\frac{[\text{FeOH}^+]}{[\text{Fe}^{2+}]} = \frac{(K)(\gamma_{\text{Fe}^{2+}})(a_w)}{(\gamma_{\text{FeOH}^+})}. \tag{3.71}$$

All the terms on the right-hand side of the equation are known. By substituting Fe_T (known) $= [\text{Fe}^{2+}] + [\text{FeOH}^+]$ into the left-hand side of Eq. 3.71, we can solve for $[\text{Fe}^{2+}]$ and $[\text{FeOH}^+]$.

Brent's method (Press et al. 1992) is used to find the root of Eq. 3.38 (V_r). Some terms in the latter equation have V_r raised to the fifth power, which means that there could be as many as five roots to such an equation. This root-finding method does not require a derivative as the following Newton–Raphson method does, but it does require a lower and upper bracket within which the true solution must lie. In this particular case with

$$Z = \frac{P_r V_r}{T_r}, \tag{3.72}$$

we bracket the true solution with

$$X_{\text{lower}} = 0.001 \left(\frac{T_r}{P_r}\right) \tag{3.73}$$

and

$$X_{\text{upper}} = 1000 \left(\frac{T_r}{P_r}\right). \tag{3.74}$$

This is equivalent to allowing Z to vary between 0.001 and 1000. From our experience and the paper by Duan et al. (1992b), this range should be sufficient to bracket the true solution up to 2000 bars of pressure. From "Numerical Recipes" (Press et al. 1992), we used ZBRENT (implemented as a subroutine) to find the root and the function FUNC, which lists the function whose root we need (Eq. 3.38). Occasionally this mathematical algorithm has failed to converge properly.

The Newton–Raphson method (NRM) is used to find the roots of three chemical equilibria: Eq. 3.29 for finding the H^+ concentration of carbonate systems (one unknown), Eq. 3.32 for finding the H^+ concentration of acidic systems where the partial pressure of acids controls equilibria (one unknown), and the equations governing dolomite equilibria where there are four unknowns [H^+, Mg^{2+}, $MgCO_3^0$, and $CaMg(CO_3)_2(cr)$]. In the dolomite case, Ca^{2+} is calculated indirectly.

For the 1-D cases

$$X_{i+1} = X_i - \frac{f(X_i)}{f'(X_i)}, \tag{3.75}$$

where f is the function containing the unknown (X_i) [e.g., Eq. 3.29 with all terms moved to the left-hand side of the equality, which leads to $f(H^+) = 0$]. $f'(X_i)$ is the derivative of this function with respect to the unknown. The

NRM requires an initial estimate (X_0) of the unknown. If this initial estimate is far removed from the true value, then this method may not converge. In the case of the carbonate systems, the program requires an initial estimate of pH; if this guesstimate is not sufficiently close (≈ 0.5 to 1.0 pH units), then an error message will be printed or the program may diverge into a computer "black hole" and never exit because the program is stuck in an infinite loop.

If partial pressures of acids are specified as input, then Eq. 3.32 is solved for [H$^+$]. If the initial specification of [H$^+$] as input is far removed from an equilibrium calculation based on Eq. 3.32, then the NRM algorithm may not converge. For example, the calculated $P_{H_2SO_4}$ above a 1 m H_2SO_4 solution at 25 °C is 4.8×10^{-20} bars. Specifying 1.0×10^{-10} bars $P_{H_2SO_4}$ for an initial 1 m H_2SO_4 solution leads to an error message: "The H routine is not converging, acid." Specifying 1.0×10^{-20} bars $P_{H_2SO_4}$ for an initial 1 m H_2SO_4 solution leads to a new equilibrium concentration of 0.4908 m H_2SO_4. It is always important in the application of this and other models that the input properties be realistic.

For the multidimensional NRM case (dolomite),

$$\mathbf{J} \cdot \mathbf{dx} = -\mathbf{F}, \tag{3.76}$$

where the bolding represents vectors. $\mathbf{J}$ is the $n \times n$ Jacobian matrix where

$$J_{ij} = \frac{\partial F_i}{\partial x_j}, \tag{3.77}$$

$\mathbf{dx}$ is the $n \times 1$ solution matrix from which we obtain

$$x_{i+1} = x_i + \mathrm{d}x_i, \tag{3.78}$$

and $\mathbf{F}$ is an $n \times 1$ matrix of functions containing the n unknowns. As $\mathrm{d}x_i \to 0$, x_i approaches a solution and $F_i \to 0$. In this multidimensional case, Eq. 3.76 is a set of linear equations that are solved using the subroutines LUDCMP and LUBKSB from "Numerical Recipes" (Press et al. 1992). As was the case for the previous two NRM examples, there are no guarantees that this approach will always work. Instructions for coping with these potential problems are discussed in Appendix A.

The algorithm used to equilibrate ice, $CO_2 \cdot 6H_2O$, and $CH_4 \cdot 6H_2O$ with the solution phase uses aspects of several of the above techniques. First, the model calculates the fugacity coefficients for $CO_2(g)$ and $CH_4(g)$ using the approach outlined in Eqs. 3.36 to 3.48 and 3.72 to 3.74. Then the model calculates which of the phases – ice, $CO_2 \cdot 6H_2O$, or $CH_4 \cdot 6H_2O$ – is thermodynamically most stable by selecting the phase that minimizes the activity of water (a_w). The reaction,

$$CO_2 \cdot 6H_2O \leftrightarrow (\phi_{CO_2})(P_{CO_2})(a_w)^6, \tag{3.79}$$

can be arranged to yield

$$a_{w,\text{equilibrium}} = \left[\frac{K_{CO_2 \cdot 6H_2O}}{(\phi_{CO_2})(P_{CO_2})}\right]^{\frac{1}{6}}. \tag{3.80}$$

A similar equation can be written for $CH_4 \cdot 6H_2O$, and bearing in mind that at equilibrium with ice, $a_{w,\text{equilibrium}} = K_{\text{ice}}$ (Eq. 3.11). Then the solid phase that minimizes $a_{w,\text{equilibrium}}$ is selected. If the solution is supersaturated with this phase (i.e., $a_{w,\text{equilibrium}} < a_{w,\text{solution}}$), then the solution is equilibrated with the selected solid phase.

Note that this is an "either-or" decision. The model does not allow the simultaneous precipitation of more than one of these three phases, except in the case of a mixed gas hydrate. To estimate where two or more of these phases are, in fact, coprecipitating, you need to run the model across a range of temperatures and pressures to select the simultaneous equilibrium points. For example, for a solution where NaCl = 0.1 m with P_{CO2} = 10 bars, ice is the stable phase between -1.9 and $-2.2\,°C$; at $-2.3\,°C$, $CO_2 \cdot 6H_2O$ becomes the stable phase. More precisely, the transition occurs at $-2.26\,°C$, which is the simultaneous equilibrium point.

Assuming that the solution is supersaturated with respect to one of these three phases, it is necessary to increase the solute concentrations until $a_{w,\text{equilibrium}} = a_{w,\text{solution}}$. Increasing solute concentrations will cause a decrease in $a_{w,\text{solution}}$. The solute concentrations (X_j) are increased in a "DO" loop $(J = 1, n)$ by

$$X_{j,i+1} = X_{j,i} + X_{j,i} 0.01 \tag{3.81}$$

for all "n" solute species until $a_{w,\text{equilibrium}} = a_{w,\text{solution}}$. This technique is similar to that used for mineral equilibria (Eq. 3.66). A declining multiplier (0.01 to 0.00001) is used in Eq. 3.81 as the solution approaches equilibrium. At equilibrium, a new estimate of solution water is made:

$$H_2O_f = H_2O_i * \frac{\sum X_i}{\sum X_f}, \tag{3.82}$$

where f = the final and i = the initial concentration. Then ice moles = $H_2O_i - H_2O_f$ or hydrate moles = $(H_2O_i - H_2O_f)/6$.

3.5 Validation

The fit of model calculations to the experimental data (Figs. 3.2–3.5, 3.8, 3.11, 3.12) and the ability of the model to simulate critical equilibria such as peritectics and eutectics are a test of the internal consistency of the experimental data and the derived model. Because the special focus of this work is to develop a model for subzero temperatures, the fit of the model to data at subzero temperatures is especially important (Figs. 3.2 to 3.5). While these fits are encouraging, they are not, for the most part, validation. Validating the model requires a comparison of the model output to independent data. Of the data presented so far, only Figs. 3.2, 3.8, and 3.12 actually compare the model to some independent data not explicitly used in the parameterization. In what follows, we will examine several cases that compare model calculations to independent data and/or independent models. These examples will include seawater cases, an activity coefficient comparison, an iron chemistry comparison, volumetric (pressure) cases, and gas hydrate cases. For a more complete discussion of model validation, one can peruse the published papers (Marion and Grant 1994; Mironenko et al. 1997; Marion and Farren 1999; Marion 2001, 2002; Marion et al. 2003a, 2005, 2006).

Figure 3.16a depicts seawater freezing along the Ringer–Nelson–Thompson pathway, where calcium precipitates as antarcticite and the eutectic temperature is $-53.8\,°C$. Figure 3.16b depicts seawater freezing along the Gitterman pathway, where calcium precipitates as gypsum and the eutectic temperature is $-36.2\,°C$. In Chap. 5 on biogeochemical applications, these two alternative pathways for seawater freezing will be discussed in detail. For now, the point of these figures is how well the model can simulate the freezing of seawater (Fig. 3.16). There is also good agreement between model-calculated unfrozen water and experimental measurements during seawater freezing (Richardson 1976) at least down to $-36\,°C$ (Fig. 3.17). Below $-36\,°C$ the experimental and model estimates diverge but come back together at $-50\,°C$, where the model predicts 3.15 g water and the experimental measurement indicates 3.53 g water. The two curves diverge at the point where $MgCl_2 \cdot 12H_2O(cr)$ begins to precipitate at $-36\,°C$ (Fig. 3.16). Given major uncertainties in both model parameterization and experimental measurements, we cannot determine if the model or experimental measurements are more accurate at low temperatures. Nevertheless, in these cases, the experimental data are independent of model parameterization, which constitutes validation of the model. Given that a major objective of the FREZCHEM model is to simulate aqueous geochemical processes at subzero temperatures, Figs. 3.16 and 3.17 demonstrate that this can be done accurately.

Gitterman (1937) examined the solubility of $CaCO_3$ in seawater at subzero temperatures. Comparison of the Gitterman data with model-calcite predictions shows a major discrepancy (Fig. 3.18). Both the data and the model show an increasing solubility up to the point where ice begins to form at $-1.9\,°C$, then a steadily declining concentration at colder temperatures; the latter trend is caused by "salting out" of the CO_2 in solution due to increasing salinity as ice forms at colder temperatures. There are some problems with the Gitterman study. The $CaCO_3$ concentrations (alkalinity) of his first two data points are 0.110 and 0.114 g $CaCO_3$/kg (soln.), which are similar to the initial seawater concentration of $CaCO_3$ generally used in our model [0.112 g $CaCO_3$/kg (soln.) = 0.00232 m (alkalinity)]. Gitterman added nucleating crystals of $CaCO_3$ to seawater. The crystalline state of the $CaCO_3$

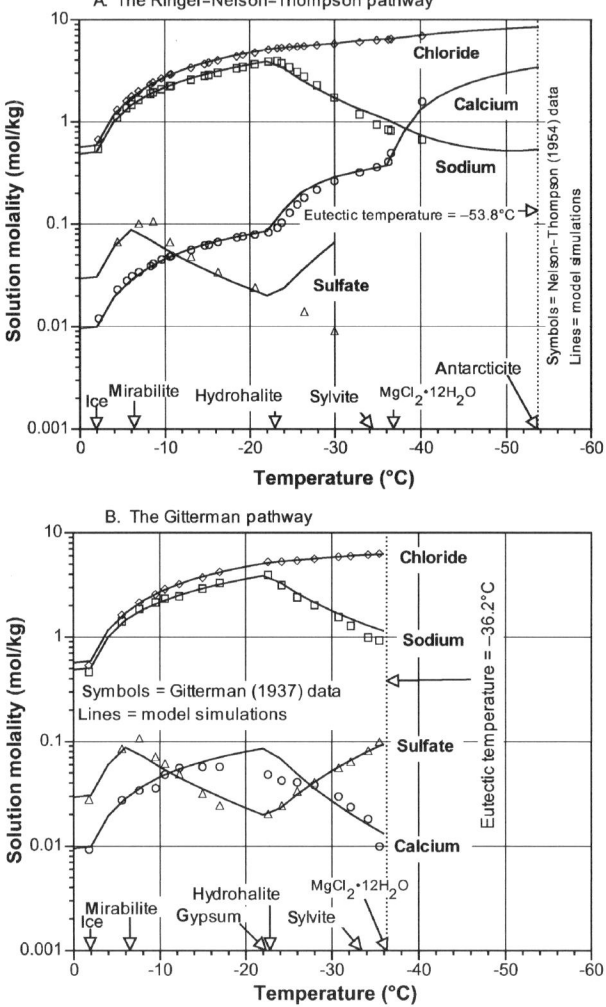

Fig. 3.16. **a** Ringer–Nelson–Thompson and **b** Gitterman pathways for seawater freezing. *Symbols* represent experimental data; *lines* are model calculations. Reprinted from Marion and Farren (1999) with permission

added (e.g., powder or large crystals), or even if the mineral was calcite, is not mentioned. The results suggest that this spiking had no effect at 0 and $-1.8\,°\mathrm{C}$; only at colder temperatures was there evidence of loss of $CaCO_3$ from solution (Fig. 3.18).

It is well known that seawater is supersaturated with respect to calcite in surface waters (Morse and Mackenzie, 1990). For example, at a seawater salinity = 35, $P_{CO_2} = 3.3\mathrm{e}{-4}$ atm, and alkalinity = 0.00240 equivalents kg^{-1}, Morse and Mackenzie (1990) estimated that seawater is 2.8- and 6.5-fold su-

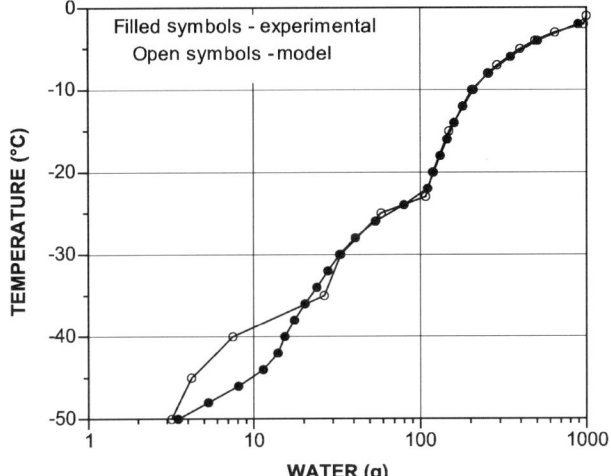

Fig. 3.17. Amount of unfrozen water during seawater freezing. Experimental measurements are from Richardson (1976). Reprinted from Marion and Grant (1994) with permission

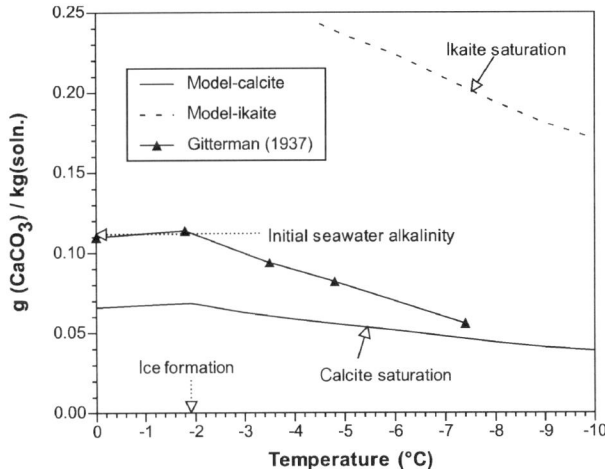

Fig. 3.18. Model and experimental estimates of $CaCO_3$ solubility at subzero temperatures. Reprinted from Marion (2001) with permission

persaturated with respect to calcite at 0 and 25 °C, respectively. Our model estimates that calcite in seawater at 0 and 25 °C are 2.8- and 6.7-fold supersaturated, in excellent agreement with the Morse–Mackenzie estimates. It is likely that in the Gitterman experiments seawater remained supersaturated with respect to calcite, at least during the initial stages of the freezing process. But as the freezing process continues, the experimental data approach the theoretical calcite saturation calculation (Fig. 3.18). Extending the Gitterman data points with a straight line leads to intersection with the model calcite saturation at $-8.8\,°C$. A fuller discussion of this equilibria is given in Chap. 5 on biogeochemical applications. As was the case for the previous examples (Figs. 3.16 and 3.17), the Gitterman data (Fig. 3.18) provide validation for the model.

Table 3.5 compares mean activity coefficients calculated with the FREZCHEM model with independent experimental measurements compiled by Robinson and Stokes (1970). "Independent" means that these specific data were not used in our model parameterizations, but similar data from other sources were used. The comparisons are in excellent agreement at 0.1 and 1.0 m, with the exception of $MgSO_4$. In that case, the trends are similar but the magnitudes are off. A footnote to Table 3.5 gives $\gamma\pm$ for $MgSO_4$ according to a recent paper by Archer and Rard (1998), which is in better agreement with our model calculations. The discrepancy at 0.1 m may reflect that solubility data were exclusively used for $MgSO_4$ parameterization (Marion and Farren, 1999). In general, both activity coefficient and solubility data were used in model parameterizations. In the $MgSO_4$ case, adding activity coefficient data to the parameterization database led to a poorer fit to the solubility data. As a consequence, we opted to optimize the solubility fit by discarding the $MgSO_4$ activity coefficient data (Marion and Farren 1999).

Table 3.5. Comparison of selected mean activity coefficients ($\gamma\pm$) between the experimental measurements compiled by Robinson and Stokes (1970) and the FREZCHEM model

Salt	Molality	Experimental[a] $\gamma\pm$	Model $\gamma\pm$
KCl	0.1	0.770	0.767
	1.0	0.604	0.603
HNO_3	0.1	0.791	0.788
	1.0	0.724	0.727
Na_2SO_4	0.1	0.452	0.452
	1.0	0.204	0.205
$Ca(NO_3)_2$	0.1	0.488	0.493
	1.0	0.338	0.341
$FeCl_2$	0.1	0.520	0.517
	1.0	0.508	0.506
$MgSO_4$[b]	0.1	0.150	0.184
	1.0	0.0485	0.0545

[a] From Robinson and Stokes (1970).
[b] The experimental measurements of Archer and Rard (1998) for $MgSO_4$ $\gamma\pm$ are 0.165 and 0.0544 at 0.1 and 1.0 molality, respectively.

The FREZCHEM model was basically designed for estimating the aqueous properties of concentrated electrolyte solutions, which is why we used the Pitzer approach. Nevertheless, it is still necessary that these models accurately describe dilute solutions. The comparisons in Table 3.5 demonstrate that the FREZCHEM model is reasonably accurate for both dilute and concentrated electrolyte solutions.

The paucity of ferrous iron data, especially of natural systems, made it difficult to validate the iron model (Marion et al. 2003a). However, Bernard and Symonds (1989) reported solution concentrations on the bottom of Lake Nyos (-200 m), where siderite ($FeCO_3$) is forming in sediments due to high concentrations of CO_2 and Fe^{2+} (Table 3.6). Note the unusually high P_{CO_2} of 4.17 atm; this accounts for the occasional catastrophic CO_2 gas bursts from Lake Nyos (Bernard and Symonds, 1989). Concentrations of Na^+, K^+, Mg^{2+}, Ca^{2+}, Fe^{2+}, Cl^-, SO_4^{2-}, alkalinity, and P_{CO_2} were used as input to the FREZCHEM model. With this input and assuming no solid phases (pure solution-phase model), the model predicts a carbonic acid concentration of 0.1485 m H_2CO_3 and a pH of 5.12 at equilibrium (Table 3.6). These values are in excellent agreement with independent model calculations of 0.1479 m H_2CO_3 and a pH of 5.13 made by Bernard and Symonds (1989) using the SOLVEQ program (Reed, 1982; Reed and Spycher, 1984). The FREZCHEM model predicts that this solution (Table 3.6) is supersaturated with siderite

Table 3.6. Chemical composition of Lake Nyos water at -200 m (Bernard and Symonds 1989). Reprinted from Marion et al. (2003a) with permission

Element	Concentration
Na^+ (m)	0.000776
K^+ (m)	0.000167
Mg^{2+} (m)	0.00257
Ca^{2+} (m)	0.000776
Fe^{2+} (m)	0.000977
Cl^- (m)	0.000017
SO_4^{2-} (m)	<0.000001
Alkalinity (m)	0.009572[a]
P_{CO_2} (atm)	4.17
H_2CO_3 (m)(calc.)[b]	0.1485
pH (calc.)[b]	5.12
Ionic strength (m)	0.0139
Temperature (°C)	23.3

[a] Adjusted from 0.01 to 0.009572 m to give a perfect solution charge balance. This ignores the small contribution of Mn^{2+} (0.000025).

[b] Calculated assuming a pure solution phase without minerals.

and no other carbonates, which agrees with Bernard and Symonds (1989). Allowing the solution to equilibrate with siderite causes the calculated pH to drop from 5.12 to 5.06 due to a small reduction in solution alkalinity as siderite precipitates.

Introducing a pressure dependence into the FREZCHEM model (Marion et al. 2005) necessitated the incorporation of volumetric properties, which allowed the model for the first time to estimate the density of aqueous solutions. Table 3.7 compares model calculated densities of mixed salt solutions at 1.01 bar total pressure derived in this work with experimental measurements and another model developed by Krumgalz et al. (1995). In our model calculations, we lumped the minor constituents Ba and Sr with Mg and Br with Cl, because Ba, Sr, and Br are not now part of the FREZCHEM model (Table 3.1). Our model estimates are in reasonably good agreement with the experimental measurements and the Krumgalz model estimates, with the exception of the Dead Sea comparison (Table 3.7). In that case, our model estimate is off by $0.004 \, \text{g cm}^{-3}$ from both the experimental value and the Krumgalz model value. It appears that this discrepancy is largely due to lumping of Ba and Sr with Mg and Br with Cl. If this lumping is done with the Krumgalz model, then their predicted density is $1.2238 \, \text{g cm}^{-3}$ (B. Krumgalz, pers. comm., 2003), in reasonably good agreement with our model estimate of $1.2224 \, \text{g cm}^{-3}$. The average discrepancy between our model and the experimental measurements is 0.21% (sign ignored) or 0.001% (sign considered). Errors of this magnitude are acceptable in the broad-scale FREZCHEM model.

Earlier we presented a model for calculating the density of salt solutions as a function of temperature and pressure (Eqs. 2.88, 3.22–3.24). Parameterization of this TP-dependent density model is based entirely on single-salt solutions (Table B.11). As validation, we compared the FREZCHEM model predictions of the density of seawater (mixed salt solution) as a function of temperature (0–25 °C) and gauge pressure (0–1000 bars) to a Gibbs thermodynamic potential model for seawater (Feistel 2003) (Fig. 3.19). The six data points at each pressure represent temperatures of 0, 5, 10, 15, 20, and 25 °C. The average discrepancies (sign ignored) between the two models at 0, 100, 200, 400, and 1000 bars of gauge pressure are 50, 42, 50, 56, and 280 ppm (Fig. 3.19). Only the fits at 1000 bars are significantly at variance. These discrepancies at 1000 bars may simply reflect the limited compositions, temperature, and pressure databases used to parameterize these chemistries for the FREZCHEM model; for example, only three of the six chemistries are parameterized with pressure data beyond 407 bars (Table B.11).

There is a small database where chemical equilibria have been measured experimentally as a function of pressure. For example, Hogenboom et al. (1995, 1999) have measured the eutectic temperatures as a function of pressure for the systems: $MgSO_4 \cdot 12H_2O$-Ice and $Na_2SO_4 \cdot 10H_2O$-Ice, respectively. For the Na_2SO_4 system, the model calculated eutectic temperatures

3.5 Validation 63

Table 3.7. Comparison of experimental and model calculated densities. Data are from Krumgalz et al. (1995), except for ρ_{model}(FREZCHEM). Reprinted from Marion et al. (2005) with permission

Solution property	Kara-Kul Lake	Orca Basin brine	Red Sea brine	Bannock II brine	Dead Sea
Na (moles kg^{-1})	1.0217	5.2955	5.4718	4.7150	1.8347
K (moles kg^{-1})	–	0.0214	0.0750	0.1420	0.2125
Mg(+Ba+Sr) (moles kg^{-1})	0.1765	0.0574	0.0546	0.7232	1.9480
Ca (moles kg^{-1})	0.0159	0.0362	0.1591	0.0183	0.4586
Cl(+Br) (moles kg^{-1})	0.7000	5.6110	5.9365	6.0475	6.8241
SO$_4$ (moles kg^{-1})	0.3090	0.0528	0.0113	0.1518	0.00558
Alkalinity (equiv. kg^{-1})	0.0117	–	0.0031	–	0.00599
Temperature (°C)	25	20	20	20	25
ρ_{expt} (g cm^{-3})	–	1.185	1.196	1.213	1.2262
ρ_{model} (g cm^{-3})	1.0625	1.1871	1.1991	1.2117	1.2263
ρ_{model} (FREZCHEM)(g cm^{-3})	1.0622	1.1876	1.1985	1.2116	1.2224

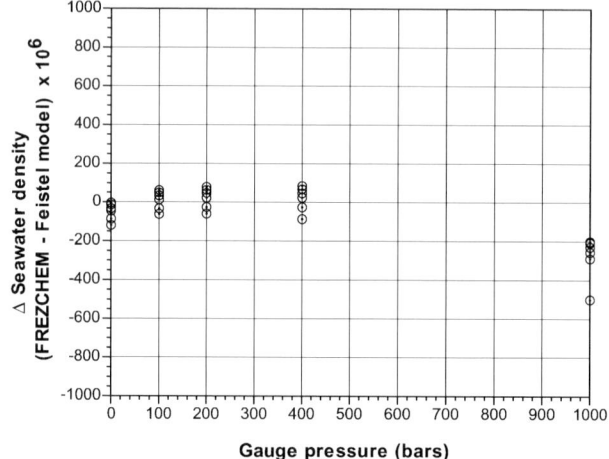

Fig. 3.19. Comparison of FREZCHEM model and Feistel (2003) model estimates of seawater density as a function of temperature and pressure

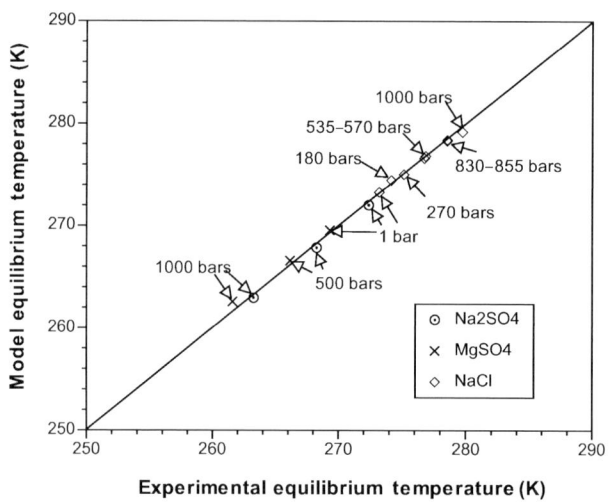

Fig. 3.20. Comparison of model-calculated and experimental equilibrium temperatures over a pressure range of 1 to 1000 bars. Experimental data are from Hogenboom et al. (1995, 1999) and Adams and Gibson (1930)

were off from the experimental measurements by -0.37, -0.31, and $-0.30\,°\mathrm{C}$ at 1 bar, 500 bars, and 1000 bars, respectively (Fig. 3.20). Similarly for the $MgSO_4$ system, the model calculated eutectic temperatures were off from the experimental measurements by 0.23, 0.43, and $1.00\,°\mathrm{C}$ at 1 bar, 500 bars, and 1000 bars, respectively (Fig. 3.20).

The NaCl example (Fig. 3.20) refers to the equilibrium

$$\text{NaCl·2H}_2\text{O(cr)} \leftrightarrow \text{NaCl(cr)} + 2\text{H}_2\text{O(l)} \quad (3.83)$$

and is based on the experimental work of Adams and Gibson (1930). In this particular case, the discrepancies between model and experimental measurements varied between -0.57 and $0.30\,°C$ over a pressure range of 1 to 1000 bars (Fig. 3.20). In contrast to the eutectic examples where increasing pressure led to a lowering of the temperature, the temperature increased with increasing pressure for the NaCl equilibrium (Eq. 3.83). These model fits to the experimental data are a slight improvement over a previous comparison (Marion et al. 2005) because we included a pressure dependence to volumetric parameters in this case (Eqs. 3.23 and 3.24), which was lacking in our previous comparison. For example, the maximum discrepancy in this case is $1.00\,°C$, while the maximum discrepancy before was $1.24\,°C$ (Marion et al. 2005). There is a tendency for the temperature discrepancies to increase with increasing pressure (Fig. 3.20). This probably reflects the limitations of the model (see discussion below). However, with the exception of one data point for $MgSO_4$, the ΔT values agreed to within $\pm 0.6\,°C$, which is sufficiently accurate for most geochemical applications.

In Fig. 3.21, we compare the percent difference between model and experimental estimates of carbon dioxide gas concentrations for NaCl solutions in equilibrium with $CO_2 \cdot 6H_2O$. Only 9% of the carbon dioxide comparisons are off by $> 10\%$ (Fig. 3.21). A large part of this variability is clearly attributable to variation in measured gas concentrations. For example in the Dholabhai et

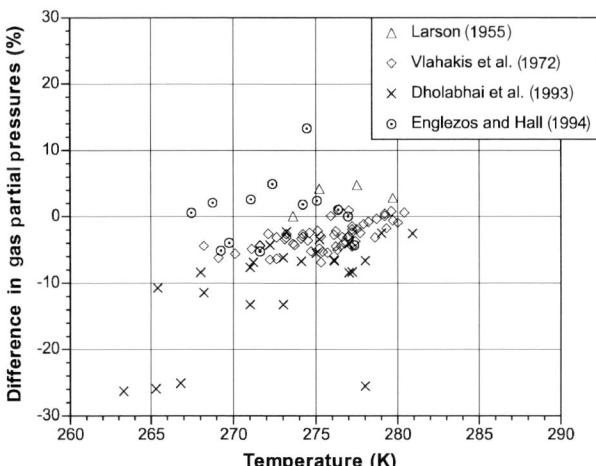

Fig. 3.21. Percent difference between model-calculated and experimental carbon dioxide gas concentrations for NaCl systems in equilibrium with $CO_2 \cdot 6H_2O$, where % Diff. $=$ [model $-$ expt./expt.] $\times$ 100. Reprinted from Marion et al. (2006) with permission

al. (1993) dataset at 278 K and NaCl = 5 wt.%, measurements in two cases gave carbon dioxide gas concentrations of 30.0 and 37.7 bars; our model predicts 28.0 bars at this temperature and NaCl concentration, which leads to "apparent" errors of -6.6 and -25.5%. The best fit of our model is to the Englezos and Hall (1994) dataset, where the percent diffferences, with one exception, are within $\pm 5\%$ (Fig. 3.21). See Marion et al. (2006) for a similar comparison of a NaCl-$CH_4 \cdot 6H_2O$ system.

In Fig. 3.22, we compare equilibrium carbon dioxide gas partial pressures with temperature for four datasets from Englezos and Hall (1994) that include pure water, 10.0 wt.% NaCl (1.901 m), 15.2 wt.% NaCl (3.067 m), and 10.57 wt.% $CaCl_2$ (1.065 m). The average absolute values of the error between our model and experimental measurements for these four datasets are 1.4, 2.4, 5.6, and 2.7%, respectively [average overall error = $3.0\,(\pm 1.8)\%$]. These errors in model fits to this experimental dataset are an improvement over the model errors of Englezos and Hall (1994) of 2.4, 6.2, 12.2, and 8.0 %, respectively [average overall error = $7.2(\pm 4.1)\%$], and the model errors of Bakker et al. (1996) of 1.2, 4.8, 9.2, and 5.7%, respectively [average overall error = $5.2(\,\pm 3.3)\%$]. In general, our model smoothes the experimental bumps, with the possible exception of one data point at 15.2 wt.% NaCl and 274.45 K (Fig. 3.22), which is also an outlier in Fig. 3.21. Both the Englezos and Hall (1994) and the Bakker et al. (1996) models use a simpler version of the Pitzer equation for defining the activity of water in electrolyte solutions than our model, which incorporates interactions between neutral species (dissolved gases) and salts (Eqs. 2.40–2.42) and corrects for pressure (Eq. 2.90).

Increasing errors with increasing salinities (see patterns in the discussion above) were observed in earlier work (e.g., Englezos and Hall 1994; Bakker et al. 1996). Equation 3.36 is the model used in this work to estimate equi-

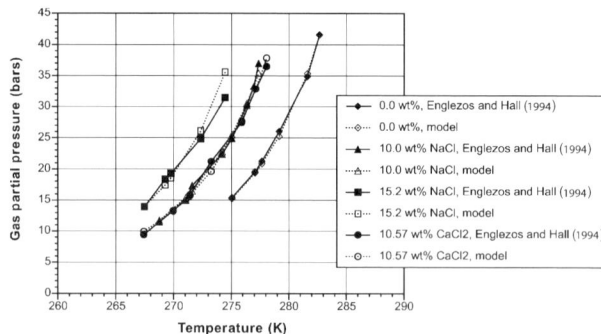

Fig. 3.22. Model and experimental gas partial pressures of CO_2 in equilibrium with $CO_2 \cdot 6H_2O$ as a function of salt concentration and temperature. *Filled symbols* and *solid lines* represent data from Englezos and Hall (1994); *open symbols* and *dashed lines* represent model fits. Reprinted from Marion et al. (2006) with permission

librium gas pressures. Since the errors in estimating gas partial pressures for pure water systems are minor (1.4% in Fig. 3.22), we can assume that the equilibrium constants are well defined in Eq. 3.36. The one property in Eq. 3.36 that is most directly related to salinity is the activity of water (a_w), which in our model is estimated from the osmotic coefficient (ϕ) (Eqs. 2.37 and 2.39). From the latter equations, as solution concentrations (m_i) increase, the potential for error in a_w increases. However, a direct test of the FREZCHEM model to accurately simulate a_w between 0.1 and 6.0 m NaCl at 298 K compared to an experimental database from Robinson and Stokes (1970) led to a maximum error of 0.040% at 1.01 bar pressure. The calculated a_w at 6.0 NaCl and 100 bars of pressure is only 0.13% smaller than at 1.01 bars. An error in a_w of 1% raised to the sixth power (Eq. 3.36) would cause an error of 6% in the calculated equilibrium gas concentration. It is possible that errors in calculated a_w may contribute to the salinity effect on errors. However, the problem is more likely due to difficulties in establishing gas equilibria at high salt concentrations. There could be interactions between gas hydrates and salts that are not directly captured by Eqs. 3.25 and 3.36. For example, this could be due to "salting out" of dissolved gases from solution (removal) as salinity increases. This phenomenon would affect the connection between soluble and atmospheric gas concentrations (Eqs. 3.25 and 3.36), which could affect both ϕ and $P_{(g)}$ in Eq. 3.36.

Figure 3.22 documents the role of salinity as an inhibitor of $CO_2 \cdot 6H_2O$ formation. At 275 K, the equilibrium carbon dioxide gas pressures for pure water, 10.0 wt.% NaCl (1.901 m) and 15.2 wt.% NaCl (3.067 m), are approximately 15, 25, and 39 bars. As salinity increases, a_w decreases, and $P_{(g)}$ must increase to satisfy the solubility product (Eq. 3.36).

The gas hydrate model developed in this work is based on an existing electrolyte model for calculating a_w in Eq. 3.36 (FREZCHEM model). The gas fugacity coefficient (ϕ) of Eq. 3.36 is based on the published Duan et al. (1992b) model. And finally, the equilibrium constants for $CH_4 \cdot 6H_2O$ and $CO_2 \cdot 6H_2O$ are based on model fits to pure gas/water data (Figs. 3.10 and 3.11). None of the salt data displayed in Figs. 3.21 and 3.22 were used to parameterize our gas hydrate model. The fact that our model can reproduce reasonably accurately these independent experimental data is validation for the approach used in developing this gas hydrate model.

3.6 Limitations

A number of limitations of the FREZCHEM model can be broadly grouped under Pitzer-equation parameterization, modeling (mathematics, convergence, and coding), and applications. The first two limitations are discussed in this chapter. Application limitations are discussed in Chap. 5 after presentation of multiple applications.

3.6.1 Pitzer-Equation Parameterization Limitations

As an interpretative research tool, every geochemical model is limited by the chemical species that constitute the model (Table 3.1). In this respect, the chemical species database is analogous to a spectroscopist's spectral library. If jarosite is missing from your spectral library, then jarosite will never be selected as a component of the spectral signature. Recent findings from the Mars Exploration Rover (MER) missions indicate that jarosite [$K(Fe^{3+})_3(OH)_6(SO_4)_2$)] is present at the Meridianai Planum site as well as "high" concentrations of Br. Neither jarosite nor Br is currently part of the FREZCHEM model, which is a limitation that is currently being rectified. One of the strengths of the FREZCHEM model is the relative ease of adding new chemistries into the sequential chain (subroutines, Fig. 3.14).

Earlier, when discussing historical development, we mentioned that different workers have used different equations to describe the Debye–Hückel constant (A_ϕ, Eq. 2.35) as a function of temperature. For example, at 0 °C, the value of this constant is 0.3781, 0.3764, and 0.3767 $kg^{1/2}\,mol^{-1/2}$ for the FREZCHEM, Archer and Wang (1990), and Pitzer (1991) models, respectively. At NaCl = 5 m and 0 °C, the calculated mean activity coefficients using these three parameters evaluated with the FREZCHEM model are 0.7957, 0.7995, and 0.7988, respectively. The largest discrepancy is 0.48%, which is within the range of model errors for activity coefficients (Table 3.5).

To replace the A_ϕ value currently in the FREZCHEM model with the most accurate A_ϕ value (Archer and Wang 1990) would require man-years of work to reestimate Pitzer-eqution parameters and solubility products derived in our work using our value of A_ϕ. This would, at best, lead to a marginal improvement in model accuracy. While the error in the example above at 0 °C is marginal, this is not necessarily the case across a broader range of temperatures. What can happen if one blindly substitutes a different equation for A_ϕ into the FREZCHEM model can be illustrated with the Archer and Wang (1990) database. We fit an equation to their 1 bar data for A_ϕ and extrapolated this equation to 210 K. Substitution of this equation into the FREZCHEM model predicts a temperature of −60.1 °C and H_2SO_4 = 5.64 m at the eutectic for $H_2SO_4 \cdot 6.5H_2O$-Ice. While the H_2SO_4 concentration (5.64 m) is close to the experimental value (5.68 m), the temperature is off by 1.9 °C (−60.1 vs. −62.0 °C). The FREZCHEM model using our equation for A_ϕ (Appendix B) predicts 5.68 m H_2SO_4 and −62.0 °C at the eutectic, in perfect agreement with the experimental data (Linke 1965). Our A_ϕ equation works because this equation is compatible with the Pitzer-equation parameters and solubility products of the FREZCHEM model; the Archer/Wang A_ϕ equation does not work because their equation is incompatible with other components of the FREZCHEM model. This example illustrates why all components of a Pitzer model must be compatible.

One of the major limitations of the FREZCHEM model is the lack of key parameters and equilibrium relationships at subzero temperatures. Several examples have already been discussed. For example, in Fig. 3.2, estimation of the molar volume of "equilibrium water" ' at 1 bar pressure at subzero temperatures requires extrapolation to subzero temperatures of a water compressibility equation (Eq. 3.17). Application of volumetric calculations at subzero temperatures necessitates extrapolation of molar volumes (Fig. 3.6) and isothermal compressibilities (Fig. 3.7). Several equilibrium constants dealing with acidic and alkaline systems require extrapolation to subzero temperatures (Fig. 3.9). In Tables 3.2 and 3.3, the temperature and concentration ranges used in parameterization and validation of the model are listed by chemistries. Extrapolation outside these ranges must be done with caution.

At a temperature of $0\,°C$, the fugacity coefficients of CH_4 (g) and CO_2 (g) vary significantly from the ideal ($\phi = 1.0$) as a function of pressure (Fig. 3.23). One potential limiting factor in using the Duan et al. (1992b) model to define fugacity coefficients (Eqs. 3.37–3.48) is that the lower temperature limit for this gas model is $0\,°C$ (273 K). In Fig. 3.24a, we depict how the fugacity coefficient for methane changes according to the Duan model between 273 and 1473 K at 1 and 20 bars of pressure. Also included in Fig. 3.24a are FREZCHEM model estimates of the methane gas fugacity coefficient at subzero temperatures for systems in equilibrium with $CH_4·6H_2O$ at 1, 5, 10, 15, 20, and 26.45 bars of gas pressure; these calculations are an extrapolation of the Duan model into the subzero temperature range. Figure 3.24b depicts similar calculations for CO_2 fugacity coefficients in equilibrium with $CO_2·6H_2O$. The equilibrium temperatures of methane and carbon dioxide hydrate at 1 bar pressure in these two cases are 193 and 218 K, respectively. The

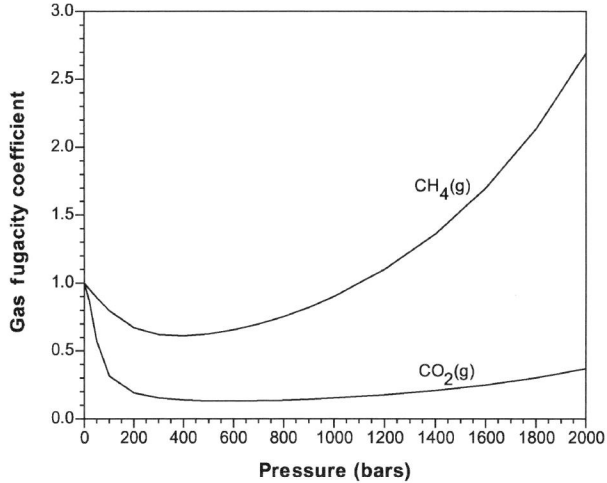

Fig. 3.23. Gas fugacity coefficients for methane and carbon dioxide at $0\,°C$ as a function of pressure using the Duan et al. (1992b) model. Reprinted from Marion et al. (2006) with permission

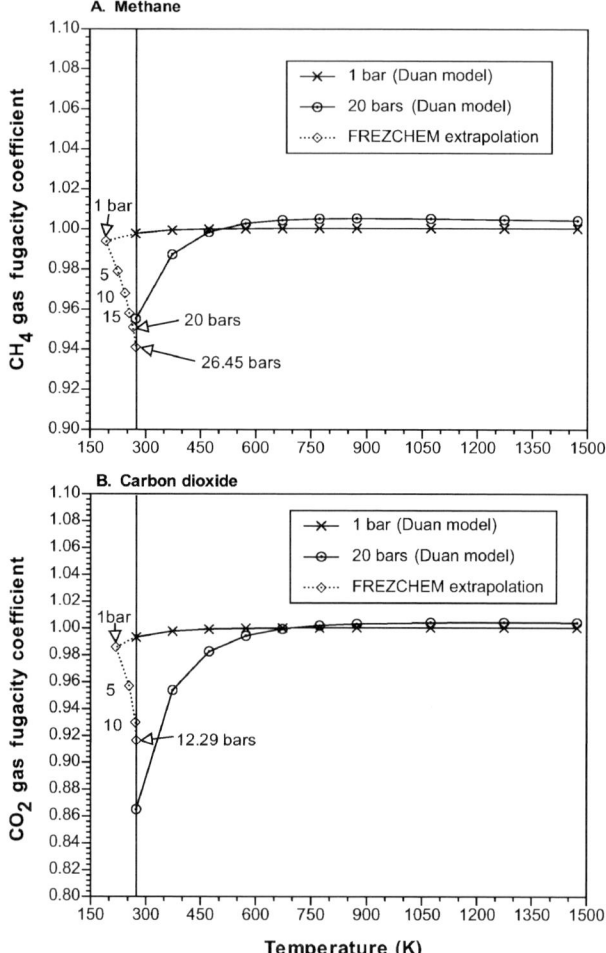

Fig. 3.24. The Duan et al. (1992b) model estimates of gas fugacity coefficients at 1 and 20 bars between 273 and 1473 K compared to extrapolations below 273 K generated by the FREZCHEM model for **a** methane and **b** carbon dioxide. Reprinted from Marion et al. (2006) with permission

point of these diagrams is that they demonstrate that these subzero extrapolations are consistent with the trends at higher temperatures. Furthermore, the magnitude of the fugacity coefficients at subzero temperatures ranges from 0.916 to 0.994 for systems in equilibrium with gas hydrates (Fig. 3.24). Compared to the broader range of fugacity coefficients (Fig. 3.23), these subzero systems are close to ideal ($\phi=1.0$). For these reasons, we feel that the subzero temperature extrapolations of the Duan et al. (1992b) model are unlikely to cause significant error in equilibrium gas hydrate calculations.

Carbon dioxide is important in governing bicarbonate/carbonate solution and mineral equilibria as well as controlling $CO_2 \cdot 6H_2O$ formation. Many of the equations used for bicarbonate/carbonate equilibria are based on Plummer and Busenberg (1982); see Marion (2001) for a complete description of bicarbonate/carbonate equilibria and model parameterizations. The maximum CO_2 gas pressure for these bicarbonate/carbonate relationships is 1.01 bar (1 atm). Introducing CO_2 gas hydrates into the FREZCHEM model necessitates increasing CO_2 gas pressures up to 45 bars at 283 K (Fig. 3.10). For a hypothetical 1×10^{-3} m $NaHCO_3$ solution, the FREZCHEM model predicts pH values of 8.16, 4.72, and 3.17 at 3.6×10^{-4}, 1.01, and 45 bars of CO_2 pressure, respectively. While the first two points are within the range of our model, 45 bars of CO_2 pressure is well beyond the range of the bicarbonate/carbonate model. So caution is necessary in interpreting mixed bicarbonate/carbonate and gas hydrate chemistries at high CO_2 gas pressures. It may even be necessary at times, when working with alkaline solutions such as seawater, to remove alkalinity from the model when simulating high $CO_2(g)$ pressures for gas hydrates.

An example not previously discussed is the Pitzer–Debye–Hückel slope for apparent molar volume (A_v) that is required in Eqs. 2.76, 2.80, and 2.81. A numerical equation for A_V as a function of temperature and pressure was derived from the database of Ananthaswamy and Atkinson (1984) over a temperature range of 273 to 298 K and over a pressure range of 1 to 1000 bars:

$$A_V = 3.73387 - 0.0289662T + 1.29461 \times 10^{-3}P - 5.62291 \times 10^{-6}TP \\ + 7.62143 \times 10^{-5}T^2 + 4.09944 \times 10^{-8}P^2. \qquad (3.84)$$

The fit of Eq. 3.84 to the database is excellent (Fig. 3.25, $R^2 = 0.99995$) and considerably simpler than the fundamental definition of A_V (Ananthaswamy and Atkinson 1984; Archer and Wang 1990; Pierrot and Millero 2000). For work at subzero temperatures, we simply extrapolated Eq. 3.84 into the subzero temperature range. The smoothness of the relationship suggests that this is a reasonable approximation. However, these curves coalesce at 236.15 K ($-37\,°C$) at $A_V = 1.14$. At temperatures below 236.15 K, we assigned A_v a value of 1.14. In the extrapolation cases examined (Figs. 3.2, 3.6–3.9, 3.24, and 3.25), the relationships change smoothly. Extrapolation of these relationships, at least down to 250 K, is probably reasonable. There are, however, a few relationships that do not extrapolate well. Those that have been identified are footnoted in the appendix tables, where alternatives to extrapolation are provided. When using the FREZCHEM model in cases that involve extreme extrapolation ($>25\,°C$), watch for activity coefficients that become either vanishingly small or absurdly large. This is a clear sign that the parameterization is failing.

One of the inherent limitations of the Pitzer approach is the necessity that all significant interactions among ions and neutral species must be quantified. In appendix Tables B.4 to B.6, we quantify 291 binary and ternary interaction

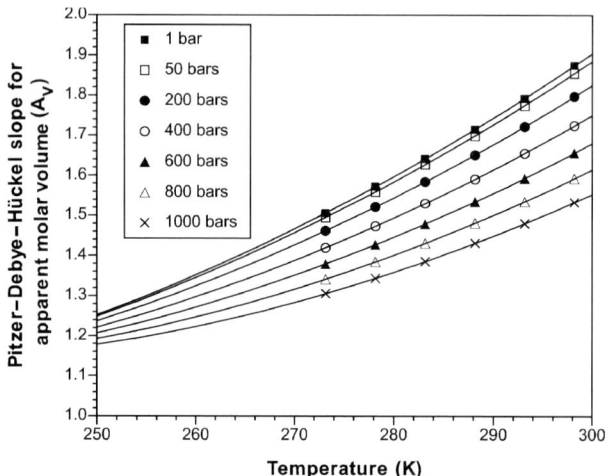

Fig. 3.25. The Pitzer–Debye–Hückel slope for apparent molar volume (A_v) as a function of temperature and pressure. *Symbols* are from Ananthaswamy and Atkinson (1984); *lines* represent model estimates

parameters among ions and neutral species at 1 bar pressure. Table B.10, in contrast, contains 74 binary volumetric parameters that are needed to quantify pressure effects. Obviously the latter volumetric database is much less complete than our primary database at 1 bar. In applying the FREZCHEM model to pressure cases, simplifications were necessary.

The pressure dependence of equilibrium constants in this work are estimated with Eq. 2.29, which requires knowledge of the partial molar volumes and compressibilities for ions, water, and solid phases. For ions and water, molar volumes and compressibilities are known as a function of temperature (Table B.8; Eqs. 3.14 to 3.19). Molar volumes for solid phases are also known (Table B.9); unfortunately, the isothermal compressibilities for many solid phases are lacking (Millero 1983; Krumgalz et al. 1999).

If we ignore compressibilities entirely in estimating the pressure dependence of the solubility product of NaCl (halite), then Eq. 2.29 predicts a K^P/K^{P0} ratio of 1.521 at 1000 bars of pressure and 25 °C. Using compressibility data from this study for ions plus a compressibility of 1.1×10^{-4} cm^3/(molbar) for NaCl(cr) (Millero 1983), the K^P/K^{P0} ratio is 1.381 at 1000 bars and 25 °C. Ignoring compressibilities leads to an error of 10.1% in estimating the NaCl solubility product at 1000 bars. Using compressibility data for Na and Cl ions but without the NaCl(cr) compressibility in Eq. 2.29 gives a K^P/K^{P0} ratio of 1.377, which is only off from the true value of 1.381 by 0.3%. This error is small because the compressiblity of NaCl(cr) is small [1.1×10^{-4} cm^3/(molbar)] relative to the Na [-3.94×10^{-3} cm^3/(molbar)] and Cl [-7.40×10^{-4} cm^3/(molbar)] ions. For NaCl(cr), Eq. 3.20 with $\bar{V}_T^0 = 27.02$ cm^3/mol (Table B.9) and $\bar{K}_T$

= 1.1×10^{-4} cm^3/(molbar) predicts a molar volume of 26.91 cm^3/mol at 1000 bars of pressure, which is a 0.41% decrease in volume. Similar calculations for calcite and siderite from data compiled by Holland and Powell (1998) predict 0.13 and 0.08% decreases in volume at 1000 bars, respectively. Because the compressibilities of solid-phase minerals are generally small (Millero 1983; Holland and Powell 1998), in the programming of the FREZCHEM model, we ignored the compressibility of solid phases (except for ice) in using Eq. 2.29 to estimate the pressure dependence of equilibrium constants.

In the FREZCHEM model, the pressure dependence for activity coefficients was calculated without a compressibility term (Eq. 2.87). This compressibility was ignored in our model because of the sparse data for this property in the literature (Millero 1983; Krumgalz et al. 1999). However, there is some compressibility data that allow us to compare pressure-dependent activity coefficient ratios, with and without the compressibility term. Table 3.8 compares the mean activity coefficient ratios at 1000 bars/1 bar for several electrolytes at I (ionic strength) = 1.0 m and 25 °C. The ratios in our work were calculated without the compressibility term; the corresponding ratios from Millero (1983) were calculated with the compressibility term. In general, the errors between these two estimates are around 1–2% (Table 3.8) at I = 1.0 m. However, the magnitude of this error increases with increasing pressure and concentration. For example, Krumgalz et al. (1999) estimated that the errors in calculating the mean activity coefficient ratios at 1000 bars, with and without compressibility terms, were 3.7 and 3.5% for NaCl

Table 3.8. Mean activity coefficients at I = 1.0 m and 25 °C calculated in this work ignoring compressibility compared to mean activity coefficients considering compressibility (Millero 1983). Values in parentheses are from Krumgalz et al. (1999). Reprinted from Marion et al. (2005) with permission

Electrolyte	$\gamma\pm$ (1 bar) (this work)	$\gamma\pm$ (1000 bars) (this work)	$\gamma\pm$ (1000 bars)/ $\gamma\pm$ (1 bar) (this work)	$\gamma\pm$ (1000 bars)/ $\gamma\pm$ (1 bar) (Millero, 1983)	% Diff.
NaCl	0.6590	0.6946	1.054 (1.053)	1.046 (1.039)	0.8
KCl	0.6027	0.6365	1.056	1.045	1.1
MgCl$_2$	0.4861	0.5158	1.061 (1.061)	1.057 (1.049)	0.4
CaCl$_2$	0.4549	0.4843	1.065	1.058	0.7
HCl	0.7986	0.8202	1.027	1.029	−0.2
NaNO$_3$	0.5456	0.5859	1.074	1.057	1.6
Na$_2$SO$_4$	0.3138	0.3554	1.132	1.115	1.5
MgSO$_4$	0.1170	0.1441	1.232	1.212	1.7

and $MgCl_2$ at $I = 5\,m$, respectively. On the other hand, these errors are insignificant at lower pressures [up to 300 bars, see Fig. 1 in Krumgalz et al. 1999)].

The magnitude of the effect of pressure on activity coefficients is much less than is the case for solubility products (see previous discussion). The errors in ignoring the compressibility term for activity coefficients are, at most, 2–5% in a pressure range up to 1000 bars (Millero 1983; Krumgalz et al. 1999). For the broad-scale FREZCHEM model, these errors are acceptable.

In the development of many geochemical models, the pressure dependence of the activity of water (a_w) is often ignored because this effect is small (Monnin 1990; Krumgalz et al. 1999). The means for calculating this pressure dependence are, however, inherent in our model algorithms (Eqs. 2.76, 2.88–2.90); therefore, this pressure dependence was incorporated into the FREZCHEM model. For NaCl at $25\,°C$ and $I = 6\,m$ (near saturation with halite at 6.1 m), the a_w (1000 bars)/a_w (1 bar) ratio is 0.9923, indicating that 1000 bars of pressure causes less than a 1% change in a_w. For $CaCl_2$, the a_w ratio is 0.9674 at $I = 18\,m$. For a more dilute brine such as seawater at $I = 0.72\,m$, this ratio is 0.99965. It is only for concentrated brines or gas hydrates that this pressure correction for a_w is needed in most geochemical applications.

Equation 2.90 for a_w ignores the compressibility of water, as did Eq. 2.87 for activity coefficients. Had we included a compressibility term in calculating the ratio of a_w^P/a_w^{P0}, then presumably this ratio would be slightly higher than our calculated ratios that excluded compressibility. However, a small correction to a small correction is insignificant in most geochemical applications.

On a percent basis, the relative effect of pressure on solution properties falls in the order: equilibrium constants (K) > activity coefficients (γ) > activity of water (a_w). The errors (%) in our model associated with these properties, however, fall in the order: $\gamma > K > a_w$. The transposition in the relative order between K and γ is because our model for calculating the pressure dependence of K (Eq. 2.29) includes the compressibility term and is therefore more accurate than the equation for calculating the pressure dependence of γ (Eq. 2.87) that ignores compressibility (see above discussions). Only activity coefficients are likely to be significantly in error. However, even in this case, the errors are likely to be only in the range of 2 to 5% up to 1000 bars of pressure (Millero 1983; Krumgalz et al. 1999). To improve on this model for activity coefficients would require adding a compressibility term to Eq. 2.87, which would make it similar in structure to Eq. 2.29. Unfortunately, the data are insufficient for developing a comprehensive set of compressibility parameters for broad ranges of composition, except at infinite dilution (Eq. 2.29) (Krumgalz et al. 1999).

Eqs. 2.76, 2.80, and 2.81 ignore volumetric ternary interaction terms (Θ_{ij}^V, Ψ_{ijk}^V) that could play important roles in mixed salt solutions (Monnin 1989;

Krumgalz et al. 1995, 1999). However, to estimate the contribution of ternary parameters, it is necessary that experimental densities be accurate to within $\pm 0.0001\,\text{g cm}^{-3}$ (Monnin 1989). Our current model is not sufficiently accurate to warrant an inclusion of ternary effects. Estimates of solution densities, with and without ternary terms, of the saline solutions of Table 3.7 differed by at most $1.8 \times 10^{-4}\,\text{g cm}^{-3}$ (Krumgalz et al. 1995), which is minor relative to other sources of error.

In several cases, key parameters are completely unknown, which requires substituting a surrogate for the unknown. In the FREZCHEM model, this is most significantly a problem for Fe^{2+} chemistry. In many cases, we substituted Mg^{2+} parameters for analogous Fe^{2+} relationships; these substitutions are identified with footnotes in the appendices (Tables B.4 to B.6 and B.8). Of the binary Mg for Fe substitutions, the most critical are the Mg-HCO_3 parameters as surrogates for Fe-HCO_3 (Marion et al. 2003a). In the latter paper, we demonstrated that this substitution leads to a constant $FeCO_3$ solubility product across a wide range of NaCl and Na_2SO_4 solutions. Furthermore, there is direct evidence that the Mg^{2+} for Fe^{2+} substitution is reasonable. For example, at 25 °C and 1.0 m, model-calculated mean activity coefficients, a direct function of binary Pitzer-equation parameters, for $FeCl_2$ and $MgCl_2$ are 0.506 and 0.589, respectively; similarly, at 25 °C and 1.0 m, the mean activity coefficients for $FeSO_4$ and $MgSO_4$ are 0.0537 and 0.0545, respectively. The experimental eutectic for a pure $FeCl_2$-H_2O system occurs at -36.5 °C and 3.45 m $FeCl_2$, which is similar to a pure $MgCl_2$-H_2O system where the eutectic occurs at -33.6 °C and 2.79 m (Linke 1965).

There are cases where the existing database appears to be poor, especially at subzero temperatures. For example, in our parameterization of HCl chemistry, we concluded that most of the pure HCl data at subzero temperatures were fundamentally flawed and unreliable (Marion 2002), a conclusion also reached earlier by Carslaw et al. (1995). There were also a few cases where it was impossible to derive Pitzer-equation parameters that fit the experimental data across the entire compositional range for mixed salt solutions. This occurred for NaCl-$NaNO_3$, KNO_3-HNO_3, K_2SO_4-H_2SO_4, and $MgSO_4$-H_2SO_4 (Marion, 2002). We attributed these discrepancies to rapid changes in activity coefficients (or, conversely, solubilities) within certain ranges that are not well predicted by our model. We suspect that the underlying problem in these cases is that the associated binaries are not adequately parameterized across a broad enough concentration range, which must extend beyond mineral saturation.

3.6.2 Modeling Limitations

There are four broad reasons why the FORTRAN program fails to converge or calculate properly at times. These include mathematical limitations, complete drying of the solution, oscillatory convergence problems, and poor coding.

Some limitations are attributable to the mathematics of the model. For example, applications of the Newton–Raphson algorithm for calculating the acidity of alkaline (Eq. 3.29) or acid (Eq. 3.32) systems necessitates an initial approximation of pH or acid [H^+]. Convergence is guaranteed provided the initial approximations are sufficiently close. Otherwise, the algorithm can diverge into a computer "black hole," never to emerge. Nonconvergence can be rectified by assuming a new, better initial approximation for pH or acid concentration. For unfamiliar systems, this may take several restarts. We mentioned previously that Brent's method, a part of the mathematical algorithm used for estimating gas fugacity coefficients, occasionally fails to converge properly (Sect. 3.4.3).

The FREZCHEM model assumes the presence of an aqueous phase. As a consequence, there must be liquid water at all times during model calculations. Evaporation or freezing to the eutectic can lead to complete drying of the system. For the evaporation pathway, always specify a lower bound on water that is above 0.0 g. For example, specifying 1.0 g of water could allow a 1000-fold concentration of the aqueous solution; by that point, the evaporite salts are sufficiently well defined. Because of the importance of eutectics in the FREZCHEM model, a special algorithm was designed to identify when the eutectic is reached. The number of independent components (IC) in an aqueous solution is given by

$$IC = (\text{cations} + \text{anions} - 1) + 1 \,(\text{water}). \qquad (3.85)$$

For the case of a Na-K-Mg-Cl-SO_4-H_2O solution, there are five independent components. What happens at the eutectic is that ice formation concentrates solutes until equilibrium is reached for the ice phase (Fig. 3.14). Then salt equilibrium precipitates salts until equilibrium is reached for salts. But now, the system is out of equilibrium with the ice phase. So the program cycles repeatedly through the ice phase and the salt phases. At each step, water is removed as ice or hydrated salts until literally no water remains in solution. Below the eutectic, no thermodynamic equilibrium is possible with an aqueous phase. After 5000 iterations (you can change this number if desired), the model checks the number of salts + ice that are precipitating. If that number is equal to the number of independent components, then the program prints a message indicating that the system is at or beyond the eutectic. Another message may also be printed that indicates when the system is running out of water ($H_2O < 0.1$ g).

Occasionally the program gets stuck oscillating indefinitely between two equilibria, typically near a junction where there is a transition between two hydrated salts. As an example, using the equilibrium crystallization pathway (Sect. 3.2.3) and an initial specification of $MgSO_4 = 2.0$ m and $H_2SO_4 = 1.0$ m, the program predicts that $MgSO_4 \cdot 7H_2O$ is stable down to $-4.0\,°C$, fails to converge at all between -4.1 and $-4.6\,°C$ (oscillates indefinitely between the 7- and 12-hydrates), and converges onto $MgSO_4 \cdot 12H_2O$ at $-4.7\,°C$.

For the same initial specification of $MgSO_4$ and H_2SO_4, this problem can be solved using the fractional crystallization pathway. In that case, the program predicts $MgSO_4 \cdot 7H_2O$ is stable at $-4.0\,°C$ and $MgSO_4 \cdot 12H_2O$ is stable at $-4.1\,°C$. Why is there a difference between these two crystallization pathways? At each temperature step for equilibrium crystallization, equilibrium is reestablished from the initial specification; if the latter is far removed from the current equilibrium, then significant precipitation of highly hydrated salts causes a substantial increase in the concentration of nonprecipitating constituents such as H^+ and HSO_4^-, which in turn affects the ionic strength, activity coefficients, and, ultimately, the equilibrium distribution of species. Finding the equilibrium point where so many factors are simultaneously changing in the vicinity of a transition is tricky. This is less of a problem with fractional crystallization because the starting point for equilibrium calculations at a given temperature is the composition at the previous temperature and not the initial composition. The difference between the equilibrium compositions at two close temperatures is generally small. Attempts to directly fix this oscillatory problem have been unsuccessful to date. But, as demonstrated above, there are generally means for dealing with these inconveniences.

The last issue to discuss with respect to computer coding is porting of the code between computers. All FORTRAN compilers have, in addition to standard options and commands, extensions that facilitate coding or perform additional tasks. The problem is that these extensions differ from compiler to compiler, which makes porting of code between computers potentially difficult. Also, some compilers are more forgiving of "poor" code. Under poor code, we can lump poorly written (by FORTRAN standards) and logical mistakes. This has become a more serious concern with the rapid expansion of the model in recent years. Major pathways that are frequently used are probably relatively error free; on the other hand, seldom-used pathways are more likely to contain poor code. Over the years, several scientists that I had given the code to have worked with me in rectifying poor code. They are thanked in the preface. Hopefully, the major problems with coding are fixed; but any future problems that users might find should be addressed to *giles.marion@dri.edu*.

4 Limits for Life

In the past few decades, our concept of a potentially habitable world has broadened from the narrow range between Venus and Mars to literally cover most of the Solar System (Fig. 4.1a,b). Three primary, interrelated developments have contributed to broadening our concept of a habitable world. First, there has been an explosion of research on the ability of microorganisms, especially bacteria and archaea, to grow in extreme environments with respect to temperature, salinity, acidity, radiation, and pressure (Tables 4.1–4.4). Environments that would have been considered inhospitable for life a few decades ago have been found to be thriving with life (e.g., hydrothermal vents and seemingly solid, frozen permafrost and sea ice). Second is a better understanding of a wide range of environments within which liquid water can exist (Fritsen and Priscu 1998; Psenner and Sattler 1998; Gaidos et al. 1999; Vogel 1999; Price 2000; Marion et al. 2003b). Third, our knowledge of the metabolic basis of terrestrial life has broadened to include many biochemical schemes that are completely independent of photosynthesis (Gaidos et al. 1999; Chyba 2000; Chyba and Hand 2001). When the Space Age began almost half a century ago, our understanding of potential extraterrestrial habitats was tied to planetary surface environments where photosynthesis is possible. The third development has decoupled our concept of the limits on potential life from Sun-warmed planetary surface conditions (Fig. 4.1); now we perceive possible habitats in any subsurface porosity, solid/fluid interface, or watery medium having the right aqueous conditions and a possible source of metabolic chemical energy.

Today in the planetary sciences, the first step in the search for life is often the search for liquid water. The current mantra of astrobiology and both NASA and ESA's broader exploration programs is "follow the water," which is as much a matter of following the heat as it is of following the water. Geophysics has become a cornerstone of astrobiology, as there has been increased recognition of geothermal and tidal heating in maintaining warm, wet conditions in planetary crusts and oceans that otherwise would be frozen (compare Fig. 4.1a and 4.1b).

The two Solar System bodies beyond Earth that have elicited the most interest as potential habitats for life are Mars and Europa because both have clearly been impacted by aqueous processes. Considering the expanse of time subsequent to the Solar System's origin, conditions conducive to the occur-

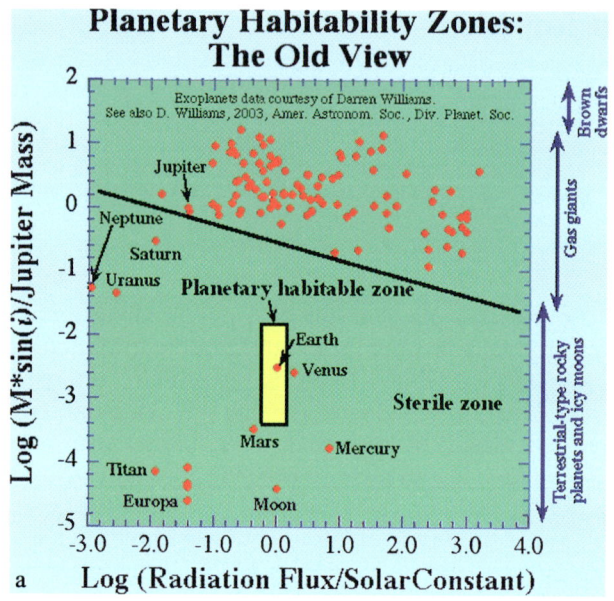

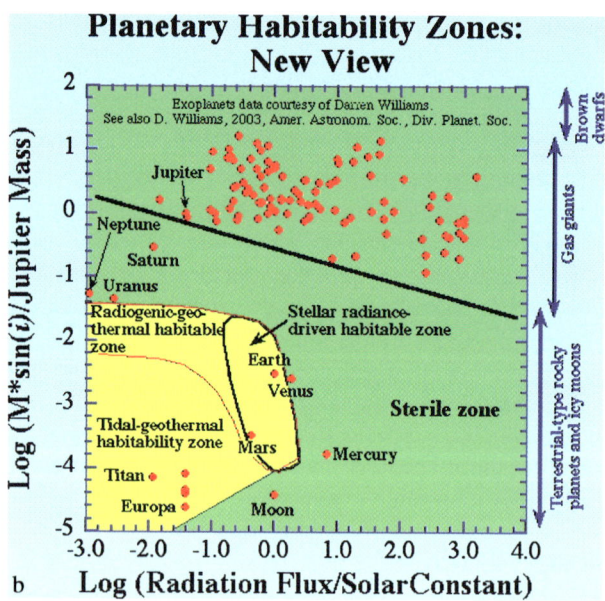

Fig. 4.1. Planetary habitability zones. **a** Old view. **b** New view. Reprinted from Kargel (2004) with permission

rence of liquid water and aqueous solutions have existed in some asteroids, many icy satellites, comets, and the gaseous envelopes of the giant planets, in addition to Venus, Earth, and Mars. Nobody would argue that a perihelion

Table 4.1. Summary of limits for biological activity at high and low temperatures. Reprinted from Marion et al. (2003b) with permission

Factor	Type of environment	Limits	References
High temperatures	submarine hydrothermal vents	110 to 121 °C	Pledger and Baross 1991; Segerer et al. 1993; Blöchl et al. 1997; Huber and Stetter 1998; Stetter 1999; Madigan and Oren 1999; Stetter 2002; Kashefi and Lovley 2003
	subterranean deep biosphere	110 °C	Pedersen 1993; Fyfe 1996; Stetter 1996
	terrestrial hot springs	103 °C	Kushner 1981; Cometta et al. 1982; Smith 1982; Farmer 1998; Stetter 1999; Farmer 2000; Rothschild and Mancinelli 2001
Low temperatures	ice	−17 to −20 °C	Vogel 1999; Rivkina et al. 2000; Junge et al. 2001; Rothschild and Mancinelli 2001; Gilichinsky 2002; Junge 2002; Junge et al. 2004
	terrestrial	−17 °C	Friedmann and Ocampo 1976; Friedmann and Ocampo-Friedmann 1984; Schroeter and Scheidegger 1995; Gilichinsky 2002
	deep sea	≈ 2 °C	Yayanos 1995; Sassen et al. 1999; Krumgalz et al. 1999
	troposphere	< 0 °C	Sattler et al. 2001

passage of a comet, which might yield transient aqueous solutions, would yield a habitable niche, and almost as improbable are certain transient aqueous environments on Earth, such as stratospheric clouds of aqueous acid aerosols. Likewise, a brief flash of wet conditions on the surface of Mars is not likely to be one where life could arise, though it could be an environment where dormant life could be reactivated. So far there has been comparatively little consideration given, outside of science fiction and some loose but serious science, to potential habitats in the gas giants, though this could be more due to our ignorance of these extremely alien environments than for solidly

Table 4.2. Tolerances to salinity for microbial activity (adapted from Mazur 1980; Kushner 1981). Reprinted from Marion et al. (2003b) with permission

Tolerances	
Most bacteria/archaea	$a_w > 0.9$
Extreme bacteria/archaea	$a_w \approx 0.70$
Most fungi	$a_w > 0.85$
Extreme fungi	$a_w \approx 0.60$

Table 4.3. Summary of extreme pH tolerances. Reprinted from Marion et al. (2003b) with permission

Type of Environment	pH limits	Selected references

Acidic systems
  acid mine drainage
  volcanic springs
	pH = −0.06 to 1.0
	organisms: bacteria, archaea, fungi, algae
	record pH = −0.06
	  Picrophilus (archaea)
		Bachofen 1986
		Schleper et al. 1995
		Johnson 1998
		Huber and Stetter 1998
		Schrenk et al. 1998
		Edwards et al. 1999
		Robbins et al. 2000

Alkaline systems soda lakes	pH > 11 organisms: bacteria	Bachofen 1986 Zhilina and Zavarzin 1994 Duckworth et al. 1996

Table 4.4. Radiation dose giving ≈ 37% survival for UV and ionizing radiation (adapted from Kushner, 1981; Baumstark-Khan and Facius, 2002). Reprinted from Marion et al. (2003b) with permission

Microorganism	UV radiation (J/m^2)	Ionizing radiation (Gy)
T1-phage (virus)	–	2600
Escherichia coli (bacteria)	50	20–30
Bacillus subtitis (bacteria)	–	33
Deinococcus radiodurans (bacteria)	600	1500–6000
Saccharomyces cerevisiae (yeast)	80	30–150
Chlamydomonas (algae)	–	24
Bodo marina (eukaryote: heterotrophic flagellate)	5000	–
Humans (eukaryote)	–	1.4

based reasons. However, most other places and times of aqueous environments in the Solar System are not out of reason for consideration as life's niches. For example, there is evidence that the moon of Pluto, Charon, may be hot enough to produce liquid water (Vogel, 1999), and there are multiple strong indications that a tidally heated, wet environment exists inside Saturn's moon, Enceladus (Porco et al. 2006; Kargel 2006); these are places where the possibility of life has not been discussed much, but they are as plausible as Mars or Europa as a potential habitat.

A key to life is the stability of liquid water, at least intermittently, for an ill-defined but long period of time; thus, we have to consider places where there is or was adequate heat to maintain H_2O in the liquid state. Due to the chemical activity of water and its amphoteric and solvent properties, wherever there is introduction of liquid water into previously anhydrous or slightly hydrous systems (such as fresh igneous rocks), there tends to be chemical reequilibration of associated solids; hydrolysis breaks down and transforms silicates and produces water-soluble salts, water-mediated oxidation-reduction reactions transform minerals bearing iron, manganese, copper, sulfur, and other elements; and hydration reactions add crystalline water of hydration to salts and other phases. During that period of phase transitions there is either a state of disequilibrium (diminishing as a stable state is approached) or metastable equilibrium (a thermodynamically unstable assemblage that is kinetically unable to reach a stable state or does so very slowly). It is a remarkable fact that on Earth many such disequilibrated or metastable aqueous environments are exploited by biochemical systems as a source of metabolic energy (Kelley et al. 2002; Huber and Wächtershäuser, 2006). Thus, "follow the water" is not a sure-fire route to finding life or habitable environments, but there is considerable science behind this approach (Furfaro et al. 2007).

The cold, dry surface of Mars today is not an ideal habitat for life. Nevertheless, evidence for a warmer, wetter early Mars has stimulated considerable speculation about the prospects for life on Mars (McKay and Stoker, 1989; Klein et al., 1992; McKay et al., 1992; McKay et al., 1996; Gibson et al., 1997; Shock, 1997; Jakosky and Shock, 1998; Fisk and Giovannoni, 1999; Max and Clifford, 2000; Cabrol et al., 2001; Kargel, 2004).

Europa is a cold, ice-covered moon of Jupiter that might at first seem inhospitable for life. But there is abundant evidence for the presence of a subsurface briny ocean (Khurana et al., 1998; Pappalardo et al., 1999; Kargel et al., 2000; Stevenson, 2000). The putative ocean of Europa has focused considerable attention on the possible habitats for life on Europa (Reynolds et al., 1983; Jakosky and Shock, 1998; Gaidos et al., 1999; McCollom, 1999; Chyba, 2000; Kargel et al., 2000; Chyba and Hand, 2001; Chyba and Phillips, 2001; Navarro-Gonzalez et al., 2002; Pierazzo and Chyba, 2002; Schulze-Makuch and Irwin, 2002; Marion et al., 2003b).

By generally clement Earth standards, the habitats for life on Mars and Europa are likely to be extreme environments. On the other hand, extraterrestrial life, if it exists, may be well adapted to its environments and would find Earth environments extreme. Nevertheless, we have adapted an Earth-centric perspective and will judge environments by terrestrial life standards ("life as we know it"). In this chapter, we will examine the limits for life on Earth as standards for life beyond Earth. We will examine temperature, salinity, acidity, desiccation, radiation, pressure, and time as potential limiting factors for life as we know it. We are implicitly assuming that the supplies of liquid water, energy, and nutrients are (were), at least, present at some minimal levels to support life on Mars (McKay and Stoker, 1989; McKay et al., 1992; Shock, 1997; Jakosky and Shock, 1998; Fisk and Giovannoni, 1999; Max and Clifford, 2000; Cabrol et al., 2001) and Europa (Reynolds et al., 1983; Jakosky and Shock, 1998; Gaidos et al., 1999; McCollom, 1999; Chyba, 2000; Kargel et al., 2000; Chyba and Hand, 2001; Chyba and Phillips, 2001; Navarro-Gonzalez et al., 2002; Pierazzo and Chyba, 2002; Schulze-Makuch and Irwin, 2002). Other potential limiting factors such as toxic metals and the presence or absence of oxygen or other oxidants are only peripherally examined in this work. For most of the factors examined (temperature, salinity, acidity, desiccation, radiation, and pressure), we focus on "normal" biological activity (i.e., respiration, growth, and reproduction, more or less as we know it); for the factor of time, we examine how much time is required for life to develop on a planet and how long organisms can survive in the dormant state. Then, in Chap. 5, we examine biogeochemical applications of the FREZCHEM model; our treatment includes the geochemistry and potential habitats for life on Mars and Europa. And, finally, in Chap. 6, we examine the search for and future of life in the Universe.

4.1 Temperature

There are three environments on Earth where microbes have been identified with temperature tolerances in a range of 100 °C to 121 °C, namely, submarine hydrothermal vents, the subterranean deep biosphere, and terrestrial hot springs (Table 4.1). The highest temperature tolerances (110–121 °C) are found in microbes from marine hydrothermal vents and the subterranean deep biosphere; high pressures prevent these waters from boiling at 100 °C, the normal boiling point of water at 1.01 bar (1 atm) pressure. From terrestrial hot springs, microbes have been isolated that can tolerate temperatures up to 103 °C (Table 4.1).

Hyperthermophiles are invariably either bacteria or archaea. Eukaryotes have an upper temperature range of ∼50–60 °C (Madigan and Marrs, 1997; Nealson, 1997; Nealson and Conrad, 1999; Rothschild and Mancinelli, 2001). Until recently, *Pyrolobus fumarii* (an archaea) had the highest known temperature tolerance of 113 °C (Blöchl et al., 1997); this organism has a minimum temperature for growth of 90 °C and an optimum temperature of 106 °C and is a strict hyperthermophile (Stetter, 1999). Recently, an archaea was isolated from a hydrothermal vent with a temperature tolerance of 121 °C; this organism is closely related to *Pyrodictium occultum* and *Pyrobaculum aerophilum* and doubled in cell number after 24 h at 121 °C (Kashefi and Lovley, 2003).

There are several types of environments on Earth where significant water exists at prevalent low temperatures such that ice and liquid aqueous solutions commonly coexist: permafrost, snow, glaciers, lake and river ice, sea ice, and parts of the atmosphere (polar troposphere, global upper troposphere, and stratosphere). In addition, the deep sea floor occurs at temperatures very close to the freezing point of water. For example, temperatures in the oceanic abysses hover around 2 °C at a maximum hydrostatic pressure of 1100 bars (10,660 m) in the Mariana Trench (Yayanos, 1995). Table 4.1 summarizes some of these environments. Furthermore, in some permafrost and sea-floor environments, the presence of nonpolar gases under pressure can stabilize a modified form of ice known as gas hydrates even where temperatures are not quite low enough for ordinary ice to form.

There have been a number of reports in recent years demonstrating that some microbes can metabolize, albeit slowly, at temperatures in a range of -17 °C to -20 °C (Table 4.1). These organisms include bacteria, lichens (a symbiotic association of algae and fungi), and fungi (yeasts). Many of these ecosystems are in protected environments such as in aqueous pockets in ice (Priscu et al., 1998; Psenner and Sattler, 1998; Thomas and Dieckmann, 2002) and within rocks (cryoendoliths) (Friedmann and Ocampo, 1976; Friedmann and Ocampo-Friedmann, 1984), where climate is more hospitable than in exposed areas. On the other hand, Price and Sowers (2004) have argued that there is no evidence of a minimum temperature for metabolism (growth, maintenance, or survival). Their extrapolated (from -20 °C) rate at -40 °C in ice corresponds to ≈ 10 turnovers of cellular carbon per billion years. If these arguments are correct, then microbes could survive indefinitely, provided they are protected from especially destructive forces such as high temperatures and radiation.

Temperature as well as salinity and other compositional variables have a profound influence on the viscosity of aqueous solutions and many aspects of life. As temperature declines and salinity increases, viscosity increases, diffusion rates decline, metabolic rates decline, motility of motile microorganisms and feeding rates decline, life spans may increase, and rates of evolution decline. All other things being equal, aqueous systems and planets allowing only cryogenic life will have less evolved life than chemical systems and planets that have permitted life to occur for substantial periods at higher temperatures.

An intriguing possibility we raise here for the first time is that there must be particular brine compositions where the net volume change during freezing is zero. That is, the combined effects of ice expansion and contraction effects due to salt precipitation and brine compositional migration toward a more salt-rich residual brine all average out to zero net volume change.

Whereas this would be a very special case to be exactly zero, a rough approximation of zero volume change has been observed in the high-pressure phase equilibrium/volumetric studies of brines by this book's second author and collaborators at Lafayette College (Hogenboom et al., 1995; Dougherty et al., 2007). This condition requires a system to be either at a eutectic or a cotectic crystallization path involving ice plus at least one salt phase. The intrigue here is that these special brine compositions may be uniquely habitable in conditions of intense freezing and thawing of psychrophilic microbial life. At least cellular damage due to expansion and contraction of a cell's environment would be minimized for these compositions, and if other causes of freeze damage (crystal punctures of cells, for instance) can be minimized, these brines might favor psychrophilic life. The same experiments have shown that under some circumstances, brines freeze by production of impressive daggerlike needles (each a potential threat to any cell wall that intervenes), whereas in other cases such quench crystals do not form (Dougherty et al., 2007). Crystal morphology is controlled both by brine composition and the rate of cooling and extent of supercooling. Ultimately, all of these issues come down to thermodynamics, phase equilibria, and kinetic factors related to crystallization and solute diffusion.

4.2 Salinity

Salinity affects microbial activity, in part, because it controls water availability. The higher the salinity, the more energy an organism must expend to maintain a favorable osmotic balance. Salts, of course, have effects on living organisms beyond water availability. For example, salts can be both a source of essential nutrients as well as a source of toxic heavy metals. Also, sulfate salts appear to be more favorable for life than chloride salts; see the discussion in Sect. 5.1.2 (Aqueous Saline Environments). However, in this section on salinity, the focus will be on salinity as a control on water availability.

Because salinity has been studied by scientists representing many disciplines, measures of salinity vary widely among disciplines. Chemists tend to prefer concentration units of molality [mol/kg (water)], while physiological microbiologists often report salinities as salt% (wt/wt) [g/100 g (soln)], salt% (wt/vol) (g/100 ml), or as the activity of water (a_w). The activity of water is probably the best measure of salinity as it relates directly to the osmotic gradient controlling flows of salts and water into and out of organisms.

This diversity of measures of salinity is especially a problem in cross-disciplinary work, where alternative measures from other disciplines are often unfamiliar. The equations needed to convert from one measure to another are

$$m \left(\frac{\text{kg(water)}}{\text{kg(soln.)}} \right) \rho = M, \qquad (4.1)$$

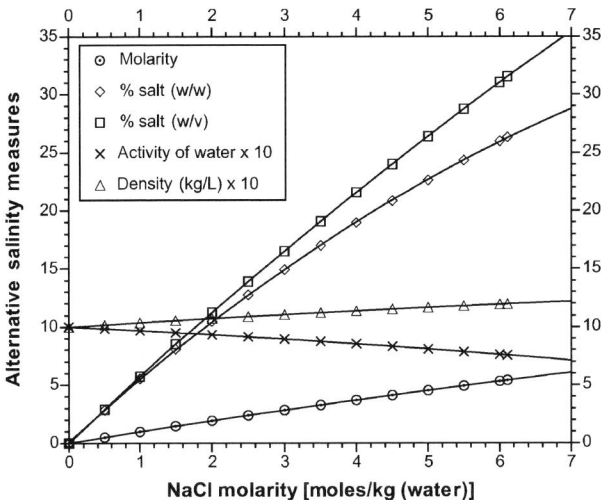

Fig. 4.2. Alternative measures of salinity for pure NaCl solutions at 25 °C and 1 atm pressure. Reprinted from Marion et al. (2003b) with permission

$$\frac{\sum_i m_i MW_i}{10} \left(\frac{\text{kg(water)}}{\text{kg(soln)}}\right) = \%\text{salt}(w/w), \tag{4.2}$$

$$\%\text{salt}(w/w)\rho = \%\text{salt}(w/v), \tag{4.3}$$

$$\frac{\text{kg(water)}}{\text{kg(soln.)}} = \frac{1000}{1000 + \sum_i m_i MW_i}, \tag{4.4}$$

where m is the molality [= mol/kg (water)], M is the molarity [= mol/L], ρ is the density (= kg (soln.)/L), and MW is the molecular weight (g/mol). The density of solutions (ρ), which is needed for some of the above conversions, can either be measured experimentally or estimated with a model (e.g., FREZCHEM, Eq. 2.88).

Figure 4.2 depicts several measures of salinity for pure NaCl solutions at 25 °C and 1.01 bar total pressure calculated using the FREZCHEM model and Eqs. 4.1 to 4.4. NaCl solutions were used in this example because this salt is often used as a background salt in physiological studies (Madigan and Oren, 1999; Kaye and Baross, 2000). For salts other than NaCl or at temperatures other than 25 °C, Eq. 4.1 to 4.4 must be relied upon to make conversions or see Marion (2007).

Table 4.2 outlines the approximate salinity limits for biological activity of bacteria/archaea and fungi. Salinity is one of the few limiting factors for life where a eukaryote (fungi) has a higher tolerance than prokaryotes (bacteria/archaea). To place these limiting a_w values in perspective, the a_w for seawater is 0.98. Most prokaryotes and fungi can tolerate much higher salinities than seawater (Table 4.2).

4.3 Acidity

Acidity is typically quantified using the pH scale

$$\mathrm{pH} = -\log_{10}(a_{\mathrm{H}^+}), \tag{4.5}$$

where a_{H^+} is the hydrogen ion activity. The activity of a single ion cannot be measured unambiguously (Pitzer, 1995); therefore, assumptions must be made in defining pH. One convention is to use a chemical thermodynamic model to estimate the activity coefficient of H^+ (γ_{H^+}). Then, given an experimental measurement of the molal concentration (m), one can calculate the activity as follows: $a = \gamma m$. This is how most geochemical models work. Another assumption, most frequently used in calibrating pH standards, is the MacInnis convention (Harvie et al., 1984). In this case, the assumption is made that $\gamma_{\mathrm{K}^+} = \gamma_{\mathrm{Cl}^-}$ in all solutions of the same ionic strength. This allows one, indirectly, to estimate γ_{H^+} and to define a_{H^+}. The reason for raising this issue is that pH values calculated with these two conventions can lead to very different pH values in extremely high acidities (pH <1.0). [See Marion (2002) for a fuller discussion of this issue.]

There are many studies demonstrating that a wide range of organisms can tolerate pH values <1.0. For example, bacteria, archaea, fungi, and algae have all been demonstrated to tolerate pH values ≤ 1.0 (Table 4.3). The current record holders are *Picrophilus oshimae* and *Picrophilus torridus* (archaea), which can grow at a pH of -0.06 (Schleper et al., 1995). Unfortunately, in these acid studies, it is not always clear which of the above two pH conventions was used. For example, the Schleper et al. (1995) study, which reported the lowest pH value of -0.06, only indicates that $1M$ HCl was used as a reference for pH 0; but there is no clear indication which convention was used to calibrate their pH electrode. On the other hand, the Iron Mountain acid studies (e.g., Schrenk et al. 1998; Edwards et al. 1999; Robbins et al. 2000) clearly use the MacInnis convention in calibrating electrodes at low pH (Nordstrom et al. 2000).

Invariably, at least on Earth, high acidities are associated with high concentrations of heavy metals because strong acids are highly effective in dissolving primary minerals and releasing heavy metals into the environment (Krishnaswamy and Hanger 1998; Robbins et al. 2000; Lopez-Archilla et al. 2001; Fernandez-Remolar et al. 2003). Therefore organisms that tolerate strong acidity also tolerate high levels of heavy metals.

There are fewer studies of high alkalinities (pH > 10) than of high acidities (pH < 1.0) probably because high alkalinities are more rare in nature. Nevertheless, there are reports of organisms tolerating pH values >11 (Table 4.3), and maybe even as high as 12.5–13 (Bachofen 1986; Duckworth et al. 1996).

4.4 Desiccation

The desiccating power of the atmosphere is generally measured by relative humidity (RH), which is related to the activity of water (a_w) by

$$a_w = RH/100. \tag{4.6}$$

Just as $a_w = 0.6$ is considered the lower limit for biological activity in saline solutions (Table 4.2), $RH = 60\%$ is considered the lower limit for biological activity under dry atmospheric conditions (Kushner 1981; Dose et al. 2001).

A clear distinction must be made between biological activity and survival under desiccating conditions. Some organisms can survive 99% loss of water with $a_w \sim 0$ (Mazur 1980). *Bacillus sphaericus* spores survived 25 million years of desiccation in amber through a process called anhydrobiosis (Fischman 1995). Bacteria, fungi, plants, and insects have been shown to survive extensive periods of dehydration (Rothschild and Mancinelli 2001). A clear distinction must also be made between conditions allowing biological activity and conditions allowing the origin of biological activity. If this distinction exists, the boundaries of these conditions are not known because the genesis of life is inadequately understood.

Examples of especially dry environments on Earth include the Atacama Desert of northern Chile and the Dry Valleys of Antarctica. Dose et al. (2001) exposed spores, conidia, and cells of several microbes to 15 months of desiccation in the dark at two locations of the Atacama Desert. *Bacillus subtilis* (bacteria) spores (survival ~15%) and *Aspergillus niger* (fungi) conidia (survival ~30%) outlived other species. *Deinococcus radiodurans* (bacteria) did not survive the desert exposure because they were readily killed at RH between 40% and 80%, which occurred during desert nights (Dose et al. 2001).

4.5 Radiation

Two types of ionizing radiation can limit life, and either can do its deadly work in long-term, low doses or short-term, intense doses: (1) short-wave electromagnetic radiation (ultraviolet, x-ray, and gamma-ray parts of the spectrum); and (2) high-energy corpuscular radiation (energetic electrons and protons especially, and heavier ions). Major sources of lethal radiation doses include (a) cosmic phenomena (gamma ray bursts, nucleosynthesis due to supernovae, stellar/solar fusion, and stellar/solar photospheric reradiation), (b) planetary magnetospheric phenomena (including trapped solar wind and sputtered, ionized material released from planetary surfaces by interaction with cosmic EM and corpuscular radiations), and (c) radioactive decay in planetary objects due to long-lived radionuclide decay and short-lived cosmogenic nuclide decay (those nuclides generated partly by neutron activation). Ultimately, all of these radiation sources trace back to nucleosynthetic processes in stars and heavier objects. The surface of Europa, for instance, would be deadly to frozen life on time scales of seconds to days (due to energetic magnetospheric ions), while the cosmic-ray-shielded deep permafrost of Earth would be deadly to frozen organisms on time scales of hundreds of thousands to millions of years due to gamma radiation caused by long-lived radionuclide decay.

Resistance to one form of radiation does not necessarily convey protection from other forms. Almost all organisms are prone to UV damage because the macromolecules that propagate genetic information (DNA) absorb UV radiation. For example, the experiments in the Atacama Desert cited in the previous section were done in the dark (shade). Direct exposure to UV radiation in these experiments killed all organisms within hours (Dose et al. 2001).

Table 4.4 depicts resistance to UV and ionizing radiation for several microbes. *D. radiodurans* is well known to have a high resistance to ionizing radiation. This resistance to radiation is thought to have evolved initially as a resistance to desiccation. The mechanism for conveying this resistance is believed to be due to their ability to quickly repair DNA damage (Kushner 1981; Smith 1982; Bachofen 1986; Jawad et al. 1998; Rothschild and Mancinelli 2001). Other mechanisms to protect organisms from UV radiation include the development of iron-enriched silica crusts (Phoenix et al. 2001) and self-shading (Smith 1982). Also, both water and ice are effective in absorbing UV radiation (Baumstark-Khan and Facius 2002).

4.6 Pressure

Pressure affects physics, chemistry, and biology. Chemical reactions that lead to a decrease in volume are favored by pressure. For example, the dissolution of gypsum at $0\,°C$ is as follows:

$$CaSO_4 \cdot 2H_2O\,(V^0 = 74.69) \leftrightarrow Ca^{2-}\,(V^0 = -19.69) + SO_4^{2-}\,(V^0 = 9.26)$$
$$+ 2H_2O\,(V^0 = 18.02)\,, \qquad (4.7)$$

where V^0 is the molar volume (cm^3/mol) at infinite dilution:

$$\Delta V_r^0 = V_{Ca}^0 + V_{SO_4}^0 + 2V_{H_2O}^0 - V_{CaSO_4 \cdot 2H_2O}^0 = -49.08\,\text{cm}^3\,\text{mole}^{-1}. \qquad (4.8)$$

In this case, ΔV_r^0 is negative, which implies that pressure will cause the reaction to shift to the right causing a dissolution of gypsum. Another important reaction is the stability of water ice and liquid water at subzero temperatures under pressure:

$$H_2O(I,cr) \leftrightarrow H_2O(aq)\,, \qquad (4.9)$$

which has a $\Delta V_r^0 = -1.63\,\text{cm}^3\,\text{mol}^{-1}$ at $0\,°C$. In this case, pressure will cause a melting of ice (to reduce the volume) with a consequent lowering of the freezing point (Fig. 3.3). Another reaction that will be considered in Chap. 5 is

$$MgSO_4 \cdot 12H_2O \leftrightarrow MgSO_4 \cdot 7H_2O + 5H_2O(aq), \quad (4.10)$$

which has a $\Delta V_r^0 = 16.31\,\mathrm{cm^3\,mol^{-1}}$ at $0\,°C$. In this case, ΔV_r^0 is positive, which will cause the reaction to shift to the left, favoring the precipitation of $MgSO_4 \cdot 12H_2O$ at a higher temperature. Also, we showed previously that the equilibrium temperature for the reaction

$$NaCl \cdot 2H_2O(cr) \leftrightarrow NaCl(cr) + 2H_2O(l) \quad (4.11)$$

increases with pressure ($\Delta V_r^0 = 5.10\,\mathrm{cm^3\,mole^{-1}}$; Fig. 3.20). These four examples of pressure on chemical reactions are sufficient to demonstrate that the effect of pressure is dependent on the volumetric properties of the individual constituents, which makes every chemical reaction highly individualistic.

High pressures can occur in both deep-earth and deep-sea environments, but there are some fundamental differences between these two systems. In the deep sea, hydrostatic pressures on organisms is simply $p = \rho g h + p_a$, where p_a is ambient atmospheric pressure at the surface (generally about $1\,\mathrm{atm} = 1.010325\,\mathrm{bars} = 0.101325\,\mathrm{MPa}$), ρ is the density of seawater (about $1026\,\mathrm{kg\,m^{-3}}$), g is surface gravitational acceleration (about $9.8\,\mathrm{m\,s^{-2}}$), and h is the depth (m). For example, for 2 atm pressure (1 atm above sea level pressure), the depth is 10 m. In the deep earth, the confining pressure could be atmospheric with organisms growing in air pockets, such as caves or soil pore spaces, or in the water-filled pore spaces of deep-sea sediments, where the pressure then is hydrostatic pressure for that depth below sea level. For depths beneath the land surface or sea floor greater than about 1 to 2 km, porosity tends to become choked off, and pressure equilibrium with the atmosphere or ocean is disrupted, so hydrostatic pressure tends increasingly to become lithostatic. Life may occur at even higher pressures within deep aquifers and crustal brine pockets, where the organisms may be subjected to both hydrostatic and lithostatic pressures. Unfortunately, the actual pressures under which these deep-earth microbes grow are poorly documented (Pedersen 1993). Another fundamental difference is that deep-sea environments decrease in temperature with increasing depth (within the sea itself), while deep-earth (continental and oceanic crustal) environments increase in temperature with increasing depth.

Microorganisms have been isolated from the Mariana Trench in the Pacific (10,660 m depth) where pressures reach 1100 bars (110 MPa; Yayanos 1995; Kato et al. 1998; Abe et al. 1999). Two bacteria similar to *Moritella* and *Shewanella* are apparently obligately barophilic with optimum pressures for growth occurring at 700 bars (70 MPa) and no growth below 500 bars (50 MPa; Kato et al. 1998). These Mariana Trench organisms grow at a temperature of $2\,°C$.

There are some archaea associated with deep-sea hydrothermal vents that can survive at pressures as high as 890 bars (89 MPa; Pledger et al. 1994). The high pressure of hydrothermal vents has a compensatory effect that allows stabilization of molecules, which allows growth at elevated temperatures up to 121 °C (Table 4.1).

Microorganisms have been found growing at depths of 2.8–4.2 km beneath the land surface (Pedersen 1993; Fyfe 1996; Kerr 1997). Microbes at 4.2 km grow at a temperature of 110 °C. Temperature, rather than pressure, is probably the most important growth-limiting factor for deep-earth microbes (Pedersen 1993; Fyfe 1996). Organic biopolymers and complex cellular structures tend to be destroyed at elevated temperatures, and apparently elevated metabolism and cellular repair activity does not compensate for the rates at which critical bonds are broken; hence, cells cannot repair thermal damage beyond a point.

Organic-rich shale-hosted zinc-lead sulfide and barite Red Dog deposits in the Brooks Range (Alaska) contain fluid inclusion bearing sphalerite veins (Leach et al. 2004). Many of the inclusions have salinities ranging from 14 to 19% NaCl equivalent, based on melting analysis. It is thought that these brines originated by formation of a hypersaline ($\sim$30% NaCl) evaporitic marine brine on tidal carbonate flats; then, after burial, this brine mixed with more dilute aquifer waters. This saline brine then presumably participated in the origin of the ore rocks. The fluids became trapped in the sphalerite at depths of 2.4 to 7.4 km (pressures of 650 to 2100 bars) and temperatures ranging from 383 to 453 K (Leach et al. 2004). Only the lowest formation temperatures of these inclusions are consistent with any form of hyperthermophilic life. The rocks later were subjected to pressures of up to 3400 bars and temperatures of up to 487 K, according to analysis of a different set of aqueous and coexisting methane fluid inclusions.

These findings for the Red Dog ore deposits, typical of similar zinc ore deposits around the world, would indicate that the original evaporitic marine environment was followed progressively by burial and increases in pressure and temperature, with a sequence of microbial life forms possible until eventually only life's organic, broiled remains could exist in the rocks. This type of burial-induced biological succession and biochemical metamorphism, starting with common evaporitic deposits in basins of high primary productivity and hypersaline conditions, is very common throughout Earths history and is a key aspect of petroleum and gas genesis. In fact, indigeneous ecological communities of thermophilic bacteria and hyperthermophilic archaea are thought to inhabit hot petroleum reservoirs (L'Haridon et al. 1995; Magot et al. 2000). Culture experiments have shown that *Ferroglobus placidus*, an obligate hyperthermophile, is able to reduce Fe^{3+} in anaerobic oxidation of a wide range or aromatic hydrocarbons (Tor and Lovley 2001).

Most petroliferous deposits originally were laid down under warm to hot arid climates and tectonic situations conducive to rapid burial, such as rift basins. One can envisage a complete ecological succession during burial starting with the inhabitants of the original depositional waters and evolving to these bizarre petroleum hosted, deep subsurface ecosystems.

Inland cold-climate basins that are both evaporitic and biologically productive are much rarer than warm-climate evaporitic basins in Earth history. Such basins now exist in western China. Cold- and warm-climate evaporitic basins are expected to be very different in several respects. Key differences should affect both the phase equilibria of precipitated and metamorphosed salts and the sustainability of chemolithoautotrophic biological activity at depth during burial. Primary precipitated salts in cold basins would tend to include a rich variety of highly hydrated salts, which then would undergo a cascading sequence of incongruent dissolution and dehydration events as the assemblage is buried and heated. In hot-climate basins, many of these salt hydrates are omitted right from the start. Primary productivity may be lower in the colder basins, but if there is a subsurface biochemical energy source, life may be maintained to greater depths owing to cooler surface temperatures and overall cooler geotherms. Rather than characteristically burning out at about 4 km burial depth and 383 K, microbial life in cold-climate depositional basins on Earth might be sustained to as deep as 6 km, if thermal conductivity, heat flow, and other factors besides surface temperature are the same.

On Mars, where surface temperatures are much lower than on Earth and temperature gradients as a function of pressure are comparable (reduced heat flow is compensated by reduced gravity), life probably can exist to much higher pressures and slightly higher temperatures and far greater depths than in Earth's crust. The lower gravity also means that, all other things being equal, pore space compaction occurs less rapidly as a function of depth on Mars than on Earth. Thus, we would not be surprised if hyperthermophilic life existed to depths of >10 km on Mars, if it exists there at all. Because of the slow cycling of water, energy, and nutrients at depth, metabolic activity is believed to be extremely slow at great depths on Earth (Kerr 1997), and this would be a limitation deep in Mars' crust as well.

Another phenomenon found under pressure at depth in terrestrial oceans and in permafrost is gas hydrate deposits. Hydrates of natural gases such as methane ($CH_4 \cdot 6H_2O$) and carbon dioxide ($CO_2 \cdot 6H_2O$) form on Earth beneath low-permeability strata under high pressure and low temperature (Kvenvolden 1993; Sloan 1998; Blunier 2000):

$$CH_4(g) \leftrightarrow CH_4(aq) + 6H_2O(l) \leftrightarrow CH_4 \cdot 6H_2O(cr). \qquad (4.12)$$

The stability of these solid-phase compounds is a function of pressure, temperature, and matrix salt composition (Sloan 1998; Marion et al. 2006). Gas hydrates could be important sources of high-energy carbon (Carney 1994). On Earth, gas hydrate deposits can sustain complex chemosynthetic communities (Sassen et al. 1999; Fisher et al. 2000). There is speculation that gas hydrates may be present on Mars and Europa (Kargel et al. 2000; Max and

Clifford 2000). Many metallic ore and evaporite minerals contain methane, though not in hydrate form. For instance, the Red Dog zinc ores mentioned above contain methane inclusions in association with saline inclusions. In many instances, microscopic methane and carbon-dioxide-rich fluid inclusions form clathrates within their tiny hydrous envelopes. Some salt deposits are so enriched in high-pressure free gas inclusions and/or gas hydrates that mining, crushing under boot, or other means of physical disturbance causes popping or even explosive decrepitation.

Recently, it was demonstrated in a diamond anvil cell that *Shewanella oneidensis* and *Escherichia coli* strains remain physiologically and metabolically active at pressures of 680 to 16,800 bars for up to 30 h (Sharma et al. 2002). At pressures of 12,000 to 16,000 bars, living bacteria resided in fluid inclusions in Ice VI crystals and continued to be viable when pressure returned to 1 bar. However, only 1% remained alive; whether this constitutes viability or survival under pressure is contentious (Couzin 2002). Nevertheless, it demonstrates that pressure may not be much of an impediment for some life forms, and that even the deep ocean of Ganymede might be suitable for life.

What do first principles say about the limits to life imposed by pressure? Life needs a polar aqueous solution. The polarity of water goes to zero beyond megabar pressures; it eventually undergoes metallization (Fig. 4.3). That kind of extreme "megabar water" is almost surely unsuited for life, though it may be liquid at the right temperature conditions. It is not known what value of polarity is needed for water to act as a life-giving medium or whether some other polar volatile, such as ammonia, might take on properties at very high pressures that would help it substitute for water as a life-giving medium. The physics of water at hundreds of kilobars to megabar pressures and the high-pressure suitability of water for life over part of this pressure range is almost completely unexplored territory. Whether there exists some range of depths and pressures within the gas giants to serve as an abode of exotic life is really unclear; it would appear that Jupiter and Saturn probably have adiabats that are too hot and atmospheric dynamics that are too strongly convective (Fig. 4.4). Uranus and Neptune might potentially be of greater astrobiological interest because the pressure conditions in the thermally acceptable zones are not exorbitant (Fig. 4.4), and these planets' atmospheres are probably more stable than the larger gas giants. The deep molecular "oceans" (or ice layers) of water, ammonia, and methane believed to exist in Uranus and Neptune are subject to pressures in the range of several hundred kilobars to several megabars (Hubbard et al. 1995). The top of this icy domain could be a habitat for unfamiliar types of life, but conditions there are so uncertain and biochemistry so unknown that we have to leave this as a speculation. Much more likely, temperatures are so high that complex organic molecules do not exist, and instead these layers truly are simple molecular mixtures. We

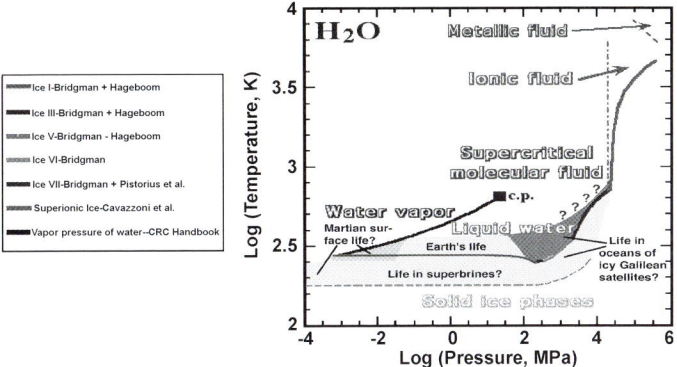

Fig. 4.3. The stability of water as a function of temperature and pressure. Reprinted from Kargel (2004) with permission

can make more definitive statements about the possibilities of life at a few tens of kilobars and lower pressures relevant to icy satellites.

Water at tens of kilobars resembles water at ordinary pressures in many respects. It has a strongly polar molecular structure, solvent and amphoteric acid/base properties, and thermal expansivity and bulk modulus properties similar to water at ordinary pressures, although boiling points and freezing

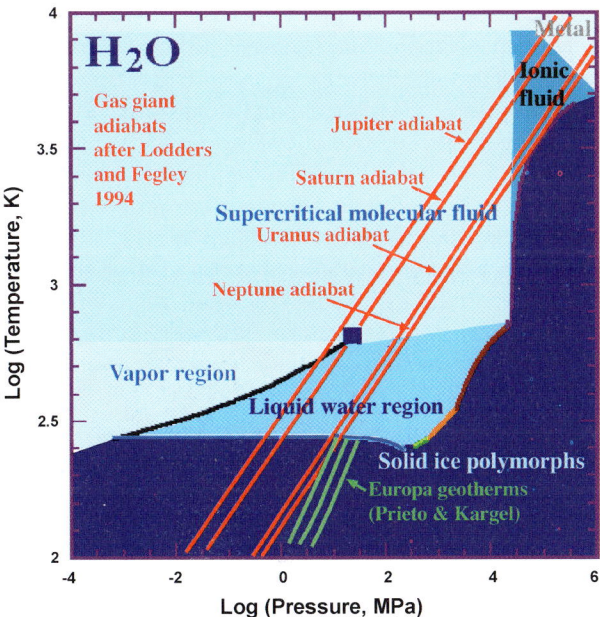

Fig. 4.4. Searching for life by mapping water

points shift considerably and the structure of equilibrium ice phases and the sign of the volume change upon freezing all are different from the case of water at low pressure. For pressures of ordinary relevance to icy satellites, asteroids, and planetary crusts (Fig. 4.3), excluding impact shock pressures, it would seem that pressure itself is not a life-limiting factor. In fact, pressure extends the conditions where liquid aqueous solutions may exist (by elevating boiling points and depressing melting points of ice 1 h); in the range of hundreds of bars, this is a key to the viability of hyperthermophiles, and we expect this aspect to become even more significant into the range of several kilobars. However, the basic properties of water do change with pressure, and these changes affect thermodynamic phase equilibria, reaction rates, and the functioning of biological systems. Gas-filled vacuoles also change their volume and density, and gas solubilities in cellular fluids change with pressure, and with it, mineral equilibria (such as carbonate precipitation boundaries) shift. For many microbial species, it is just necessary to have life adapted and biochemistry "calibrated" to the relevant pressures. This happens all the time in hot springs, hydrothermal systems, and marine environments, where a given individual may be subjected to changes in pressure ranging from a few bars to a few hundred bars. The ability of individuals to adapt to pressure changes has been investigated in the laboratory, and this adaptability has its limits (Couzin 2002; Sharma et al. 2002), but pressure itself does not appear in any way to be a life-limiting factor over the range of pressures encountered in Earth's hydrosphere and upper crust.

Adaptations over pressure ranges of many kilobars or tens of kilobars may be difficult for many individuals because of large thermodynamic shifts, but the properties of water probably do not change so much that biology would be impossible. For a given set of conditions where a liquid-water phase exists, organic chemistry probably will tend to find the right equilibria where biochemistry may get rooted. If one biochemical reaction pathway is not stable for a certain set of conditions, then another probably will be. If so, the type of life that emerges and its major biochemical pathways will be suited to the prebiotic chemistry that prevails for a certain set or range of conditions. We expect that there are limits to this biochemical flexibility to adapt to or evolve with the pressure variable, but the pressure limits to life at least are broad and probably extend beyond the science community's observational and experimental limits so far explored.

One way that pressure must affect life is by its control of the freezing point of water and the viscosity of water and aqueous solutions near the freezing point (or at any temperature) and, hence, pressure control on ionic mobility and reaction rates, including metabolism. This in itself does not limit life, but it helps to set the pace of life's genesis, evolution, and life span, and that may affect the viability of life in the face of unstable or dynamic conditions.

A major discontinuity in the behavior of water occurs at about 2100 bars, where the equilibrium ice phase changes from ordinary ice Ih, which is less dense than water, to ice III, which is more dense. The volume expansion on freezing below 2100 bars is a severe problem for many species because growing ice crystals in freezing cells rupture cell walls both by the expansion of cell contents and by crystal penetration through the cell walls. Cellular damage during freezing ought to be reduced or eliminated at pressures above the ice I/ice III transition. Freezing, however, may still kill cells by rupturing their cell walls with growing crystallites. Thawing at these elevated pressures, on the other hand, may be traumatic due to the sudden expansion. Freezing at any pressure may cause cell damage by exclusion and forced precipitation of solutes, which may isolate nutrients away from the cellular domains where metabolism would utilize them once thaw conditions occur.

4.7 Time

Two aspects of time have a bearing on the possibility of extraterrestrial life. (1) How much time is required for life to develop on a planet? (2) How long can life survive in the dormant stage in isolation from conditions normally considered vital for life such as cycling of liquid water, energy, and nutrients? There are further higher-order issues that pertain to time. Does the time required for biogenesis and evolution depend on the composition of the chemical system? Do these critical time spans depend also on the temporal and spatial stability of composition? How do the physical nature and biochemical processes of life depend on the stability (dynamical time constants pertinent to changing conditions) of a chemical system? For instance, are acidic solutions characteristic of stratospheric clouds fundamentally hostile to any type of life, or are they hostile because of the dynamic state? Where a salt pan's composition admits certain types of life, might frequent flushing by fresh water render that life nonviable? How do diurnal fluctuations in solar ultraviolet radiation affect the viability of habitats in surface soils? These questions are clearly relevant to the day-to-day viability of life in certain terrestrial habitats, but even more so it seems that dynamic conditions would affect biogenesis. The effects are not necessarily all deleterious to life, as dynamic conditions, including oscillating compositions, can drive biochemical energy gradients, which may be useful to life; the life simply has to avoid being killed by a zeroing or reversal of a favorable gradient.

The Earth began forming about 4.6 billion years (Ga) ago. The first 600 to 800 million years (Ma) of Earth's existence have been erased by the constant early bombardment of asteroids and comets (Arrhenius and Lepland 2000; Delsemme 2001; Ehrenfreund and Menten 2002; Wharton 2002). The earliest geologic evidence for life on Earth dates to 3.5 to 3.8 Ga (Schopf and Packer 1987; Mojzsis et al. 1996; Ehrenfreund and Menten 2002; Stetter 2002; Wharton 2002). Based on this evidence, it has been argued that life on Earth developed rapidly within about 200 to 300 Ma. During this interval, the Earth evolved from a hot dry rock to a cool wet world. Evidence suggests that the rain of asteroids and comets brought to Earth water, organic molecules, and gases that are key ingredients for the establishment of life (Delsemme 2001; Horneck and Baumstark-Khan 2002). According to present knowledge, the time necessary for life to develop might require hundreds of millions of years.

This model of the evolution of life on Earth does not preclude the possibility that life arrived on Earth fully formed from another body [the Panspermia Hypothesis (Horneck and Baumstark-Khan 2002; Wharton 2002; Napier 2004; Wallis and Wickramasinghe 2004)]. In this case, only a short-term temporary abode would be necessary for life to become established.

Another, perhaps more important, question is: How long can life survive in the dormant state on a planet even under hostile conditions? A number of reports in recent years have suggested that microbes can survive in the frozen state (in ice or permafrost) for periods ranging from thousands to 3 million years (Soina et al. 1995; Stone 1999; Christner et al. 2000; Gilichinsky 2002). On an even longer time scale (25 to 40 Ma), viable microbes, similar to *B. sphaericus*, have been isolated from bees encased in amber (Fischman 1995; Cano and Borucki 1995). The longest reputed record for survival goes to a *Bacillus* spp. that has been isolated from halite crystals believed to be 250 Ma (Vreeland et al. 2000). In that study, only 2 of 53 salt crystals had viable bacteria, suggesting that survival is a rare occurrence. However, Hazen and Roedder (2001) have argued that in the absence of primary growth features in the specific halite crystals studied, the age of these crystals and their fluids must remain in doubt. In a reply to these concerns, Powers et al. (2001) defended their crystal and fluid inclusion ages.

Earlier, under "Temperature," we discussed the work of Price and Sowers (2004), who argued that there is no lower temperature limit for microbial activity, which implies that microbes could survive indefinitely, provided they are protected from especially destructive forces such as high temperature and radiation. All claims of exceptionally long-dormant life in frozen terrestrial environments have been challenged on the basis of low-temperature metabolism, which must continue, very slowly, despite the microbe's state of suspended animation. A further challenge has been raised on the basis of radiation damage due to radiolysis from long-lived radionuclides; without functioning means to repair damage to DNA and other cellular molecular apparatus, the mounting toll of radiolytic damage would render any life dead within the amount of time claimed for these dormant periods. Hence, cell debris resembling cells might survive in a frozen state for millions or billions of years, but the cells themselves would have resisted resuscitation after hundreds of thousands of years even under optimum conditions. One possibility, however, is that sluggish metabolism in frozen microorganisms, coupled with resupply of nutrients and reductants or oxidants, might take place in unfrozen brine films, and possibly even molecular repair activities might occur slowly in frozen microorganisms. However, it seems to us that the challenges to frozen life are pretty severe.

Life in our Solar System could have started rapidly if the Panspermia Hypothesis is correct and life was seeded to Earth (and Mars and Europa?) from outside, or it may have taken hundreds of millions of years. At this time, only the crudest boundaries can be placed on the time for life to develop or the survival time for life after environmental conditions become hostile, but it may be on the order of hundreds of millions of years.

5 Biogeochemical Applications to Solar System Bodies

In this chapter, we examine biogeochemical applications of the FREZCHEM model to Earth, Mars, and Europa, where cold aqueous environments played and continue to play a critical role in defining surficial geochemistry. Interpretations include the potential for life in these environments. These simulations cover applications to seawaters, saline lakes, regoliths, aerosols, and ice cores and covers. These examples are the proverbial "tip of the iceberg" in terms of the potential of this model to describe cold aqueous geochemical processes. At the end of the chapter, we discuss application limitations, cases where the underlying thermodynamic assumptions are at variance with real-world situations.

One of the great virtues of the FREZCHEM model is its ability to examine complex chemistries. The number of independent components for the systems examined in this chapter range from four to eight. Earth seawater consisting of Na^+, K^+, Mg^{2+}, Ca^{2+}, Cl^-, SO_4^{2-}, and alkalinity has seven independent components (six salts and water). The most complex system evaluated is the snowball Earth seawater (eight independent components), which in addition to the above seven components also includes Fe^{2+}. This ability to cope with complexity makes models like FREZCHEM more realistic in describing natural systems than simpler binary and ternary diagrams; we demonstrate this point with data from Don Juan Pond, the most saline body of water on Earth.

The FREZCHEM model is also highly flexible in the range of compositions (Table 3.1) and their temperature (< -70 to $25\,^\circ$C) and pressure (1 to 1000 bars) dependencies that can be examined. This flexibility allows one to examine a broad range of hypothetical scenarios. However, a major limitation in applying the FREZCHEM model are the poor constraints, at times, for many planetary situations. For example, we examine three hypothetical cases for a Europan ocean: neutral, alkaline, and acidic scenarios. All of these scenarios for a Europan ocean have been suggested by past work. Obviously, these scenarios are mutually exclusive. Geochemical models such as FREZCHEM benefit tremendously from space missions that constrain potential chemistries as well as laboratory studies that define chemical parameters. In turn, thermodynamic models can be used to constrain interpretations of planetary geochemistries; some speculations are simply not thermodynamically feasible. Space missions and geochemical models are routinely used to channel laboratory studies into the most meaningful planetary chemistries.

There is a strong synergy among space missions, laboratory studies, and geochemical models in furthering our understanding of planetary biogeochemistry. One thing space exploration has taught us is that planets and satellites in the Solar System are extremely diverse geologically and geochemically. Among these worlds, dry ones (such as the Moon, Mercury, and Venus) are the exception. Because oxygen is so abundant in the Universe, H_2O also is abundant. This fact is very well reflected in the solid bodies of our Solar System. Among planets and satellites that are geologically evolved, a striking number are or have in the past been "water worlds," as indicated by the landscape-shaping roles of liquid water and ice on rocky surfaces; the role of cryovolcanism as a major process on some outer planet satellites; the widespread occurrence of aqueous chemical alteration and mineral products of that alteration on Earth, Mars, many asteroids/meteorite parent bodies, and several outer planet satellites; and geophysical or geochemical indications of ice-crusted oceans on Europa, Ganymede, Callisto, and Enceladus.

Salts seem to be nearly as widespread as ice (though not as abundant) in the Solar System, being present on nearly every object where ice has also been observed and where indications of a watery past are in evidence (Enceladus is the key exception so far). There are very few instances (Venus is the most notable one) where salts occur on worlds where H_2O is absent or nearly absent. This is a striking correlation, given the terrestrial situation of most evaporites, which are mostly associated with hot and arid areas of Earth. Even so, notable instances of ice-associated evaporites also occur on Earth. In detail, each world and the paragenesis of its salts tells a different story. The key point here is that in this Solar System, salts commonly are associated directly with ice. Hence, phase equilibria generally are described by FREZCHEM, even if on Earth some of the most important evaporite deposits were not formed in the presence of ice or icy-cold conditions.

5.1 Earth

5.1.1 Seawater Freezing

Seawater is the most abundant aqueous solution on Earth, and, as a consequence, has been the subject of countless studies (Millero 2001). The occurrence of marine evaporites has spurred much of the work on seawater solidification, which can occur by either evaporation or freezing. Early experimental work on the evaporation of seawater was done by Usiglio (1849). Whereas most evaporites on Earth probably involved mainly or only evaporation (particularly under hot, arid climatic conditions, 25–50 °C), cold-climate evaporation and freezing without evaporation are also common means by which some evaporites and similar salty deposits form. Freeze-induced precipitation certainly occurs in sea ice and in cold-climate tide pools, it occurs with modified seawater in Antarctic dry-valley lakes, and it was probably a major process during the Neoproterozoic snowball Earth episode. Despite this long-term seawater experimental history, controversy still exists with respect to the pathway for seawater freezing. Until recently, the paradigm for seawater freezing was largely based on the experimental work of Ringer (1906) and Nelson and Thompson (1954). According to this model, the solid phases (excluding trace carbonates) that precipitate during seawater freezing and the temperatures at initial formation are as follows: ice at $-1.9\,°C$, mirabilite at $-8.2\,°C$, hydrohalite ($NaCl \cdot 2H_2O$) at $-22.9\,°C$, sylvite (KCl) and $MgCl_2 \cdot 12H_2O$ at $-36\,°C$, and antarcticite ($CaCl_2 \cdot 6H_2O$) at $-54\,°C$, which is the eutectic temperature for seawater freezing along the Ringer–Nelson–Thompson pathway.. Other experimental (Richardson 1976; Herut et al. 1990) and theoretical work (Spencer et al. 1990; Marion and Grant 1994; Mironenko et al. 1997) support this pathway. However, the experimental work of Gitterman (1937) suggested an alternative pathway that has gypsum precipitating during seawater freezing. Along this pathway, the last salt to precipitate during seawater freezing is $MgCl_2 \cdot 12H_2O$ at $-36\,°C$, which is the eutectic temperature for seawater freezing along the Gitterman pathway.

The original versions (1 and 2) of the FREZCHEM model lacked gypsum in the minerals database. Our original validation of the model compared model predictions (sans gypsum) with the experimental data of Nelson and Thompson (1954) (Fig. 3.16a). The predicted solution concentrations are, in general, in good agreement with the Nelson–Thompson data. Calcium concentrations continue to increase as temperature decreases until antarcticite begins precipitating at $-53.8\,°C$, which defines the eutectic temperature for seawater freezing along the Ringer–Nelson–Thompson pathway. However, a major discrepancy exists between the model predictions and the Nelson–Thompson data for sulfate at temperatures $<-23\,°C$ (Fig. 3.16a). The model predicts the dissolution of mirabilite following hydrohalite precipitation, which causes sulfate to increase to high concentrations in this simulation. In fact, our numerical model (Marion and Grant 1994; Mironenko et al. 1997; Marion and Farren 1999) fails to converge at temperatures $<-36\,°C$ in the presence of high sulfate; to overcome this problem, simulations below $-30\,°C$ were done by removing the remaining soluble sulfate as mirabilite and continuing the simulation as a pure chloride system. Similarly, Spencer et al. (1990) simulated seawater freezing below $-37\,°C$.

One of the earliest additions to the original FREZCHEM model was gypsum (Table 3.1). When seawater freezing is simulated with gypsum in the minerals database, mirabilite begins precipitating at $-6.3\,°C$ and is the dominant sink for sulfate (Fig. 3.16b). Initiation of mirabilite precipitation at $-6.3\,°C$ is at a higher temperature than the often-cited experimental value of $-8.2\,°C$ based on the Ringer (1906) and Nelson–Thompson (1954) experiments. Gitterman (1937) seeded his solutions with mirabilite crystals during short-term freezing experiments and found a temperature of $-7.3\,°C$ for initiation of mirabilite precipitation. Since solutions require some degree of supersaturation before minerals can precipitate, the true equilibrium temperature for mirabilite precipitation is probably $\geq -7.3\,°C$.

Contrary to some experimental evidence, the model also predicts the precipitation of gypsum, which begins at $-22.2\,°C$. A reversible reaction for mirabilite is indirectly involved in gypsum precipitation (see below). First hydrohalite begins precipitating at $-22.9\,°C$, lowering the Na concentration (Fig. 3.16b), which causes the dissolution of mirabilite. The increasing sulfate concentration, in turn, causes gypsum to precipitate. At temperatures $<-23\,°C$, the sulfate concentration gain in solution, resulting from mirabilite dissolution and ice concentration, is greater than that lost from solution by gypsum precipitation. As a consequence, the soluble sulfate concentration increases at temperatures $<-23\,°C$ (Fig. 3.16b). Note the good agreement between model and experimental sulfate concentrations at temperatures $<-23\,°C$ for the Gitterman pathway (Fig. 3.16b) compared to the poor agreement for the Ringer–Nelson–Thompson pathway (Fig. 3.16a).

The only minor discrepancy between the Gitterman data and the model predictions is the temperature at which gypsum first begins to precipitate. According to Gitterman, this begins at $-15\,°C$, while our model predicts that this occurs at $-22\,°C$. To test whether seawater between $-15\,°C$ and $-22\,°C$ is undersaturated, saturated, or supersaturated with respect to gypsum, we equilibrated two seawater samples, with and without gypsum crystals, at $-15\,°C$ for 366 d and similarly at $-20\,°C$ for 43 d. These data (not presented) indicate a dissolution of the introduced gypsum crystals, indicating that the seawater samples were undersaturated with respect to gypsum at $-15\,°C$ and $-20\,°C$, in agreement with our model. Despite this minor discrepancy, the model predictions closely follow the Gitterman data, especially the complicated sulfate pathway (Fig. 3.16b). According to the theoretical model, the last salt to precipitate along the Gitterman pathway is $MgCl_2·12H_2O$ at $-36.2\,°C$, which defines the eutectic temperature in agreement with Gitterman (1937).

The differences between the two alternative models (Fig. 3.16a and b) are likely caused by differences in experimental protocols. Gitterman's experimental protocols differed in two major ways from those used by others who have examined seawater freezing. First, the Gitterman data presented in Fig. 3.16b are based on static methods with equilibration periods of up to 4 weeks, whereas other studies are based on short-term dynamic methods with equilibration times measured in hours (Ringer 1906; Nelson and Thompson 1954; Richardson 1976; Herut et al. 1990). Insufficient reaction times for mirabilite dissolution and gypsum precipitation, known to be sluggish even at $25\,°C$ (Harvie et al. 1982; McCaffrey et al. 1987), are probably major causes of the discrepancies. In general, sulfate salt equilibria are sluggish, with liquid–solid and solid–solid equilibration times ranging from a few hours up to a few days, unless there is vigorous stirring and thermal cycling (Kargel 1991; Hogenboom et al. 1995). Second, all freezing experiments at very low temperatures involve preconcentration of seawater. For these concentrated seawaters, Gitterman made certain that an adequate amount of precipitated salts was present in addition to the solution phase for work at low temperatures, whereas other workers have used only the concentrated solution phase (Ringer 1906; Nelson and Thompson 1954). According to the theoretical model, 90% of sulfate is removed from the solution phase as mirabilite at $-20\,°C$ (Fig. 5.1). Using only such a sulfate-poor solution as the base for work at temperatures $<-20\,°C$ provides insufficient sulfate for all of the Ca to precipitate as gypsum. It is clear that the sulfate for gypsum precipitation comes primarily from mirabilite dissolution (Fig. 5.1). A sufficient mass of mirabilite must be maintained in the system to serve as a sulfate source for gypsum precipitation. Removal of a phase such as mirabilite from the reaction site is known as fractional crystallization.

To experimentally validate the Gitterman model, we prepared an artificial seawater sample that had the composition of seawater's liquid partially frozen down to $-23\,°C$ (Marion et al. 1999). To this sample we added an excess of mirabilite crystals to ensure an adequate sulfate source. The sample was then placed in a $-26\,°C$ temperature-controlled bath and allowed to equilibrate with periodic sampling and analyses over a 12-week period. The precipitation of hydrohalite between $-23\,°C$ and $-26\,°C$ (Fig. 3.16) led to an initial decrease in the sodium molality. Magnesium, on the other hand, was conserved in the solution phase, as ice formed and hydrohalite precipi-

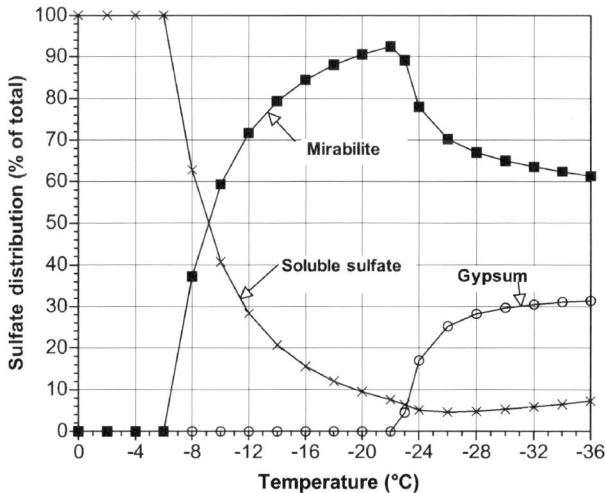

Fig. 5.1. Distribution of sulfate during seawater freezing along the Gitterman pathway. Reprinted from Marion et al. (1999) with permission

tated, and increased in molality between $-23\,°C$ and $-26\,°C$. Calcium also initially increased to a maximum molality of 0.164 m. However, in contrast to sodium and magnesium, which stabilized quickly, the calcium molality began to slowly decrease, reaching a minimum of 0.087 m at the end of 12 weeks. This loss of calcium from the solution phase is indirect evidence that gypsum can precipitate spontaneously, albeit slowly, from seawater given an adequate sulfate source and an appropriate temperature. We believe the slow approach to equilibrium is caused, in part, by the slow kinetics for mirabilite dissolution-gypsum precipitation (Fig. 5.1) and, in part, by poor mixing in the unstirred reaction chamber, where the brine is dispersed in a matrix of ice and hydrohalite. See Marion et al. (1999) for more complete details on these experiments.

These results indicating gypsum precipitation at -23 to $-26\,°C$ are in contrast to evidence for gypsum dissolution (albeit without added mirabilite) at -15 and $-20\,°C$. Hogenboom et al. (1995) provide an independent study pertaining to sluggish sulfate hydrate/solution equilibria and common metastable phenomena in these systems. Working primarily in the system $MgSO_4$-H_2O, they identified several types of metastability. These metastable factors are probably common to most sulfate-water systems. The key limitations on equilibria stem from crystal nucleation kinetics, and limited H_2O, cation, and anion diffusivity in highly viscous, high-concentration, low-temperature brines.

There is experimental evidence for two pathways for seawater freezing (Fig. 3.16). These pathways differ principally in the Ca salt that precipitates. Along the Ringer–Nelson–Thompson pathway, Ca precipitates as antarcticite, while along the Gitterman pathway, Ca precipitates as gypsum (Fig. 5.2). A consequence of this difference in Ca salt precipitation is a major difference in the eutectic temperature of seawater freezing along the two pathways ($-53.8\,°C$ vs. $-36.2\,°C$).

The experimental protocols used by Gitterman are more reliable for ascertaining thermodynamic stability, especially the longer equilibration times and the maintenance of a sulfate source. The theoretical model can simulate either pathway, depending on whether or not gypsum is included in the mineral database. However, with gypsum in the database, the theoretical model also favors the Gitterman pathway as the thermodynamically stable pathway, because the model literally selects the configuration of solid- and solution-phase constituents that minimizes the Gibbs energy of the system, which defines thermodynamic stability. The theoretical and experimental evidence argue in support of the Gitterman pathway as the thermodynamically stable pathway for seawater freezing. The metastable Ringer–Nelson–Thompson pathway is attributable to slow kinetics and fractional crystallization during the freezing process.

In our assessment of the more stable freezing pathway, we must distinguish between "more stable" and "natural." Natural crystallization processes

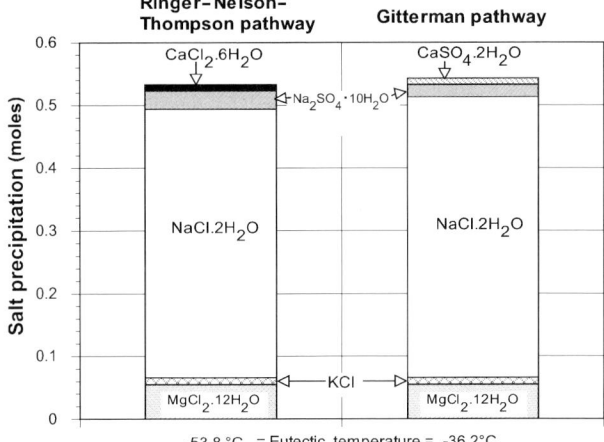

Fig. 5.2. Salt precipitation during seawater freezing to the eutectic temperatures along the Ringer–Nelson–Thompson and Gitterman pathways Reprinted from Marion et al. (1999) with permission

do not always move along the stable pathway. The natural, physical pathway for seawater freezing will depend on several key parameters describing the system, the processes, and the time allotted for these processes. Solution viscosity, the spatial scale of the system, rates of temperature change, the presence of suitable nucleation substrates, surface gravitation, and other parameters, in addition to thermodynamics, will determine what hydrate phases form, the sizes of crystals, whether the system crystallizes in equilibrium or by fractional crystallization, and other process attributes.

Rapid seawater freezing, for example nightly freezing in cold-region tidal pools or in leads within sea ice, may very well follow more closely the Ringer–Nelson–Thompson pathway due to insufficient time to achieve thermodynamic equilibrium. Slow freezing of thick sea ice on Earth or thick cryovolcanic brine flows on Europa, on the other hand, would likely be described better by the Gitterman pathway. However, slow freezing does not ensure adherence to equilibrium crystallization. In fact, very large brine bodies, for instance Europa's ocean or very deep cold-region brine lakes on Earth, are more apt to solidify by fractional crystallization, where ice floats to the top of the system, and layer after layer of hydrated salts are deposited at the bottom in a chemical sequence. Despite these uncertainties, there is supportive evidence on Earth for freezing of Antarctic saline lakes along the thermodynamically more stable Gitterman pathway (Stark et al. 2003).

An aspect of seawater freezing that will play a role in our discussion of a "snowball Earth" (Sect. 5.1.3) is the quantity of water that remains unfrozen at subzero temperatures. For validation of this facet of the model, see Fig. 3.17. For a seawater system starting with 1.0 kg water at 0 °C, ice starts

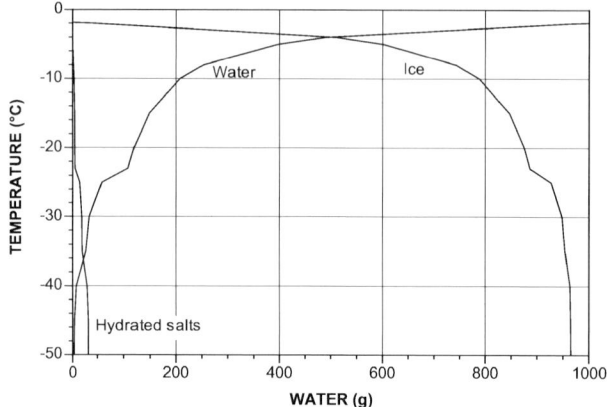

Fig. 5.3. FREZCHEM model prediction of distribution of water during seawater freezing along the Ringer–Nelson–Thompson pathway. Reprinted from Marion and Grant (1994) with permission

forming at $-1.9\,°C$ (Fig. 5.3). Approximately 50% of the water has turned to ice by the time the temperature has dropped to $-4\,°C$. By $-10\,°C$, only 20% of the original water is unfrozen. At about $-23\,°C$, hydrated salts become a significant sink for water; this is where hydrohalite begins to precipitate. At $-50\,°C$, the original 1000 g of water is redistributed as 964.7 g of ice, 32.1 g of hydrated salts, and 3.1 g of unfrozen water. The simulations of Fig. 5.3 are based on the Ringer–Nelson–Thompson pathway, the Gitterman pathway for seawater freezing would be identical down to $-22\,°C$, where gypsum starts to precipitate; below $-22\,°C$, the two seawater freezing pathways are somewhat different down to their respective eutectics (Fig. 3.16).

Seawater drying via evaporation and freezing are compared in Fig. 5.4. Seawater is supersaturated with respect to calcite at atmospheric P_{CO_2} and precipitates calcite early in both the evaporation and freezing processes (Gitterman 1937; Nelson and Thompson 1954). At $25\,°C$, gypsum begins precipitating at a seawater concentration factor (SCF) for evaporation of $3.3\times$ (Fig. 5.4a), whereas mirabilite begins precipitating during freezing at a similar SCF of $3.0\times$ (Fig. 5.4b). At close to the same SCF of $10.3\times$, halite and glauberite ($Na_2SO_4 \cdot CaSO_4$) begin precipitating during evaporation, and hydrohalite and gypsum begin precipitating during seawater freezing. The first K salt to precipitate during evaporation is polyhalite ($2CaSO_4 \cdot MgSO_4 \cdot K_2SO_4 \cdot 2H_2O$) at a SCF of $34.0\times$, and during freezing, sylvite at a SCF of $39.8\times$. The first Mg salt to precipitate during evaporation is polyhalite at a SCF of $34.0\times$, and during freezing, $MgCl_2 \cdot 12H_2O$ at a SCF of 44.4, which occurs at the eutectic for seawater freezing along the Gitterman pathway ($-36.2\,°C$).

Although there are many similarities in evaporating and freezing of seawater (Fig. 5.4), these processes generally produce different suites of minerals.

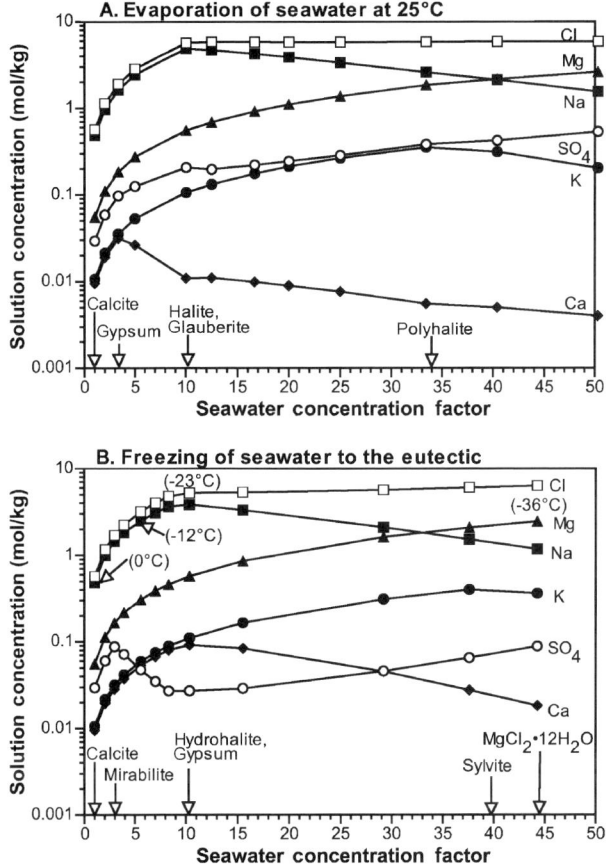

Fig. 5.4. Comparison of seawater concentration by **a** evaporation at 25 °C and **b** freezing to the eutectic. Reprinted from Marion (2001) with permission

A few generalities can be made with respect to the minerals that precipitate during drying by evaporation or freezing. The minerals that precipitate during freezing are generally mixtures of simple salts such as $Na_2SO_4 \cdot 10H_2O$ and $CaSO_4 \cdot 2H_2O$ rather than complex salts such as $Na_2SO_4 \cdot CaSO_4$ or $KMgCl_3 \cdot 6H_2O$ that precipitate during evaporation at higher temperatures. Similar but more highly hydrated salts precipitate during freezing compared to evaporation (e.g., $MgCl_2 \cdot 12H_2O$ vs. $MgCl_2 \cdot 6H_2O$ or $NaCl \cdot 2H_2O$ vs. $NaCl$). Differences of this nature can provide clues to the environmental conditions that prevailed during salt deposition on Earth and in other planetary environments, if postdepositional alteration of hydration states or partial dissolution can be assessed, constrained, or ruled out.

5.1.2 Aqueous Saline Environments

Saline environments include lakes, oceans, sea ice, soils, salterns (seawater evaporation basins), and evaporite deposits. The chemical analyses of representative saline waters demonstrate a wide range of properties (Table 5.1). The FREZCHEM model was used to estimate the a_w, ionic strength, pH, and density of these waters. The Orca Basin contains a deep-sea (2400 m) brine at the bottom of the Gulf of Mexico that formed as the result of dissolution of evaporite salt beds. Because of the high density of the Orca Basin brine (1.190 kg/l) compared with seawater (1.023 kg/l), these deep-sea brines are gravitationally stable at the bottom of the ocean. The diffusive mixing times of these submarine brines and overlying seawater are less than the dissolution timescales of the seabed salts. The Great Salt Lake is a predominantly Na-Cl brine, while the Dead Sea is a predominantly Na-Mg-Cl brine. The most dense brine is Don Juan Pond in Antarctica, which is a predominantly Ca-Cl brine. Basque Lake in Canada is a predominantly Mg-SO_4 brine. Mono Lake and Lake Magadi are examples of alkali lakes dominated by Na-Cl-CO_3 brines; note the high pH values (9.9–10.1) and the virtual absence of Mg and Ca from these alkali lakes (Table 5.1). These extreme brines, with the sole exception of Don Juan Pond, have in common a suitability for life.

We also included in Table 5.1 the calculated chemical composition of Earth seawater at $-20\,°C$ (a possible lower temperature limit for biological activity) to demonstrate the powerful influence of freezing on chemical composition. According to model calculations, by the time the temperature has dropped to $-20\,°C$, ice, calcite, and mirabilite are precipitating; of the original water, 87.6% has precipitated as ice, 11.9% is still present in the aqueous phase, and 0.5% has precipitated as the hydrate mineral mirabilite (Fig. 5.3). The salinity of this solution ($a_w = 0.823$, Table 5.1) is within the tolerance range of halophiles (Table 4.2), thus supporting our view that probably some of these microorganisms can tolerate and perhaps grow/reproduce at $-20\,°C$ (Chap. 4).

Not all measures of "salinity" convey the same degree of salinity. For example, compare Orca Basin, the Great Salt Lake, the Dead Sea, and Basque Lake (Table 5.1). All four of these waters contain about the same salinity % [25.1–26.4% salt (wt/wt)]. Note, however, that Basque Lake has a much more favorable (for life) a_w (0.919) compared with Orca Basin (0.774), Great Salt Lake (0.776), and, especially, the Dead Sea (0.690). The impact of salts on life depends on the anions and cations and their charges and molecular weight. Bacterial sulfate reduction occurs with salt concentrations up to 24% (Oren 1988), but chloride salt solutions at such concentrations deals much more harshly with life. Only the most halophilic organisms can live in the Dead Sea (Table 4.2). The Dead Sea was called "dead" because it was only in 1936 that life forms (e.g., bacteria, algae, yeast) were first isolated from this hypersaline water (Ventosa et al. 1999).

Table 5.1. Chemical analyses of selected saline waters[a]. Reprinted from Marion et al. (2003b) with permission

Property	Seawater[b]	Seawater at −20 °C	Orca Basin, Gulf of Mexico[c]	Great Salt Lake, USA[d]	Dead Sea, Israel[e]	Don Juan Pond, Antarctica[f]	Basque Lake, Canada[d]	Mono Lake, USA[d]	Lake Magadi, Kenya[g]
Na	0.48610	3.636	5.532	4.860	1.835	0.112	0.797	0.991	6.277
K	0.01058	0.089	0.021	0.139	0.212	0.008	0.054	0.032	0.054
Mg	0.05475	0.460	0.058	0.396	1.944	0.110	2.350	0.001	≈ 0
Ca	0.01065	0.080	0.036	0.008	0.459	5.830	≈ 0	0.0001	≈ 0
Cl	0.56664	4.757	5.633	5.278	6.824	12.192	0.064	0.404	2.476
SO_4	0.02927	0.023	0.051	0.228	0.006	< 0.001	2.746	0.081	0.024
Alkalinity	0.00232	0.00073	0.0063	0.0055	0.0060	≈ 0	0.067	0.458	3.899
Salinity (%, w/w)	3.51	21.4	25.1	25.2	26.4	40.2	25.8	6.86	32.0
a_w[h]	0.981	0.823	0.774	0.776	0.690	0.414	0.919	0.974	0.819
Ionic strength[h]	0.722	5.37	5.88	6.43	9.26	17.75	10.68	1.28	8.28
pH[h]	8.30 (7.92)[i]	8.09 (7.55)[i]	8.36	8.10	7.71	≈ 5.4	8.70	9.90	10.13
Density (kg/l)[h]	1.023	1.180	1.190	1.197	1.223	1.385	1.278	1.047	1.250

[a] Concentration units are mol/kg(water) except for alkalinity, which is given in units of equivalents/kg(water).
[b] Millero and Sohn 1992.
[c] Krumgalz et al. 1999.
[d] Eugster and Hardie 1978.
[e] Krumgalz et al. 2000.
[f] Marion 1997.
[g] Jones et al. 1977.
[h] Calculated with the FREZCHEM model (Marion and Farren 1999). Model runs assume temperature = 25 °C (except for seawater at −20 °C), $P_{CO_2} = 3.6e-4$ atm, and total pressure = 1.01 bars.
[i] pH values without parentheses are supersaturated with respect to calcite. pH values in parentheses are in equilibrium with calcite.

Don Juan Pond (DJP) is the only selected saline water wherein the reported a_w (= 0.414, Table 5.1) is generally below the experimentally derived lower limit for microbial activity ($a_w \sim 0.6$, Table 4.2). There was an early report that suggested that DJP had a viable microbial population (Meyer et al., 1962); but later it was found that the isolated bacteria could not grow in the pond water and were likely brought in from outside (Kushner 1981). DJP, located in the Dry Valleys of Antarctica, is unique for several reasons. It is the most saline body of water on Earth, being a near-saturated $CaCl_2$ solution. As a consequence of this high salinity, DJP generally remains unfrozen in winter, even at temperatures below $-50\,°C$. DJP is the site where antarcticite ($CaCl_2 \cdot 6H_2O$) was first identified forming naturally.

DJP pond is also nearly unique in that it appears to violate the NASA and popular mantra of "where there is water, there is life." In this case, the chief obstacles to life most likely are: (1) the high salinity generates destructively high osmotic pressure gradients across cell walls and (2) low temperatures and high solution viscosities reduce reaction and diffusion rates and low input rates of key nutrients. Also limiting life's prospects in DJP is its small size, instability and short longevity, and limited variety of physical niches and chemical processes. We wonder whether a planetary ocean, if maintained near DJP's composition and temperature overall but subjected to physical and chemical variation across the sea floor, and maintained for hundreds of millions of years, might acquire or initiate life, which might then adapt and flourish throughout the ocean. Even more extreme and inhospitable to life are the aqueous acid aerosols making up stratospheric clouds (Sect. 5.1.4). Those solutions – also a possible microscopic chemical model of Europa's ocean (Kargel et al. 2000) – tend toward multicomponent acid eutectics, undergo complete freezing and thawing, and generally are probably completely devoid of life.

For a pure $CaCl_2$ solution, the FREZCHEM model predicts a eutectic temperature of $-50.4\,°C$ with a $CaCl_2$ molality of $3.99\,\mathrm{mol\,kg^{-1}}$ (Fig. 5.5) (Marion 1997), which is in reasonable agreement with literature values of $-51\,°C$ to $-49.8\,°C$ and $CaCl_2$ molalities of 3.90 to $4.32\,\mathrm{mol\,kg^{-1}}$ (Spencer et al. 1990). A pure $NaCl$-$CaCl_2$-H_2O system more nearly matches the chemical behavior of DJP than does the pure $CaCl_2$-H_2O system. The calculated eutectic temperature for a pure $NaCl$-$CaCl_2$-H_2O system is $-51.6\,°C$, with a Ca concentration of $3.78\,\mathrm{mol\,kg^{-1}}$, a Na concentration of $0.51\,\mathrm{mol\,kg^{-1}}$, and a Cl concentration of $8.07\,\mathrm{mol\,kg^{-1}}$, which is similar but not identical to the composition of DJP at the eutectic. The small differences between this pure system and DJP are due to the minor amounts of K and Mg present in DJP. The calculated eutectic temperature for DJP is $-51.8\,°C$; the Ca concentration is $3.72\,\mathrm{mol\,kg^{-1}}$; the Na concentration is $0.50\,\mathrm{mol\,kg^{-1}}$; and the total Cl concentration is $8.08\,\mathrm{mol\,kg^{-1}}$, which leaves a residual charge of $0.14\,\mathrm{mol_c\,kg^{-1}}$ consisting of K and Mg. Compared to pure $CaCl_2$, the eutectic for DJP is displaced to a lower Ca concentration and a lower temperature

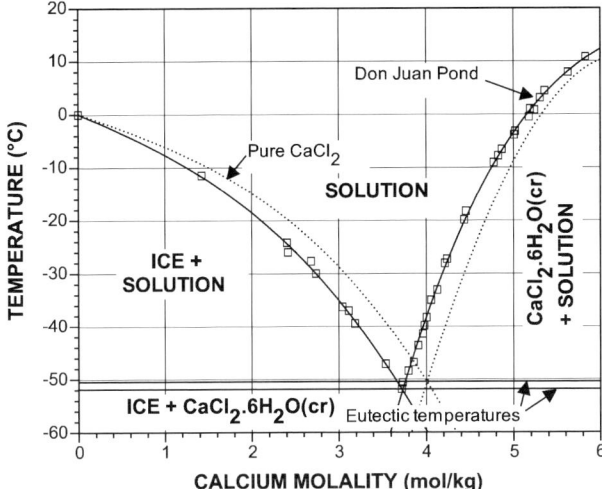

Fig. 5.5. A stability diagram for ice and antarcticite in a pure CaCl$_2$ solution and in Don Juan Pond. Reprinted from Marion (1997) with permission

(Fig. 5.5). The FREZCHEM model enables one to tailor a stability diagram to a specific saline water rather than doing what has been common practice among Earth and planetary scientists, which is approximating it with a simpler system (e.g., Kargel et al. 2000).

5.1.3 Snowball Earth-Hothouse Earth

Abundant evidence exists that during the latter part of the Neoproterozoic era from 750 to 580 million years ago (or part of this period), the Earth was intermittently covered by glaciers and sea ice into the tropics (Kirschvink 1992; Hoffman et al. 1998; Kerr 2000; Hoffman and Schrag 2002; Bodiselitsch et al. 2005). This cold phase of Earth's geologic history has come to be known as "snowball Earth," which interestingly alternated with unusually warm intervals, now called "hothouse Earth." The causes of these alternating climates are variously attributed to obliquity changes, continental drift, albedo feedbacks, weathering, changes in the carbon cycle, or some combination of these factors (Kirschvink 1992; Hoffman et al. 1998; Hyde et al. 2000; Jenkins 2000; Kerr 2000; Kirschvink et al. 2000; Pavlov et al. 2000; Jacobsen 2001; Kennedy et al. 2001b; Godderis et al. 2003; Hoffman and Schrag 2002; Baum and Crowley 2003; Jiang et al. 2003; Smith and Pickering 2003; Donnadieu et al. 2004; Ramstein et al. 2004). For example, drifting continents could lead to (1) newly exposed basalts that would trap atmospheric CO$_2$ (via weathering) or (2) changes in oceanic currents vis a vis the equator and poles leading to global cooling and ultimately glaciation.

Remnants of these geologic times are clearly imprinted in geologic formations with "cap" carbonate deposits overlying glacial tillites and occasionally banded iron formations (Hoffman et al. 1998; Kerr 2000; Kirschvink et al. 2000; Kennedy et al. 2001b; Hoffman and Schrag 2002; Jiang et al. 2003; Lorentz et al. 2004; Holland 2004; Porter et al. 2004). Banded iron formations strongly suggest that seawater was anoxic in order for ferrous iron to have accumulated to concentrations high enough to precipitate distinct iron layers (Hoffman and Schrag 2002; Holland 2004). Carbonate deposits suggest that the oceans were saturated (or supersaturated) with carbonate minerals. One of the mechanisms proposed for bringing the Earth out of these deep freezes is the release of greenhouse gases from gas hydrates that could have contributed to atmospheric warming (Kennedy et al. 2001b; Jiang et al. 2003).

In our simulations, we assumed that seawater compositions during the Neoproterozoic era would have been similar to present-day seawater, except for high Fe(II) concentrations. We arbitrarily assigned our hypothethical ocean a Fe(II) concentration of 0.01065 m, the same concentration as Ca; to maintain a charge balance, we removed an equivalent amount of Na (from 0.48610 to 0.46480 m) (Fig. 5.6).

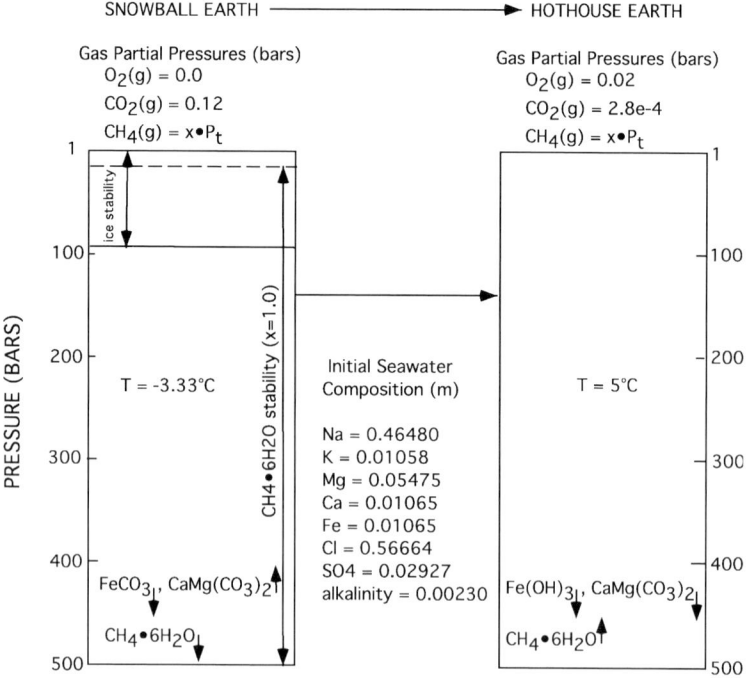

Fig. 5.6. A summary of model inputs and outputs for "snowball Earth" and "hothouse Earth' scenarios. For minerals in *boxes*, *down arrows* signify precipitation, and *up arrows* signify dissolution

According to Holland (2004), it was during the late Neoproterozoic era that seawater sulfate concentrations rose to modern values (29 mm, Fig. 5.6). However, Porter et al. (2004) have argued for oceanic sulfate concentrations of 1 mm or less during Neoproterozoic glacial periods; similarly, Hurtgen et al. (2005) have argued for oceanic sulfate concentrations as low as 10% of current oceanic levels during the Neoproterozoic. In our specific simulations, it is of no consequence whether we use 1 or 29 mm sulfate concentrations. Sulfide anions are not now part of the FREZCHEM model and, therefore, were not considered in our simulations. It is unlikely that sulfide anion concentrations in snowball Earth oceans exceeded ferrous iron concentrations because the insolubility of FeS_2 (iron pyrite) would have prevented the high solubility of ferrous iron needed for banded iron formations (Hoffman and Schrag 2002; Petsch 2004).

We also assumed equilibrium with respect to dolomite during snowball Earth and hothouse Earth phases (Fig. 5.6); this assumption implicitly assumes that dolomite is present in oceanic sediments and would equilibrate with seawater. This is a snowball mechanism (slow dissolution) for boosting the carbonate content of seawater prior to hothouse precipitation (quick). Both dolomite and calcite are typically present in "cap" carbonates (Kirschvink et al. 2000; Kennedy et al. 2001a,b; Condon et al. 2002; Hoffman and Schrag 2002; Young 2002; Jiang et al. 2003; Porter et al. 2004). Porter et al. (2004) have argued that dolomite formed in isotopic equilibrium with seawater as a primary precipitate or as a very early diagenetic replacement. Replacing dolomite solubility with calcite solubilty in our model simulations would not alter our conclusions.

We assumed fixed $O_2(g)$ partial pressures of 0.00 and 0.02 bars for snowball Earth and hothouse Earth, respectively. A low model value is necessary for snowball Earth because otherwise ferrous iron is oxidized to ferric iron, whose solubility is too low to account for banded iron formations. It is in the late Neoproterozoic era that atmospheric O_2 rose to modern values (Holland, 2004). According to Canfield and Teske (1996), between 1.05 and 0.64 Ga ago, O_2 levels were 5–18% of the modern value (0.2 bars). For hothouse Earth, we choose a level of 10% of the modern value. For our simulations, it is immaterial whether we used 0.002, 0.02, or 0.2 bars of O_2 pressure. All are sufficient to completely oxidize ferrous to ferric iron.

In the simulations, $P_{CO_2(g)}$ is fixed during a given cold or warm interval in order to keep the carbonate chemistries and the associated pH values narrowly defined. If hothouse warming was attributable to high atmospheric CO_2 concentrations, then the sources of the CO_2 on an ice-covered Earth were likely subaerial and submarine volcanism. According to Kirschvink et al. (2000), subaerial and submarine volcanism could contribute 9 and 18 mbars (CO_2)/Ma, respectively. We ran the snowball Earth case with an oceanic $P_{CO_2(g)} = 0.12$ bars (Hoffman et al. 1998) and the hothouse Earth case with an oceanic $P_{CO_2(g)} = 2.8 \times 10^{-4}$ bars (Godderis et al. 2003)

(Fig. 5.6). To elevate seawater P_{CO_2} from 2.8×10^{-4} to 0.12 bars at the rate of 18 mbars/Ma would take 6.65 Ma, which is on the time scale of these global phases (Hoffman and Schrag 2002). The concept is that CO_2 would have slowly built up to high levels beneath an ice-covered ocean, leaking slowly through cracks or open waters into the atmosphere. At some point, the atmospheric CO_2 would lead to a greenhouse warming, a rapid melting of the ice cover, and a temporary boost in atmospheric CO_2, which would then have reverted to a lower level (280 ppm) relatively rapidly. Warming of seawater and a decrease in oceanic CO_2 levels would cause the rapid precipitation of carbonate minerals.

In contrast to fixed $P_{O2(g)}$ and $P_{CO2(g)}$, we defined $P_{CH_4(g)}$ by

$$P_{CH_4(g)} = xP_t, \quad (5.1)$$

where x is the mole fraction of $CH_4(g)$ in the gas phase and P_t is the total pressure. Application of this equation for model simulations is through Henry's law (Eq. 3.25) and gas hydrate equilibrium (Eq. 3.36). In these simulations, CH_4 serves as a surrogate for all potential gas hydrates and was assigned a mole fraction of 1.0 (pure CH_4 gas). Using Eq. 5.1 as the means for estimating the $CH_4(g)$ partial pressures allows us to quantify the stability of $CH_4 \cdot 6H_2O$ as a function of total system pressure.

The model as structured specifies pressure (in bars) as input. To convert from pressure (P) to depth (D) of ice or seawater, we used the equation

$$P(\text{bars}) = \rho(\text{g/cm}^3)g(\text{cm/s}^2)D(\text{km})/10, \quad (5.2)$$

where ρ is density and g is gravitational acceleration (980.6 cm/s^2). Densities of ice and seawater were estimated using the FREZCHEM model.

The thickness of sea ice during these glacial periods is the subject of considerable controversy. The original "hard" snowball Earth hypothesis argued for a virtually complete global ice cover that ranged from tens of meters to 1500 m (Kirschvink 1992; Hoffman et al. 1998; Kirschvink et al. 2000). Other studies have argued for at least open waters in the tropics (Hyde et al. 2000; Kennedy et al. 2001a; Condon et al. 2002; Young 2002; Holland 2004). For the purposes of our simulations, we assumed an average ice thickness of 1 km. This represents a "hard" snowball assumption. To produce a 1.0-km sea-ice thickness for global oceans where the mean thickness is 3.73 km (oceans, including adjacent seas) (Millero and Sohn 1992) would require conversion of 26.8% of seawater into ice (1/3.73), which would leave 73.2% of seawater unfrozen. Such freezing also implies that seawater ionic concentrations during the snowball Earth phase would be 36.6% (3.73/2.73) higher than during the hothouse Earth phase. Using the FREZCHEM model with an elevated (36.6%) seawater composition would require freezing the seawater to $-3.33\,°C$ to produce this quantity of sea ice (1 km). The temperature of $-3.33\,°C$ is the freezing point of seawater with a 36.6% elevated composition at a pressure of 90 bars (= 1 km of sea ice, Eq. 5.2). This temperature compares to $-1.92\,°C$,

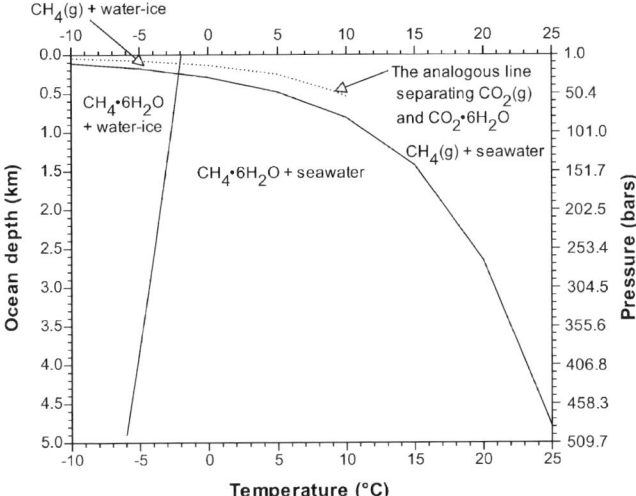

Fig. 5.7. The stability of water ice and CH$_4$·6H$_2$O in "standard" seawater (*solid lines*). The *dashed line* separates the analogous equilibrium between CO$_2$(g) and CO$_2$·6H$_2$O. In both cases, the assumption was made that the mole fraction in the gaseous phase (x) is 1.0 [i.e., pure CH$_4$(g) or CO$_2$(g)]. Reprinted from Marion et al. (2006) with permission

which is the temperature of present-day surface seawater freezing (Fig. 5.7). For the snowball Earth simulation at $-3.33\,°$C, ice is stable between 1 and 90 bars of pressure (Fig. 5.6). At $-3.33\,°$C and $x = 1.0$, CH$_4$·6H$_2$O is stable from 23.1 bars ($P_{\mathrm{CH_4(g)}}$) to the base of our simulation at 500 bars (Fig. 5.6).

Seawater temperatures in a hothouse Earth are likely to have been much more variable than those of a snowball Earth. For example, ocean temperature depth profiles at high latitudes are typically in the range of -2 to $2\,°$C; similar profiles from low and mid-latitudes can vary from 2 to $28\,°$C (Millero and Sohn 1992). Temperature plays a critical role in the stability of gas hydrates. Figure 5.7 depicts the stability of CH$_4$·6H$_2$O ($x = 1.0$) in seawater as a function of temperature and pressure (depth). For illustrative purposes, we also included the analogous stability line for CO$_2$(g)-CO$_2$·6H$_2$O in Fig. 5.7; but this CO$_2$ stability relation was not used in the snowball/hothouse simulations. This figure is based on "standard" seawater (Millero and Sohn 1992), which has the composition given in Fig. 5.6, except that Fe = 0.0 m and Na = 0.48610 m. We used "standard" seawater in this graph to broaden the applicability to present-day oceans. This "standard" seawater begins freezing at $-1.92\,°$C; the ferrous iron seawater (Fig. 5.6) begins freezing at $-1.91\,°$C, which is an insignificant difference in temperature. The depth (y-axis) is estimated with Eq. 5.2 using FREZCHEM model estimates of seawater density. In a snowball Earth ocean at $-3.33\,°$C, CH$_4$·6H$_2$O is stable from 0.227 to 5.0 km ($x = 1.0$) (Figs. 5.6 and 5.7). If the oceans are warmed, then gas hydrates at shallow depths would become unstable, dissociate, and release gas to the atmosphere, which would accelerate global warming. But this effect is likely to be confined to shallow depths. For example, at 1.0 km, seawater temperature would have to increase to above 11.5 °C before gas dissociated. This much warming at these depths is unlikely. For example, seawater temperatures at low and mid-latitudes today are about 5 °C at 1.0 km (Millero and Sohn 1992). On the other hand, gas hydrates formed at 0.5 km would only need to warm to 5 °C before gas dissociated. Also, flooding of permafrost gas hydrates with subsequent gas dissociation is clearly possible with warming (Kennedy et al. 2001b; Jiang et al. 2003).

In the "hard" snowball Earth hypothesis (Hoffman et al. 1998; Kirschvink et al. 2000; Hoffman and Schrag 2002), the assumption is made that during the glacial phase, weathering would be negligible because atmospheric CO_2 would be isolated from terrestrial and marine substrates. "Cap" carbonate deposits are attributed to rapid weathering when the ice/snow cover is removed and atmospheric CO_2 reacts with terrestrial and marine rocks. Some have questioned whether weathering rates could be high enough to create such "cap" carbonates (e.g., Kennedy et al. 2001a; Sankaran 2003). Also, Sr isotopes provide no support for reduced or enhanced weathering during these alternating cold-hot phases (Jacobsen 2001). For illustrative purposes, Table 5.2 shows solution properties at 1 bar of total pressure. Equilibrium with dolomite at -3.33 °C and $P_{CO_2} = 0.12$ bars causes dissolution of dolomite (weathering) relative to present-day seawater and a substantial increase in alkalinity as well as Ca and Mg concentrations (cf. snowball Earth in Table 5.2 with initial seawater in Fig. 5.6). Subsequent equilibration of this solution with a hothouse Earth environment ($P_{CO_2(g)} = 2.8 \times 10^{-4}$ bars, 5 °C) causes a massive precipitation of dolomite; note the decline in alkalinity, calcium, and magnesium and the change in pH from 6.59 to 7.88 (Table 5.2). This "slow dissolution–rapid precipitation" mechanism for "cap" carbonates is more plausible than the rapid weathering mechanism of the original "hard" snowball hypothesis.

It is possible to do a "back-of-the-envelope" calculation for dolomite formation between a snowball Earth and a hothouse Earth. The total volume of water in the Earth's oceans is 1.34993×10^{21} l (Millero and Sohn 1992). Adjusting this quantity of seawater for seawater density and salt content leads to an estimate of 1.3371×10^{21} kg (H_2O), which is appropriate for a hothouse Earth. For a snowball Earth with 1.0 km of ice, this seawater volume would be reduced to 9.7863×10^{20} kg (H_2O). Multiplying these quantities of water times molal concentrations (Table 5.2) gives the total quantity of this constituent in the Earth's oceans. In this simulation, we used 1 bar pressure on the assumption that these carbonates would be deposited in shallow seas. For every mole of Mg or Ca that is removed from solution, a mole of dolomite forms. Multiplying snowball and hothouse Mg concentrations [0.08570 and 0.05967 moles kg $(H_2O)^{-1}$, Table 5.2] by the quantity of water yields 8.3869×10^{19} and 7.9785×10^{19} moles of Mg, respectively, for a net loss of 0.4084×10^{19} moles of Mg. A similar calculation for Ca leads to a net loss of 0.3963×10^{19} moles of Ca. Multiplying these mole losses by the molar density of dolomite ($64.93 \, \mathrm{g \, mole^{-1}}$) and dividing by the density of dolomite ($2.85 \, \mathrm{g \, cm^{-3}}$) yields $9.304 \times 10^{19} \, \mathrm{cm}^3$ (based on Mg) and $9.029 \times 10^{19} \, \mathrm{cm}^3$ (based on Ca) of dolomite formation. Dividing these quantities by the surface area of the ocean ($3.620 \times 10^8 \, \mathrm{km}^3$; Millero and Sohn 1992) yields an average depth of dolomite formation of 0.257 and 0.249 m based on Mg and Ca, respectively. These estimates are a small fraction of the depth of cap carbonates, which range from meters to hundreds of meters (Hoffman and Schrag 2002). However, there is evidence that these cap carbonates may have formed in upwelling regions (Grotzinger and Knoll 1995; Hoffman and Schrag 2002), which would have concentrated cap carbonate deposition over a much smaller area leading to much thicker deposits. In addition to upwelling uncertainties, pressure would probably preclude dolomite formation in deep oceans because pressure increases the solubility of carbonate minerals to the point that oceans at depth become undersaturated with these minerals (Millero and Sohn 1992). Also, our calculations are based on equilibrium assumptions, while oceans typically are supersaturated with respect to calcium and magnesium carbonates (Millero and Sohn 1992). Despite the above simplifications, the main point of these calculations is to demonstrate that significant dolomite (or calcite) could have formed via precipitation during the transition from a snowball Earth to a hothouse Earth. Whether this is sufficient by itself or can only partially explain cap carbonate formation will necessitate a more comprehensive analysis of this hypothesis.

Table 5.2. Comparison of snowball Earth and hothouse Earth equilibrium oceanic compositions at 1 bar of pressure (assuming the ocean is saturated with dolomite)

Constituent	Snowball Earth	Hothouse Earth
Na (mol/kg)	0.63490	0.46480
K (mol/kg)	0.01445	0.01058
Mg (mol/kg)	0.08570	0.05967
Ca (mol/kg)	0.02535	0.01559
Fe (mol/kg)	0.00002	0.00000
Cl (mol/kg)	0.77403	0.56664
SO_4 (mol/kg)	0.03998	0.02927
Alkalinity (equi./kg)	0.01753	0.00074
pH	6.593	7.877
Temperature (°C)	-3.33	5.0
Total pressure (bars)	1	1
P_{CO_2} (bars)	0.12	2.8×10^{-4}
P_{O_2} (bars)	0.0	0.02

In our model, cap carbonates are explained as a result of high carbonate alkalinity in an ice-covered ocean, regardless of the specific mechanism for melting the snowball Earth. Ice could melt for a variety of reasons (e.g., greenhouse gases, a positive perturbation in solar forcing, a decrease in planetary albedo), but cap carbonates are expected to be produced as a result of the transition from high to low oceanic carbonate alkalinity.

Soluble Fe(II) concentrations are low for both snowball Earth and hothouse Earth (Table 5.2). In a snowball Earth environment, Fe(II) largely precipitates as siderite because of the high carbonate alkalinity; in a hothouse Earth, siderite is converted to ferrihydrite (a ferric mineral surrogate) because of the high O_2 levels (Fig. 5.6). During the transition from snowball to hothouse, either siderite changes to ferrihydrite in place or the precipitation of dolomite (decreasing carbonate alkalinity) solubilizes siderite, which then reacts with oxygen to form ferrihydrite. On the other hand, the fact that iron concentrations are theoretically low in both snowball and hothouse oceans (Table 5.2) is an argument against the commonality of banded iron formations during these periods. Young (2002) has argued that banded iron formations during these glacial periods only occurred in Red Sea rift-type basins due to the presence of Fe-charged brines. Low Fe(II) concentrations, except for localized high Fe(II) brines, could explain the uncommon occurrence of banded iron formations associated with snowball Earth—hothouse Earth periods.

While the chemical compositions of snowball Earth/hothouse Earth oceans are still poorly constrained (Holland 2004), the FREZCHEM simulations can (1) estimate that the temperature of a snowball Earth ocean was probably between $-1.92\,°C$, the present-day surface seawater freezing temperature, and $-3.33\,°C$ for a 1.0-km ice cover, (2) quantitatively explain the deposition of ferric iron, (3) make a case for why these banded iron deposits are uncommon, (4) quantitatively explain the deposition of carbonate minerals, (5) provide a more plausible mechanism for "cap" carbonate formation, (6) make a case for a high carbonate alkalinity during the snowball Earth phase, and (7) make a case for a shallow zone of gas hydrate influence on global warming during these alternating climate shifts.

5.1.4 Why Are Clouds not Green?

On Earth there exists a cold aqueous environment not generally considered a habitat for life, namely clouds. What prevents clouds from supporting life? Factors that could hinder life in clouds include cold temperatures, lack of nutrients or energy sources, toxic chemistries, and cloud life spans.

In a recent study, Sattler et al. (2001) examined cloud water collected at high altitudes (3.1 km) on Mt. Sonnblick, Austria. They found that bacteria in cloud droplets were actively growing and reproducing at temperatures at or below $0\,°C$. They concluded that temperature, nutrients, and energy sources are unlikely to limit microbial processes in cloud droplets, at least at lower altitudes. Instead, they felt that the short residence time of cloud droplets in the atmosphere was the most limiting step for microbial processes. Residence times of aerosols in the atmosphere are measured in hours to weeks (Sattler et al. 2001; Buseck and Schwartz 2004). While these residence times are suitable for global dispersal of microbes, they are inadequate for the establishment of a self-sustaining ecosystem.

The FREZCHEM model can be used to simulate what would happen theoretically to cloud droplets and their chemistries as they are lofted (convected) to higher (colder) altitudes. We used two aqueous datasets to simulate atmospheric chemistries. The first dataset consist of mean annual concentrations of ions in precipitation from the Hubbard Brook ecosystem (1.0 km altitude) (table 4 in Likens et al. 1977). The second dataset is from Mt. Sonnblick, Austria (3.1 km altitude) and is a direct measure of cloud chemistry (table II, May 1991 in Brantner et al. 1994). In both cases, the chemistries are similar in relative concentrations with H^+, NH_4^+, Na^+, Cl^-, NO_3^-, and SO_4^{2-} as the dominant ions.

The FREZCHEM algorithm for estimating pH in acidic solutions requires a perfect charge balance (Equations 3.31 and 3.32). A minor adjustment in the Na and Cl data from Hubbard Brook was made to insure a perfect charge balance. A larger adjustment was necessary for the Mt. Sonnblick data because carboxylic acids constitute about 11% of the total anion charge. Organic acids are not now part of the FREZCHEM database (Table 3.1). In this case, we replaced carboxylic acids with proportional amounts of Cl^-, NO_3^-, and SO_4^{2-} to balance the cation charges. For both datasets, we lumped NH_4^+ with K^+ because NH_4^+ is not now part of the FREZCHEM database (Table 3.1). As a check on the internal consistency of our assumptions and calculations, the model estimated pH values of the Hubbard Brook and Mt. Sonnblick solutions over a temperature range of 0 to 25 °C are 4.16 and 4.22, respectively (Fig. 5.8), which are in good agreement with the experimental measurements of 4.14 (Likens et al. 1977) and 4.2 (Brantner et al. 1994).

The bulk of cloudiness is found in the lowest level of the atmosphere, which is called the troposphere. The thickness of the troposphere is $\sim$8 km at high latitudes, $\sim$12 km at mid-latitudes, and $\sim$16 km at low latitudes (Buseck and Schwartz 2004). In our simulations, we assumed a 12 km troposphere and a lapse rate of 6.5 °C/km (Lide, 1994). We also assumed a starting point of 0.0 km and 25 °C (Fig. 5.8) for our two solutions, which were then lofted to colder (higher) altitudes.

These simulations were done assuming chemical equilibria at each temperature step ($\Delta = 5$ °C). Relatively little happens at lower elevations until the freezing point of the aqueous solutions is reached at just below 0.0 °C. At this point, ice starts forming rapidly because of the dilute starting concentrations. Consequently, all ion concentrations also increase rapidly. For illustrative purposes, we only included pH in Fig. 5.8. The pH $[= -\log_{10}(H^+)]$ decreases rapidly (H^+ increases). A pH of 0.0 is the approximate lower limit for metabolic activity (Table 4.3; Schleper et al. 1995; Marion et al. 2003b); this value is reached at about 5 km. The lower limit of metabolic activity for temperature is ca. $-20\,°C$ (Rivkina et al. 2000; Junge et al. 2004; Marion et al. 2003b); this value is reached at ca. 7 km. Active metabolic activity is probably limited to the lower 5 to 7 km of the troposphere.

In addition to ice formation, salts also precipitate as these solutions are lofted to higher altitudes. A consequence of the formation of these solid phases (ice and salts) and the low-temperature eutectics of strong acids (Fig. 3.5) is that the atmospheric solutions become more and more acidic with altitude (Fig. 5.8). For example, the final elevation (temperature) examined is 11.54 km ($-50\,°C$). At this point, the calculated concentrations of the Hubbard Brook solution are $H^+ = 7.55\,m$ with acid anions (Cl^-, NO_3^-, SO_4^{2-}, HSO_4^-) = 7.91 m. Similarly, for the Mt. Sonnblick solution, $H^+ = 6.50\,m$ and acid anions = 6.90 m. These acidic trends are in line with stratospheric chemistries, which are predominantly sulfuric/nitric acid aerosols (Carslaw et al. 1997). For example, the total acid concentration at 20.7 km in the stratosphere is 10.17 m (calculated from fig. 7 in Carslaw et al. 1997), which is in line with our lower atmospheric concentrations.

These calculations (Fig. 5.8) assumed chemical equilibria at each temperature. It is well known that atmospheric liquid aerosols are often supercooled (Carslaw et al. 1997; Sattler et al. 2001; Buseck and Schwartz, 2004). If these simulations were run without water ice as a solid phase in the minerals database (in essence, supercooling), then the predicted pH values up to 12 km would be just an extension of the lower data points (temperature = 0 to 25 °C) in Fig. 5.8. Given that stratospheric aerosols are concentrated acidic solutions (Carslaw et al. 1997), clearly the equilibrium calculations

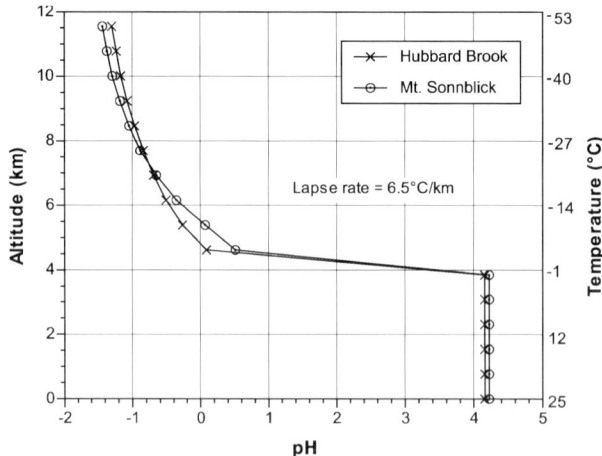

Fig. 5.8. Effect on pH of lofting (convecting) tropospheric chemistries to higher altitudes

(Fig. 5.8) more accurately reflect what is likely than assuming supercooling throughout the troposphere. Nevertheless, the model used for these cloud simulations is oversimplistic for a number of reasons. For example, equilibrium between liquid aerosols and acidic gases are largely what controls the chemical composition of liquid aerosols at high altitudes (Carslaw et al. 1997). The FREZCHEM model is structured such that aqueous compositions can be controlled by acidic gases (Sect. 3.3.3); but this pathway was not implemented in these simulations. Nevertheless, the model demonstrates the importance of increasing acidity with increasing altitude. Moreover, much of our understanding of cloud-water chemistry at high altitudes is based on thermodynamic models rather than observations (Carslaw et al. 1997; Wise et al. 2003).

Transatlantic airline flights typically fly at 39,000 feet (11.9 km), which is approximately at the top of the troposphere. For obsessive-compulsive scientists, constantly watching the airplane monitors that give location, speed, ETA, and outside temperatures is a necessity. Outside temperatures are typically in the range of -50 to $-56\,°C$ (approximately the eutectic temperature of seawater freezing according to the Ringer–Nelson–Thompson pathway, Fig. 3.16a). Liquid aerosols that might exist at such altitudes are not conducive to active life, if for no other reason than temperature ($-50\,°C$) and pH (-1.30 to -1.43) (Fig. 5.8). Active life is likely in clouds only at altitudes <5–7 km. While cloud water at lower altitudes can be considered a habitat for microbes, the short residence time (hours to weeks) (Sattler et al. 2001; Buseck and Schwartz 2004) probably keep clouds from ever becoming green or self-sustaining ecosystems.

An aspect of life in clouds that is beyond the scope of this book (cold environments) is potential life in Venusian clouds. The surface of Venus is too hot ($464\,°C$) for liquid water or carbon-based life (Cockell 1999). Atmospheric constraints include sulfuric acid clouds and high doses of ultraviolet radiation; in principle, these atmospheric constraints can be overcome (Cockell 1999; Schulze-Makuch et al. 2004), which means that Venus could be close to possessing a habitable environment. However, it still remains to be demonstrated that the residence time in Venusian clouds is sufficiently long to create a self-sustaining ecosystem.

5.1.5 Other Earth Applications

In Chap. 3, two Earth systems were discussed that provided partial validation of the model: Lake Nyos and calcite equilibrium. Lake Nyos is a system dominated by Mg, Fe(II), alkalinity, and high P_{CO_2} (Table 3.6). While such systems are rare on Earth, they could be good analogs for early Mars systems resulting from carbonic acid weathering of ferromagnesian minerals (see Mars example below and Marion et al. 2003a). In the Lake Nyos case, our model (Table 3.6) was in excellent agreement with independent model calculations (Bernard and Symonds 1989).

In the calcite case, we demonstrated that experimental measurements of $CaCO_3$ solubility approached model estimates at subzero temperatures (Fig. 3.18). Discrepancies between experimental data and model calculations at higher temperatures were attributed to calcite supersaturation. However, another possible explanation for the discrepancy is that calcite might not control carbonate alkalinity at subzero temperatures. A number of workers have suggested that ikaite ($CaCO_3 \cdot 6H_2O$) is the most likely calcium carbonate mineral that should precipitate during seawater freezing (e.g., Ringer 1906; Assur 1958; Richardson 1976; Weeks and Ackley 1982); however, no one has ever presented physicochemical support for this idea. By removing more stable, less soluble, calcium and magnesium carbonate minerals (e.g., calcite, magnesite, dolomite) from the mineral database, one can simulate the conditions necessary for ikaite to precipitate during seawater freezing. According to model calculations, seawater is grossly undersaturated with respect to ikaite at $0\,°C$ (Fig. 3.18), in agreement with Bischoff et al. (1993). During the freezing process, ikaite begins precipitating from seawater at $-4.5\,°C$, where the alkalinity has been concentrated to $0.243\,g\ CaCO_3/kg$ (soln.) (Fig. 3.18). The fact that these simulated alkalinities are considerably higher than Gitterman's experimental measurements (Fig. 3.18) does not necessarily preclude the possibility that ikaite could precipitate during seawater freezing because it is highly likely that Gitterman (1937) seeded his seawater samples with calcite, which would favor calcite precipitation. These simulations, however, do suggest a simple test to distinguish whether calcite or ikaite is the solid phase that naturally precipitates during seawater freezing. One could freeze a seawater sample to $-10\,°C$ and measure the equilibrated alkalinity; there is a 4.4-fold difference in alkalinity between calcite and ikaite equilibria at $-10\,°C$ (Fig. 3.18). According to a "standard sea ice" model developed by Assur (1958), $CaCO_3$ begins precipitating at $-2.2\,°C$, shortly after freezing begins at $-1.9\,°C$. If this is the case, then this would argue in favor of calcite as the precipitating $CaCO_3$ mineral as seawater is supersaturated with calcite at $-2.2\,°C$ and undersaturated with respect to ikaite (Fig. 3.18). This example makes a strong case for calcite, and not ikaite, as the controlling $CaCO_3$ phase during seawater freezing.

An Earth example not previously discussed deals with the roles of temperature and pressure on the density of ice cores (Marion and Jakubowski 2004). Gow (1971) has shown that the density of deep ice cores under pressure relaxes elastically as soon as the cores are extracted. In Fig. 5.9, we used our model parameters to calculate how the density of an ice core from Antarctica (Gow et al. 1968; Gow 1971) would vary with core temperature at 1 atm, which is what is measured at the surface with corrections for temperature, to the same core under both temperature and pressure constraints. At 1 atm pressure, the core density changes linearly with temperature (Fig. 5.9), in agreement with our model (Fig. 3.2) and the Gow (1971) results (see his table 1). In contrast, the density of the ice core subjected to both temperature

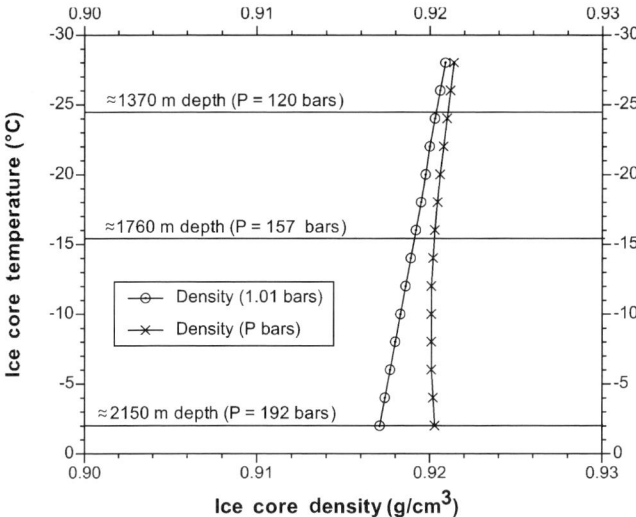

Fig. 5.9. Calculated densities of an Antarctic ice core as a function of pressure and temperatures. Reprinted from Marion and Jakubowski (2004) with permission

and pressure constraints is always higher and becomes progressively more separated from the 1 atm curve as pressure increases (Fig. 5.9). At the base of the core, the predicted densities are 0.9171 and 0.9203 g/cm^3 for 1 and 192 bars, respectively. Between ca. 1730 and 2150 m, the calculated ice density only fluctuates between 0.9201 and 0.9203 g/cm^3. Over this range, the increasing pressure raises the density and the increasing temperature lowers the density; the two forces of temperature and pressure are in relative balance over this range. Gow et al. (1968) estimated that the pressure melting point of the ice at the base of this core at 197 bars of pressure is −1.6 °C. Our model predicts −1.5 °C at 197 bars for a pure ice/water system. The presence of solutes at the ice/water junction would reduce our estimate to lower temperatures. Our parameterization allows one to estimate the effect of temperature and pressure on ice core density. This is an example where the FREZCHEM model was not directly used but only a subset of parameters dealing with properties of ice (Equations 3.15 and 3.18–3.20). Our parameterizations of equilibria and processes have value beyond the FREZCHEM model per se.

5.2 Mars

5.2.1 Surficial Aqueous Geochemical Evolution

In a recent paper (Marion et al. 2003a), we quantified a conceptual model for the surficial aqueous geochemical evolution of Mars. This model rests

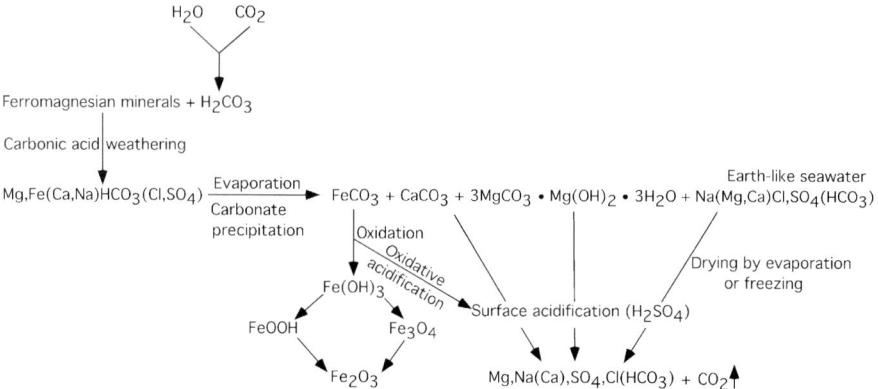

Fig. 5.10. Conceptual model of carbonic acid weathering of ferromagnesian minerals on Mars. Reprinted from Marion et al. (2003a) with permission

on a foundation of previous work (e.g., Clark and Van Hart 1981; Schaefer 1990 1993; Burns 1993; Banin et al. 1997; Catling 1999; Morse and Marion 1999; Bridges and Grady 2000; Catling and Moore 2000; Bridges et al. 2001; Catling and Moore 2003). The five stages of the conceptual model are: (1) carbonic acid weathering of primary ferromagnesian minerals to form an initial magnesium-iron-bicarbonate-rich solution; (2) evaporation and precipitation of carbonates, including siderite, with evolution of the brine to a concentrated NaCl solution; (3) ferrous/ferric iron oxidation; (4) evaporation or freezing of the brine to dryness; and (5) surface acidification (Fig. 5.10).

For the highland crust of early Mars, a mafic to ultramafic character is expected based on Martian meteorites (Gooding 1992; McSween 1994) and fundamental geochemical considerations (Toulmin et al. 1977; Baird and Clark 1981; Burns 1993). These rock types are dominated by ferromagnesian minerals that weather to produce solutions dominated by Fe^{2+}, Mg^{2+}, Ca^{2+}, K^+, and Na^+. Initial P_{CO_2} was assumed to be 2 atm, which is sufficient for greenhouse warming to >273 K with a faint early sun (Forget and Pierrehumbert 1997). Under the assumed high atmospheric CO_2 of early Mars, the dominant anion would have been bicarbonate. Therefore, based on Mars minerology and putative early high CO_2 levels, we used a hypothetical solution for early Mars water based on groundwater in terrestrial ultramafic rocks that is dominated by alkalinity (Catling 1999) (Table 5.3, Soln. A). Note that this solution is similar in composition to the CO_2-HCO_3 water of Lake Nyos, Cameroon (cf. Tables 3.6 and 5.3a). This is not surprising given that biotite, a ferromagnesian silicate, is a major constituent of the Lake Nyos sediment silt (Bernard and Symonds 1989) and Lake Nyos is situated amidst mafic to ultramafic alkaline volcanic rocks (Lockwood and Rubin 1989).

This hypothetical early Mars water was subjected to concentration by evaporation; we tracked its evolution until evaporation increased solutes by

Table 5.3. Martian and comparative solution compositions in the geochemical evolution of Mars. Reprinted from Marion et al. (2003a) with permission

Solution Constituent	A. Hypothetical early Mars water[a]	B. 1000-fold concentration of Soln. A	C. Soln. B at lowered P_{CO_2}	D. Terrestrial seawater[b]	E. Soln. C with 1.665 m H_2SO_4	F. Hypothetical Mars brine[c]
Na^+ (m)	0.0008	0.8068	0.8101	0.4870	0.8101	0.9108
K^+ (m)	0.00007	0.0706	0.0709	0.0106	0.0709	0.0599
Mg^{2+} (m)	0.001	0.3744	0.0821	0.0552	1.743[d]	1.743
Ca^{2+} (m)	0.0005	0.00068	0.0011	0.0103	0.0022[d]	0.0022
Fe^{2+} (m)	0.0008	0.00000151	—[e]	—	—[e]	—
Cl^- (m)	0.00065	0.6556	0.6582	0.5682	0.6582	0.5256
SO_4^{2-} (m)	0.00018	0.1815	0.1823	0.0294	1.847	1.958
Alkalinity (m)	0.00446	0.6090	0.0246	0.00143	0.0184	0.0184
P_{CO_2} (atm)	2	2	0.0053	0.00035	0.0053	0.0053
pH (calculated)	5.00	6.84	8.03	8.05	7.57	7.58
Ionic strength (m)	0.008	2.19	1.31	0.723	7.95	8.17
Temperature (°C)	0°C	0°C	0°C	0°C	0°C	0°C

[a] Catling, 1999.
[b] Marion and Farren 1999. This analysis assumes equilibrium with respect to calcite at 0°C with $P_{CO_2} = 3.5e - 4$ atm (current Earth concentration).
[c] Clark and Van Hart 1981. Their table III equilibrated at 0°C.
[d] Assumes that sufficient H_2SO_4 is added to solubilize Mg and Ca salts to match Soln. F.
[e] At this stage and beyond, the assumption is made that all Fe would be irreversibly precipitated as a ferric mineral.

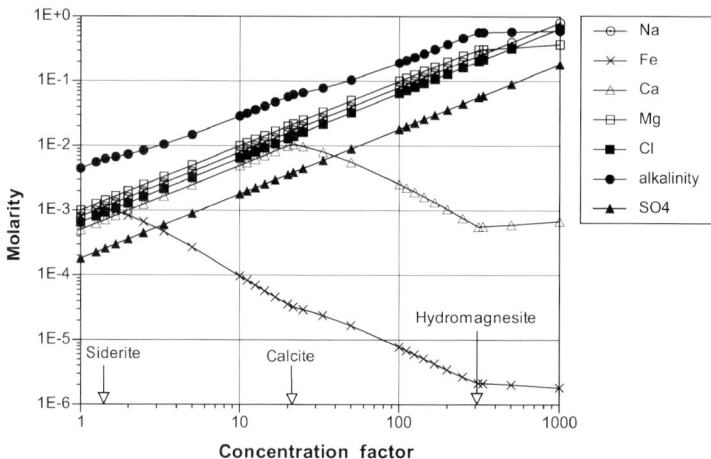

Fig. 5.11. A hypothetical early Mars water concentrated 1000-fold at $0\,°C$. Reprinted from Marion et al. (2003a) with permission

1000-fold (Soln. B, Table 5.3). Because of the dominance of carbonate alkalinity in this solution, first siderite, then calcite, and finally hydromagnesite are predicted to precipitate at $0\,°C$ (Fig. 5.11). Because of the insolubility of siderite, very little of the original Fe^{2+} remains in solution after a 1000-fold concentration. This particular simulation assumed no $O_2(g)$. Fe^{2+} is unstable relative to Fe^{3+} minerals even at low levels of $O_2(g)$ (Fig. 5.12). The presence

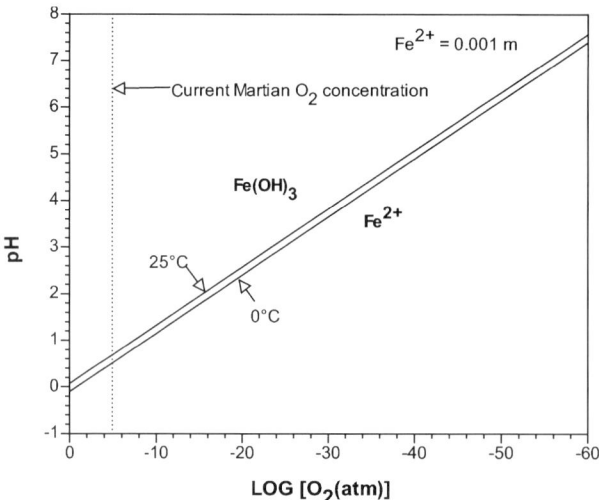

Fig. 5.12. Stability of ferrous/ferric iron as a function of pH and atmospheric oxygen concentration. Reprinted from Marion et al. (2003a) with permission

of hematite deposits on Mars – possible products of surface aqueous deposition (Catling and Moore 2000; Christensen et al. 2000a, 2001; Catling and Moore 2003) – and siderite in Martian meteorites (Clark and Van Hart 1981; Gooding 1992; Bridges et al. 2001) suggests that there must have been times when environmental conditions were sufficiently reducing for high levels of ferrous iron to have existed. However, these periods were probably as short-lived on Mars as they were on Earth (Burns 1993; Catling and Moore 2000) because equilibrium today on Mars exists well within the stability fields of ferric minerals (Fig. 5.12), as they do on Earth's surface. Today, iron oxides are the dominant form of surface iron on Mars (Toulmin et al. 1977; Burns 1993; Banin et al. 1997; Rieder et al. 1997; Catling and Moore 2000, 2003). The assumption was made based on this simulation that iron probably precipitated early in the geochemical evolution of Mars, perhaps first as the ferrous mineral siderite but ultimately oxidizing to ferric iron minerals; and once present as ferric iron, this oxidative transformation was probably irreversible (Fig. 5.10). In subsequent simulations, iron is assumed to be essentially absent because ferric iron compounds have a low solubility.

The evaporatively concentrated Mars water (Soln. 5.3B) was then allowed to drop in $CO_2(g)$ concentration from 2 atm (the assumed initial Mars concentration) to 5.3×10^{-3} atm (the current average Mars concentration, Kieffer et al. 1992). This loss of CO_2 leads to a significant increase in pH from 6.84 to 8.03 and a significant additional precipitation of hydromagnesite; note the large drops in magnesium and alkalinity concentrations (cf. Soln. B and Soln. C, Table 5.3).

The resulting Soln. C is a predominantly NaCl solution similar to terrestrial seawater (Soln. D, Table 5.3). Had we chosen a concentration factor of 600-fold, the agreement would have been even better. In any case, the concentration factor is arbitrary. The point is that simple processes, starting with a dilute Fe-Mg-HCO_3-rich solution formed by reaction of water with ultramafic and mafic rocks, evaporation, and carbonate precipitation, converted the solution into an Earth-like seawater NaCl brine. The Na/Mg ratio of solution C is 9.9, while terrestrial seawater has a Na/Mg ratio of 8.8 (Soln. 5.3D). Note also the similar pH values (8.03 and 8.05, Table 5.3). This solution did not (cannot) evolve into an alkali soda-lake composition as some have hypothesized or assumed for Mars (e.g., Kempe and Kazmierczak 1997; Morse and Marion 1999) because the mass of hypothesized soluble iron and magnesium and the low solubility of their respective carbonate minerals are sufficient to precipitate most of the initial soluble bicarbonate/carbonate ions.

The current surface of Mars is cold and dry (Fig. 5.13). Early oceans or lakes would have dried out either by evaporation or freezing. Solution C was allowed to dry by either (a) evaporation at 0 °C (Fig. 5.14a) or (b) freezing to the eutectic (Fig. 5.14b). Both equilibrium-mode evaporation and freezing lead to six precipitated salts (Fig. 5.14) because the number of independent salt components is equal to the number of cations (4) + the number of anions

Fig. 5.13. The landing site (Gusev Crater) of the Mars "Spirit" rover. Courtesy NASA/JPL/Caltech

(3) $-1 = 6$. The ensuing suite of minerals that theoretically precipitate are, however, different for these two scenarios. For example, drying by evaporation leads to precipitation of predominantly halite (NaCl), while freezing leads to predominantly hydrohalite (NaCl·2H$_2$O) (Fig. 5.14a–b). If states of mineral hydration are preserved in ocean or lacustrine deposits on Mars, then these records could provide valuable clues to the environmental history of Mars during the drying process.

The last solid phase to precipitate during equilibrium-mode freeze drying is MgSO$_4$·12H$_2$O at $-35.4\,°C$, which sets the eutectic temperature. However, at this point, the solution is also nearing saturation with respect to MgCl$_2$·12H$_2$O. If MgSO$_4$·12H$_2$O is dropped from the FREZCHEM mineral database, then MgCl$_2$·12H$_2$O begins precipitating at $-36.3\,°C$. The accuracy of the magnesium mineral solubility products at $-36\,°C$ are not adequate to distinguish which of these two magnesium salts (chloride or sulfate) is the more likely.

If drying at this stage was the last step in the aqueous geochemical evolution of Mars, then we would expect to find surficial salts present in the order of abundance: NaCl > (MgNa)SO$_4$ > (MgCa)CO$_3$ (Fig. 5.14a–b). Instead we find surficial salts present in the order of abundance: (MgNa)SO$_4$ > NaCl > (MgCa)CO$_3$ (Clark and Van Hart 1981). To reconcile this apparent discrepancy, we need to either add MgSO$_4$ to or remove NaCl from the initial solution or at any step along the way. Past work suggests that acidification of the surface by volcanic acidic volatiles such as HCl, HNO$_3$, or H$_2$SO$_4$ can explain the predominance of MgSO$_4$ salts on the surface of Mars (Settle 1979; Clark and Van Hart 1981; Clark 1993; Banin et al. 1997). For example, addition of sulfuric acid to secondary hydromagnesite

$$3MgCO_3 \cdot Mg(OH)_2 \cdot 3H_2O + 4H_2SO_4 \leftrightarrow 4Mg^{2+} + 4SO_4^{2-} + 3CO_2 \uparrow + 8H_2O \tag{5.3}$$

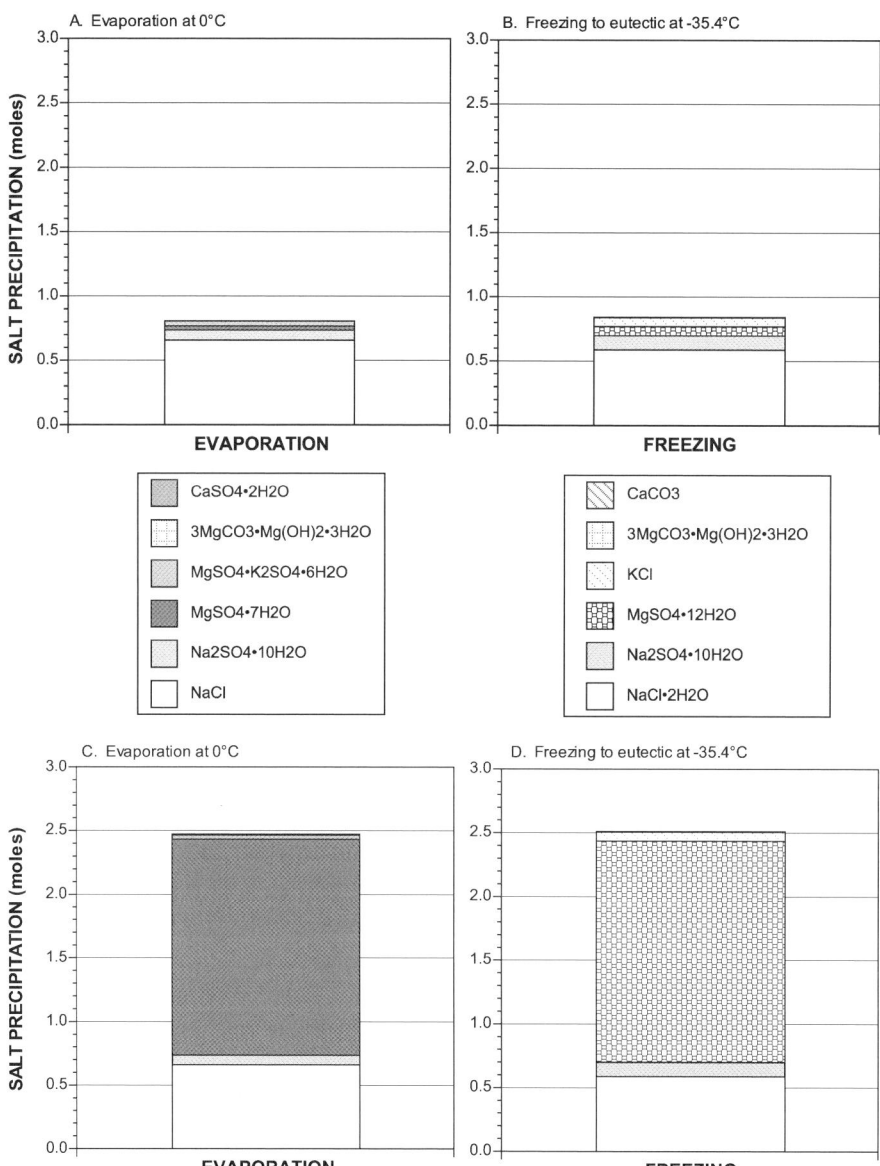

Fig. 5.14. Salt precipitation from Soln. C (Table 5.3) subjected to **a** evaporative drying at 0 °C and **b** freeze drying to the eutectic at −35.4 °C, and salt precipitation from Soln. E (Table 5.3) subjected to **c** evaporative drying at 0 °C and **d** freeze drying to the eutectic at −35.4 °C. Reprinted from Marion et al. (2003a) with permission

would result in an increase in soluble magnesium and sulfate and a loss of CO_2 to the atmosphere. Or alternatively, sulfuric acid could react with primary ferromagnesian minerals

$$2(Mg,Fe)_2SiO_4 + 2H_2SO_4 + 0.5O_2 \leftrightarrow 2Mg^{2+} + 2SO_4^{2-} + 2SiO_2 \downarrow + Fe_2O_3 \downarrow + 2H_2O. \quad (5.4)$$

The net result is the same, an increase in $MgSO_4$ salts (Eqs. 5.3 and 5.4). Banin et al. (1997) have hypothesized that these volcanic acids are a relatively recent addition (up to 10^9 years B.P.).

Clark and Van Hart (1981) hypothesized that a Mg-Na-SO_4 brine from New Mexico (their table III) is analogous to a Martian brine dominated by Mg-Na-SO_4. Their brine was equilibrated at 0 °C using the FREZCHEM model (Soln. F, Table 5.3) and was used in this study for comparative purposes.

We added sufficient sulfuric acid (1.665 m H_2SO_4) to the Martian seawater solution (Soln. C, Table 5.3) to increase the Mg and Ca molalities (Soln. E, Table 5.3) to those of the hypothetical Martian brine (Soln. F, Table 5.3). The assumption was made that for every two moles of added acidity, one mole of Mg or Ca and one mole of SO_4 were released into solution (Equations 5.3 and 5.4). As a consequence, the derived Soln. E and Soln. F agreed exactly in their Mg and Ca molalities (Table 5.3). Other constituents of our derived Soln. E, however, were not so constrained. Nevertheless, note the similar Na, K, Cl, SO_4, alkalinity, pH values, and ionic strengths of Solns. E and F (Table 5.3). Note that the final brine composition (Soln. E, Table 5.3) has little carbonate alkalinity. By this stage, any significant surficial carbonates formed during the early geochemical evolution of Mars would either be buried by more soluble seawater salts and dust or removed by acidification. This may account for the minor amounts of carbonate found on the surface of Mars (Blaney and McCord 1989; Pollack et al. 1990; Gooding 1992; Calvin et al. 1994: Griffith and Shock 1995; Christensen et al. 1998, 2000b).

In addition to acidification of the surface through volcanic acids, which we explicitly modeled (Table 5.3), there is another acidification process that is implicit in our model but that was not explicitly considered: namely, acidification resulting from oxidation of ferrous to ferric iron (Fig. 5.10). For every mole of iron that is oxidized, two moles of acid are produced:

$$4Fe^{2+} + O_2(g) + 10H_2O \Leftrightarrow 4Fe(OH)_3(cr) + 8H^+. \quad (5.5)$$

In terms of the original concentrations (Soln. A, Table 5.3), 0.0008 moles of Fe^{2+} would produce 0.0016 moles of H^+. The amount of volcanic acid added (1.665 m H_2SO_4) in terms of the original concentrations (before the 1000-fold concentration) is equivalent to 0.00333 moles of H^+. Together these two sources of acid add up to 0.00493 moles, which is more than enough to neutralize all the original alkalinity (0.00446 moles, Table 5.3). However, acids would also be neutralized by reaction with primary minerals (Eq. 5.4). Based on the prevalence of carbonates, especially siderite, in Martian meteorites

(Bridges et al. 2001) it is unlikely that acidification processes neutralized all the alkalinity on Mars or that all the ferrous iron was oxidized.

If Soln. E is allowed to dry by evaporation at $0\,°C$ (Fig. 5.14c) or freezing to the eutectic (Fig. 5.14d), then the distribution of precipitated salts is similar to Soln. C except for the large increase in $MgSO_4$ salts (cf. Figs. 5.14a,b with 5.14c,d). Exactly the same suite of salts precipitate for Solns. C and E. Also, the eutectic temperature is the same ($-35.4\,°C$). For Soln. E, the salt quantities fall in the order $(MgNa)SO_4 > NaCl > (MgCa)CO_3$, in agreement with estimates of salt distribution on the Martian surface (Clark and Van Hart 1981).

Figure 5.10 summarizes the evolution of the surface aqueous geochemistry of Mars. The process begins with the weathering of ferromagnesian minerals by carbonic acid leading to a predominantly $Mg-Fe-HCO_3$ solution. As this solution evaporates, iron, calcium, and magnesium carbonate minerals precipitate, drawing down the atmospheric CO_2, which cools the surface temperature on Mars. The resulting solution has an Earth-like seawater composition (NaCl dominated). Ferrous iron is irreversibly oxidized to ferric minerals early in the geochemical evolution of Mars. Eventually, the seawater would dry either by evaporation or freezing. And, finally, the surface is acidified by volcanic volatile acids or ferrous iron oxidation, producing a predominantly $Mg-Na-SO_4-Cl$ salt phase. The precise timing of drying and acidification is uncertain. There is evidence for enormous floods throughout Martian history (Carr, 1996). Oceans/lakes were likely ephemeral, alternating forming and drying out. Banin et al. (1997) have hypothesized that volcanic acidification was a relatively recent phenomenon (up to 10^9 years B.P.); on the other hand, oxidative acidification could have occurred early in Mars history whenever oxygen levels in the atmosphere became sufficiently high. The transformation from a NaCl into a $Mg-Na-SO_4-Cl$ brine via acidification could have been gradual with alternating wet and dry periods.

What do these simulations imply about the prospects for life on early or current Mars? Salinities are predicted to increase over time (cf. ionic strengths of Solns. A and E, Table 5.3). The calculated a_w for Soln. E is, however, only 0.926; despite the high ionic strength of this saline solution, the prevalence of sulfates presents a relatively favorable environment for life (cf. Basque Lake in Table 5.1).

Can this model published in 2003 (Marion et al. 2003a) explain all the geochemical findings of the 2004 Mars Exploration Rover (MER) missions? Not exactly! In our model we predicted that ferrous iron would precipitate as siderite ($FeCO_3$) early in the temporal sequence, and siderite would ultimately be oxidized to ferric minerals such as ferrihydrite [$Fe(OH)_3$] and hematite (Fe_2O_3) (Fig. 5.10). There is no place in this conceptual model for the precipitation of ferrous or ferric sulfate minerals as suggested by the MER missions (Squyres et al. 2004; Lane, 2004). This problem could be simply rectified by drawing an arrow from siderite through the surface acidification

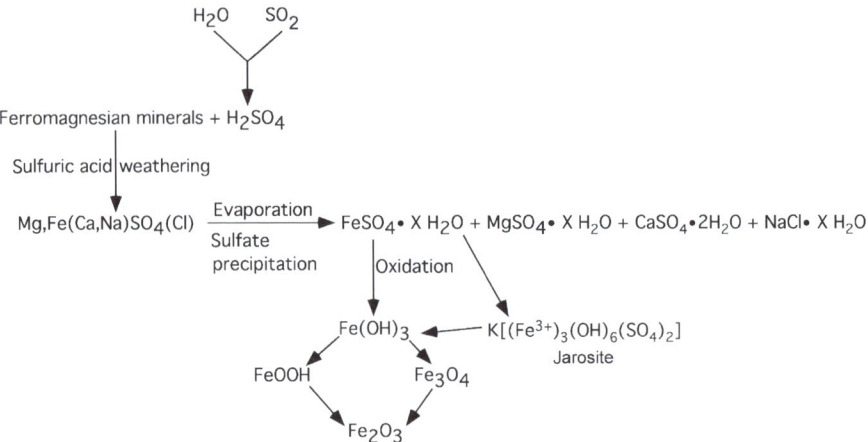

Fig. 5.15. Conceptual model of sulfuric acid weathering of ferromagnesian minerals on Mars

process to the Mg-Na-SO$_4$-dominated brine (Fig. 5.10), which would allow for the precipitation of ferrous sulfates and, ultimately, ferric sulfate minerals. This conceptual model (Fig. 5.10) presupposes that early Mars was warmer and wetter due to high CO$_2$ levels (1 or more bars) in the atmosphere, which would lead predominantly to carbonic acid weathering.

A very different conceptual model is based on the assumption that the early Martian atmosphere was dominated by acidic volatiles from volcanic activity, which leads directly to sulfuric acid weathering (Fig. 5.15). This model leads to the dominance of sulfate minerals and leaves open the possibility of both ferrous and ferric iron sulfates. In addition to atmospheric acidification,

Fig. 5.16. "Blueberries" in the stratified rocks in the landing site (Eagle Crater) of the Mars "Opportunity" rover. Courtesy NASA/JPL/Caltech

another acidification process that likely was important on Mars is iron oxidation (Figs. 5.10 and 5.15, Eq. 5.5). Figure 5.16 shows the laminated rocks at the Eagle Crater site on Mars that have high concentrations of chloride, bromide, sulfate, and the iron minerals jarosite and hematite. The hematite spherules that formed in these rocks stand out as false-color "blueberries." These MER results suggest rock formation from acidic brines. At this time, it is clear that acidification processes played an important role in the surficial geochemistry of Mars. The exact acidic processes and mechanisms are still unclear but are being actively worked on by many independent groups motivated by the MER findings.

5.2.2 Early Mars Oceans

In earlier work (Morse and Marion 1999), we also examined a hypothetical early Martian ocean by assuming that the oceanic composition would be similar to early Hadean Eon (∼4.3–3.8 Ga) oceans on Earth, which were characterized by low sulfate (Morse and Mackenzie 1998). Our hypothetical starting solution in this case was identical to current terrestrial oceans, except that sulfate was replaced by alkalinity (Fig. 5.17). While these simulations did not include iron chemistry, the starting solution was alkaline, as was our previous hypothetical early Mars water (Table 5.3a). These high alkalinities are a consequence of assuming that early Mars was warm and wet due to high atmospheric CO_2 concentrations (2 to 5 bars), which naturally leads to carbonic acid weathering and the dominance of carbonate alkalinity (Fig. 5.10).

In this simulation, we followed the removal of atmospheric CO_2 as carbonate minerals, which leads to cooling of the climate, freezing of the ocean, and, ultimately, to curtailing of the hydrologic cycle and rock weathering. Differences were found in the suite of minerals precipitating depending on atmospheric CO_2 concentrations (Fig. 5.17). Freezing of this hypothetical solution to the eutectic at $P_{CO_2} = 0.1$ atm (Fig. 5.17c) produced similar minerals to our previous case (Fig. 5.14b), except for the lack of sulfate in these simulations (Fig. 5.17a). Comparisons of the salts that hypothetically precipitate and their eutectic temperatures at 1 to 5 atm (Fig. 5.17b) and 0.1 atm (Fig. 5.17c) of CO_2 pressure demonstrate the idiosyncratic nature of the freezing process. Chemical systems and their response to temperature and pressure depend, ultimately, on the thermal and volumetric properties of individual constituents, which makes every system response highly individualistic. One of the virtues of geochemical models is that they can cope with natural, complex variability.

5.2.3 A Cold, Intermittently Wet Mars

The evolution of an early warm, wet Mars to the present-day cold, dry Mars can be explained, in principle, by assuming high atmospheric concentrations of CO_2 early in Martian history whose removal as carbonate minerals led to planetary cooling. These are underlying assumptions in both of our previous Martian simulations (Figs. 5.10–5.12, 5.14, 5.17). However, if this were the case, where are the precipitated carbonates? Explanations for lack of sufficient carbonate near the Martian surface today include surficial covering by dust and removal by subsequent acidification later in Mars history (Fig. 5.10). A very different concept of Martian climate history does not assume a warm-wet phase but proposes that Mars was always cold with only intermittent short-term wet intervals driven largely by glacial or subterranean processes (Kargel 2004). Under these conditions, pressure becomes a critical component driving the hydrologic cycle.

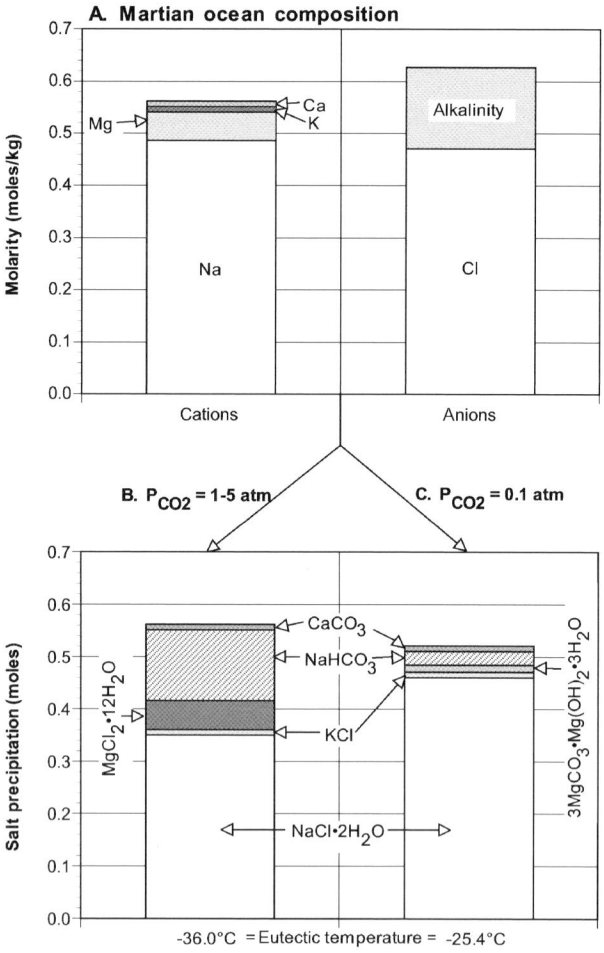

Fig. 5.17. The **a** hypothesized composition of an early Martian ocean and the composition of salts precipitating by freezing to the eutectic at **b** $P_{CO_2} = 1$ to 5 atm and **c** $P_{CO_2} = 0.1$ atm. Reprinted from Morse and Marion (1999) with permission

In a recent paper (Gaidos and Marion 2003), we used the FREZCHEM model to simulate hypothetical processes controlling hydrologic features on Mars. Geomorphic evidence previously interpreted to indicate a more temperate early climate can be explained by cold climate processes and the transient flow of liquid water at the surface rather than a terrestriallike hydrological cycle. In this scenario, freezing and confinement of crustal aquifers leads to eruption of water or brines to the surface: Hesperian events formed massive ice sheets and triggered floods that carved the outflow channels, while smaller, present-day aqueous eruptions are responsible for the seepage and gullylike landforms identified in high-resolution imaging.

We ran a series of simulations to quantify how evaporation and freezing would affect the chemical compositions of aqueous solutions on Mars. We assumed an aquifer 5 km thick, divided for modeling purposes into ten layers, each 0.5 km in thickness. The lithostatic pressure for Mars was assumed to be 102 bars km^{-1} (a mean crustal density of 2700 kg m^{-3}). The temperatures for the simulations were based on the freezing point depression of the aqueous solutions, which is a function of chemical composition and pressure, and is calculated by the FREZCHEM model. As the starting point for chemical composition, we assumed a hypothetical early Mars water based on groundwater compositions for ultramafic rocks on Earth (Catling 1999) {Na = 0.8 mM, K = 0.07 mM, Mg = 1.0 mM, Ca = 0.5 mM, Fe = 0.8 mM, Cl = 0.65 mM, SO$_4$ = 0.18 mM, and alkalinity = 4.46 mM, where mM is the molal concentration [10^{-3} moles/kg(water)]}; this is the same starting solution that we used previously (Table 5.3a). We assumed that the partial pressure of CO$_2$ on early Mars would be in a range of 50 to 250 mbars (Bridges et al. 2001). For the purposes of this study, we used an initial partial pressure of 100 mbars that declined to 50 mbars over time, and we assumed that this decline was sufficient to halt any further concentration of brine by evaporation. A decline of 50 mbars of atmospheric CO$_2$ is equivalent to precipitation of 0.06 moles of carbonate. The original alkalinity (4.46 mM) would have to be concentrated 26.9-fold, presumably by surface evaporation, in order to supply the carbon needed to precipitate 0.06 moles of carbonate. So, the original "Catling" solution (see above) was concentrated 26.9-fold, and this served as our Martian brine.

Initially, this Martian brine was equilibrated at P_{CO_2} = 50 mbars within each of the ten layers. Under these conditions, virtually all the Fe (99.99%) precipitates as siderite (FeCO$_3$), and 77.6 to 96.6% of the Ca precipitates as calcite (CaCO$_3$) (Fig. 5.18). After equilibration of each 0.5-km layer separately, we then froze the profile from the top down, layer by layer, assuming that the freezing process would be sufficiently slow that all soluble salts would be ejected into the lower, unfrozen, layers. This process leads to increasing salt concentrations with depth (Fig. 5.18). These freezing simulations were done assuming that all Fe was removed in the initial evaporative concentration. Freeze concentration leads to the additional precipitation of calcite and

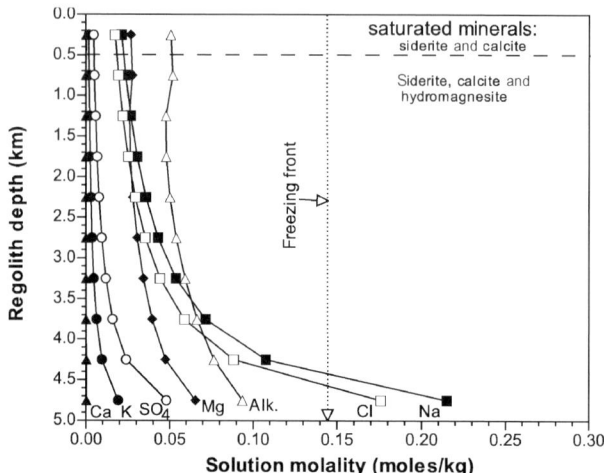

Fig. 5.18. The regolith layer (0.5 km) molal concentrations as the freezing front moves through the Martian profile. Minerals that should theoretically precipitate within the profiles either from the initial evaporative concentration or later from freezing concentration are listed

the removal of some Mg as hydromagnesite at depth (Fig. 5.18). During the early phases of this freezing process, eruptions could have come from any of the layers. But over time, as layers freeze, eruptions come from progressively deeper layers. If the equilibrated concentrations from the deepest layer (4.5–5.0 km) erupts to the surface, the lower total pressure and lower P_{CO_2} at the surface {from 50 mbars to 5 mbars [current Mars P_{CO_2} (Kieffer et al. 1992)]} would lead to significant additional precipitation of hydromagnesite at the surface. For example, 77.5% of the Mg and 88.0% of the alkalinity in the deepest layer (Fig. 5.18) would precipitate as hydromagnesite on the surface; the residual solution is dominated by NaCl. Allowing the residual erupted solution to freeze to the eutectic (237.25 K) ultimately leads to precipitation of ice, additional hydromagnesite, mirabilite ($Na_2SO_4 \cdot 10H_2O$), sylvite (KCl), hydrohalite ($NaCl \cdot 2H_2O$), and $MgCl_2 \cdot 12H_2O$.

The long-term consequence of evaporation at the surface, followed by freezing of the regolith, and brine eruptions to the surface is to first remove alkalinity, then sulfate, and finally chloride salts from the initial Martian aqueous composition. Our scenario predicts that carbonates have largely accumulated within the regolith, except for hydromagnesite, which may be present locally on the surface along with sulfate and chloride salts. A 5.2- to 5.4-mm feature in spectra obtained by spacecraft and ground-based telescopes and ascribed to hydrous carbonates has yet to be confirmed (Calvin et al. 1994). Cryogenic formation of calcite is associated with fractionation of stable carbon isotopes, an effect that could explain the observed $d^{13}C$ of carbonates in the Nakla and ALH 84001 meteorites (Socki et al. 2001, 2002).

A cold early climate and a thin CO_2 atmosphere are consistent with the ubiquity of primary igneous minerals and the apparent absence of secondary minerals and copious carbonates on the surface. Aqueous eruptions introduced soluble chloride and sulfate salts into Martian soils, leaving clays and carbonates within the crust. If the Martian crust hosts a "deep biosphere," aqueous eruptions could bring organisms or their chemical signature to the surface. The biotic component of any recent eruptions may be preserved in transient ice in cold traps on the surface.

5.2.4 Hydrate Deposits and Thermal Stratification

Low-temperature aqueous geochemistry can have far-reaching effects on the geology and geophysics of planets. Hydrates – both salt hydrates or gas hydrates – have low thermal conductivities compared to water ice Ih and most common silicates (Ross and Kargel 1998; Prieto-Ballesteros and Kargel 2005). Hydrates typically have thermal conductivities a factor of four to ten less than the conductivities of other common planetary materials, so it is not a small contrast. Thus, any thick bed of hydrates acts as a thermal insulation blanket. If the bed is on the order of 100 m thick or thicker (or a sequence of hydrate interbeds that cumulatively amounts to 100 m or more), it can have a profound influence on the subsurface thermal environment. Beds of hydrates, however, can so perturb the thermal environment of underlying crust that planetary geophysical conditions and processes and surface geology can be dramatically influenced (theoretically at least). The physics of heat flow is such that widely disseminated nodules or grains of hydrates embedded in a silicate or ice matrix would not have a large influence on the crust's thermal state since heat can readily flow around the insulating inclusion. With an insulating bed of material, however, heat flow is so impeded that temperatures beneath the insulator can rise to a point where melting or solid-state convection can take place where otherwise it would not have.

The possible geological and geophysical roles of hydrate bed insulators was raised by Prieto-Ballesteros and Kargel (2005) for the case of Europa and was modeled quantitatively by Kargel et al. (in press) for Mars. In the case of the Europa work, it was concluded that a heterogeneous patchwork of salt hydrates could localize subsurface melting and domains of thermally softened ice; thus, hydrate deposits could affect cryovolcanism, crustal melt-through, brittle lithospheric thickness, and solid-state diapirism on Europa. In the case of Mars (Fig. 5.19), salt hydrates in buried evaporitic crater lakes or clathrate hydrate beds contained in permafrost (Prieto-Ballesteros et al. 2006) can allow shallow ice melting and may explain young water-formed features Kargel et al. (in press).This horizontally stratified model (Fig. 5.19) is in contrast to an earlier model developed by Gaidos and Marion (2003), where freezing of the regolith was assumed to be horizontally uniform (Fig. 5.18). Cryovolcanic eruptions to the surface would come from deeper in

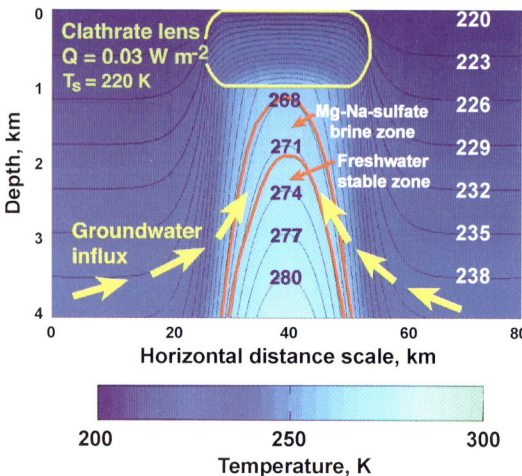

Fig. 5.19. Steady-state conductive thermal anomalies associated with insulating clathrate lens on Mars

the Martian regolith with the Gaidos/Marion model than would be the case with the Kargel et al. (in press) model.

As Kargel et al. (in press) point out, this type of thermal perturbation can be stable so long as the hydrates neither dehydrate nor disperse. This is fundamentally unlike other types of hydrothermal systems, such as those generated by impact or magmatism, which are self-dissipating. Dispersion would be prone to occur in open hydrological systems due to dissolution and transport of solutes, but reinforcement of the chemical anomaly and its geophysical effects is apt to occur in hydrologically closed basins. Dehydration may ensue if the climate becomes too warm for the hydrates to remain stable, or if the thermal anomaly produced by the hydrates becomes so extreme that the hydrates dissociate; in the latter case, the hydration state and the deposit's thermal anomaly might be regulated. Other systems might behave episodically or chaotically with cycles of dehydration/rehydration and rising and falling of the thermal anomaly as hydrothermalism also rises and falls.

We emphasize that it is not anhydrous salts but salt hydrates (and gas hydrates) that have this important thermal insulating property. Cool climates, which tend to support high hydration states, favor this type of process. Cold evaporitic basins are ideal, and so we find that Mars is where this phenomenology is most likely to have widespread relevance.

The same phenomenology must be important locally on Earth, too, where thick evaporite deposits of hydrated salts and local thick beds of methane clathrate in permafrost or seafloor sediments should influence the thermal environment of the crust. The predicted control on the crust's thermal state by hydrate deposits should have consequences for the localization of hydrothermal springs around and within evaporite basins, hydrothermal metamorphism and petroleum maturation beneath these basins, and permafrost thermokarst development in areas where methane clathrate is abundant. Thus far, this effect is a hypothetical concept, though it seems an inevitable consequence in some environments. We are not aware of work that has definitively linked Earth processes and geomorphology specifically to this effect, but we expect that they will be found.

5.3 Europa

After Earth and Mars, the ice-covered moon of Jupiter, Europa, is the most likely Solar System body that could support life. The presence of an aqueous ocean beneath the ice cover (Figs. 5.20 and 5.21) might possess environmental conditions that could be within the tolerance ranges of life as we know it (Kargel et al. 2000; Marion et al. 2003b). Many of the properties and processes on Europa are inferred based on indirect evidence. For example, estimates of the thickness of the water/ice layers range from a few kilometers to almost 200 km (McKinnon 1997; Anderson et al. 1998; Pappalardo et al. 1999; Kargel et al. 2000). The total thickness of this water/ice shell has a direct bearing on pressures that organisms might face in a Europan ocean and the cycling of critical elements for life such as oxidants and nutrients between the ocean and the surface. The surface of Europa is strongly oxidizing because of the heavy radiation load on the surface (Carlson et al. 1999a; Cooper 2001; Cooper et al. 2002; Greenberg 2002; Greenberg et al. 2002). The ocean and seafloor sediments are presumably reducing environments. Biology utilizes oxidation–reduction reactions to fuel metabolism (Gaidos et al. 1999). To maintain a long-term viable ecosystem on Europa would require cycling of oxidants and reductants. Presumably a thin crust would lead to quicker cycling between the surface ice and the subsurface ocean. If the crust is thick enough, slow cycling of surface oxidants into the ocean could limit life in the ocean (Barr et al. 2002).

The composition of a Europan ocean has a major bearing on the suitability of this ocean for life. It has been hypothesized that a Europan ocean could be (1) a neutral Na-Mg-SO_4-H_2O solution, (2) an alkaline Na-SO_4-CO_3 solution, or (3) an acidic Na-H-Mg-SO_4 system (Kargel et al. 2000; Marion 2001, 2002; Kempe and Kazmierczak 2002). Simulations of these three alternatives and their bearing on potential life on Europa are discussed below.

Kargel (2001), Thomson and Delaney (2001), and O'Brien et al. (2002) have hypothesized that a Europan ocean could become thermally and compositionally stratified. Disturbance and overturning of this ocean could lead to exsolution of gases and a massive thermal flux to the surface via cryovolcanism (Fig. 5.21). Such massive fluxes, if frequent enough, could play an important role in cycling oxidants compared with the presumably slower process of solid–phase convective flow in the soft ice. There may also be hydrothermal vents in a Europan ocean. If massive enough, these vents could play an important role in the cycling of oxidants, nutrients, and heat. Even small hydrothermal vents could serve as refugia for life in an otherwise inhospitable environment.

The variability in the surface temperature (Fig. 5.21) reflects measured diurnal fluctuations (Spencer et al. 1999). The variability in the ocean temperature reflects uncertainties in the chemical composition, the thicknesses of the solid-ice and liquid brine layers, and internal and external sources of heat. For example, the eutectic temperature (the temperature below which only solid phases are possible) for a pure Na-Mg-SO_4-H_2O system at 1.01 bar is ∼238 K (Kargel et al. 2000), while the eutectic temperature for a H-Mg-SO_4-H_2O system at 1.01 bar is 211 K (Kargel et al. 2001; Marion 2002). So, depending on the chemical composition, the ocean temperature could fall anywhere from the initial freezing point of ice for a given composition (∼270 to 255 K) down to the eutectic temperatures (250 to 210 K). Also, as we will demonstrate later, pressure may play a role in defining the temperature of an ice-covered ocean.

5.3.1 Ocean Compositions

The near-infrared mapping spectrometer (NIMS) satellite reflectance spectra from the "nonicy" regions of Europa's surface indicate the presence of highly distorted water-absorption bands, suggesting the presence of bound water in compounds such as hydrated salts or acids (McCord et al. 1998, 1999; Carlson et al. 1999b). The leading hypotheses to explain the surficial salt (or acid) assemblages on Europa are that salts (or acids) are forming from brines such as (1) $MgSO_4$-Na_2SO_4-H_2O (the neutral hypothesis), (2) Na_2SO_4-Na_2CO_3-H_2O (the alkaline hypothesis), and (3) $MgSO_4$-Na_2SO_4-H_2SO_4-H_2O (the acid hypothesis) (McCord et al. 1998, 1999; Carlson et al. 1999b; Kargel

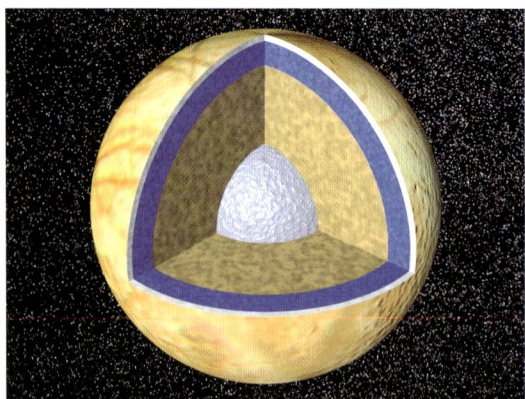

Fig. 5.20. An artist rendition of a cross-sectional view of Europa. Courtesy NASA/JPL/Caltech

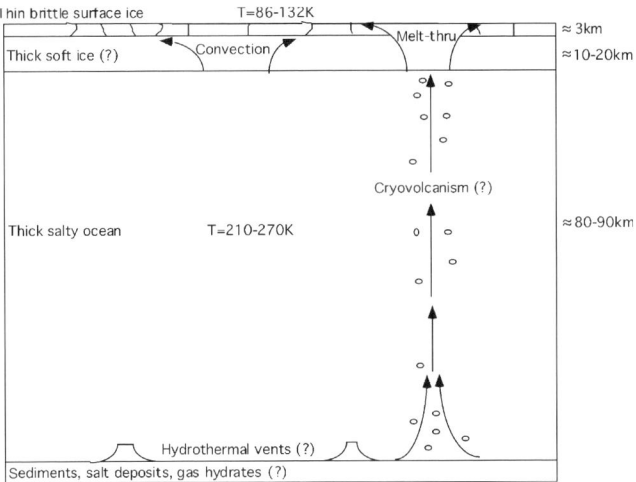

Fig. 5.21. A schematic diagram of the surface layers of Europa. Reprinted from Marion et al. (2003b) with permission

et al. 2000; Kempe and Kazmierczak 2002; Marion et al. 2003b). In what follows, we will first examine these three hypotheses at 1.01 bars of pressure, then consider the potential role of pressure on equilibrium within an Europan ocean. In all cases, we will track how these ocean compositions would have changed from an earlier warmer period (298 K) to the subzero eutectic.

The initial chemical composition for the neutral Na-Mg-Ca-SO_4-Cl-H_2O system is believed to be representative of chondritic weathering (Kargel 1991) with the addition of chlorides in amounts observed by Fanale et al. (1998) in their leachates (Kargel et al. 2000). According to model calculations, the starting solution is slightly supersaturated with respect to both gypsum and epsomite at 298 K (Fig. 5.22a). At a temperature of 271 K, $MgSO_4 \cdot 7H_2O$ is replaced by $MgSO_4 \cdot 12H_2O$; ice starts forming at 266 K. The dual effect of the precipitation of a highly hydrated sulfate salt and ice causes the precipitous drop in sulfate and the rapid rise in chloride. The colder the ultimate temperature, the higher the Cl/SO_4 ratio (Fig. 5.22a). The eutectic temperature reached by this composition is 238.65 K at 1.01 bars. If the temperature of the ocean is lower than 238.65 K, then liquid water cannot exist for this solution. The salinity can be estimated by a_w, which rises from 0.85 (298 K) to 0.93 (266 K), where ice begins to form, and then drops to 0.72 at the eutectic. By the salinity standards of Table 4.2, a_w could be a limiting factor, at times, for biological activity of many microbes, but not for extreme halophiles. Zolotov and Shock (2001), using a similar but more dilute starting salt solution and the FREZCHEM model for calculations, found a eutectic temperature of 237.05 K, which is slightly lower than we found (238.65 K). They also found that the last salt to precipitate was $MgCl_2 \cdot 12H_2O$ at the eutectic; in con-

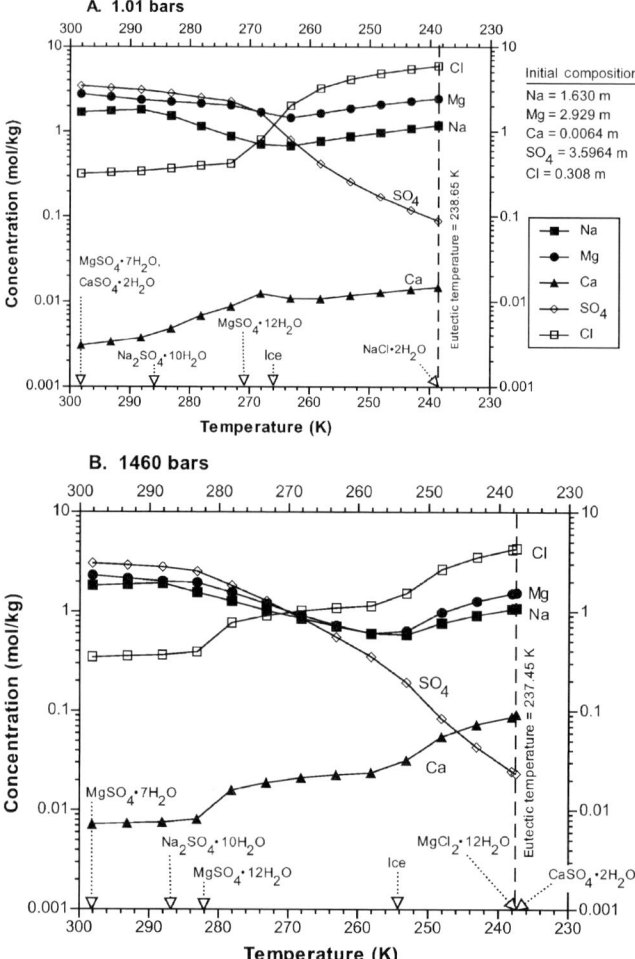

Fig. 5.22. The evolution of a Na-Mg-Ca-SO$_4$-Cl brine at **a** 1.01 bars and **b** 1460 bars of pressure as temperature decreases to the eutectic. Reprinted from Marion et al. (2005) with permission

trast, we found NaCl·2H$_2$O precipitating at the eutectic (Fig. 5.22a). These small discrepancies probably reflect variations in the relative concentrations of the starting solutions. As we point out repeatedly, system equilibrium depends on the thermal and volumetric properties of individual constituents, which makes every system response highly individualistic. One of the virtues of models like FREZCHEM is that they can easily cope with natural variability.

The initial composition for a hypothetical Europan alkaline system is taken from Alkali Valley, Oregon (Marion 2001) and represents the type of

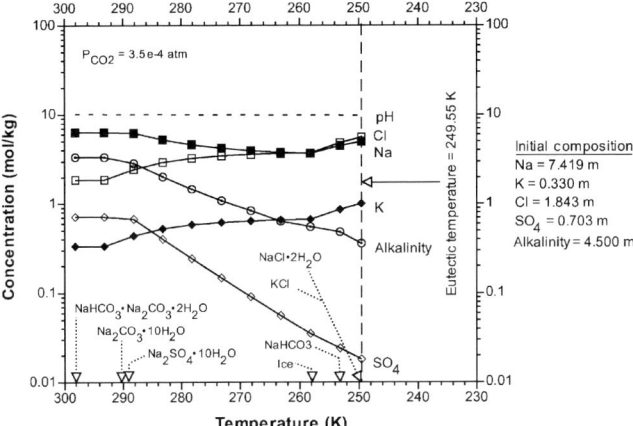

Fig. 5.23. The evolution of a Na-K-Cl-SO$_4$-alkalinity brine as temperature decreases to the eutectic. Reprinted from Marion et al. (2003b) with permission

extreme alkaline system that can lead to precipitation of minerals such as trona (NaHCO$_3$·Na$_2$CO$_3$·2H$_2$O) and natron (Na$_2$CO$_3$·10H$_2$O) (Fig. 5.23). Natron has been suggested as a possible salt on the Europan surface (McCord et al. 1999). Extreme alkaline systems are dominated by Na and K; Mg and Ca concentrations are extremely low because of the relative insolubility of Mg and Ca carbonates in high-pH, high-alkalinity solutions (see Mono Lake and Lake Magadi in Table 5.1) (Marion 2001). The starting solution is supersaturated with respect to trona (Fig. 5.23), but this salt is replaced by natron at 290 K, and eventually by nahcolite (NaHCO$_3$) at 253 K. The high solubility of these alkaline salts keeps ice from forming until the temperature drops to 258 K. The last salts to precipitate are KCl and NaCl·2H$_2$O at the eutectic temperature of 249.55 K at 1.01 bars (Fig. 5.23). Both alkalinity and sulfate drop in concentration with decreasing temperature, leading ultimately to a chloride-dominated system. The calculated pH of this alkaline system ranges from 10.11 at 298 K to 9.73 at the eutectic. But bear in mind that calculated pH is very much a function of the assumed P_{CO_2} ($= 3.5 \times 10^{-4}$ atm) in alkaline solutions. A higher P_{CO_2} would lower pH, and a lower P_{CO_2} would raise the pH. However, it is unlikely that such an alkaline system could rise to levels that might limit life (pH $>$ 11, Table 4.3). The calculated a_w values hover between 0.80 and 0.86 across the range of temperatures depicted in Fig. 5.23. This would limit life forms to extreme halophiles (Table 4.2).

The initial composition for the acidic system is based on low-temperature leaching of chondritic material to produce MgSO$_4$ and high-temperature devolatilization and venting of SO$_2$ into the ocean to produce H$_2$SO$_4$ (Kargel et al. 2001; Marion 2002). The salts that precipitate from this acidic system and the temperatures at which they precipitate (Fig. 5.24) are very similar to the initially considered Na-Mg-Cl-SO$_4$-H$_2$O system (Fig. 5.22a). In this acidic

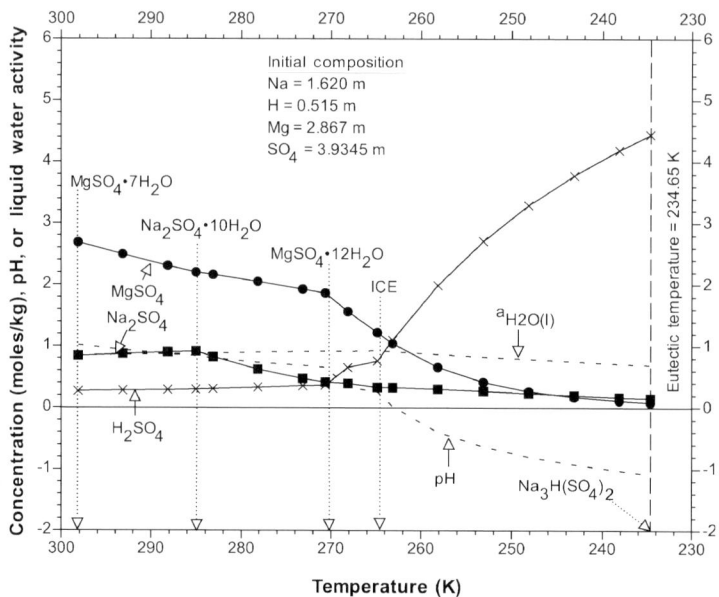

Fig. 5.24. The evolution of a Na-H-Mg-SO$_4$ brine at 1.01 bars of pressure as temperature decreases to the eutectic. Reprinted from Marion et al. (2003b) with permission

case, the last salt to precipitate is Na$_3$H(SO$_4$)$_2$ (an acid salt) at the eutectic temperature of 234.65 K at 1.01 bars (Fig. 5.24). We also ran this simulation without sodium; in that case, the sink for hydrogen was H$_2$SO$_4$·6.5H$_2$O, which precipitated at the eutectic temperature of 211 K (Marion 2002). Because of the high solubility of sulfuric acid, decreasing temperatures and freeze concentration leads to increasing H$_2$SO$_4$ concentrations (Fig. 5.24); this is in marked contrast to the earlier Europan sulfate systems where low temperatures led to the precipitous drop in sulfate concentrations (Figs. 5.22a and 5.23) and is more akin to what happens when slightly acid systems are lofted to higher altitudes (Fig. 5.8). The pH in this acid case dropped below 0.0 (the lower limit for life) at around 263 K (Fig. 5.24); the minimum pH at −1.09 was reached at the eutectic. The a_w rose from 0.85 at 298 K to 0.92 at 265 K, where ice begins to form, and then dropped to 0.69 at the eutectic. Extreme halophiles could survive the salinity; but pH values below 0.0 would place life forms under an additional extreme stress.

All of the above Europan simulations were done at 1.01 bar (1 atm) total pressure. A 100-km ice-covered ocean on Europa would have a pressure >1200 bars at the base of the ocean. What effect would such a high pressure have on chemical equilibria and the thermodynamic properties of such a Europan ocean? How might pressure affect the likely composition of seafloor sedimentary deposits, the chemistry of possible hydrothermal brines,

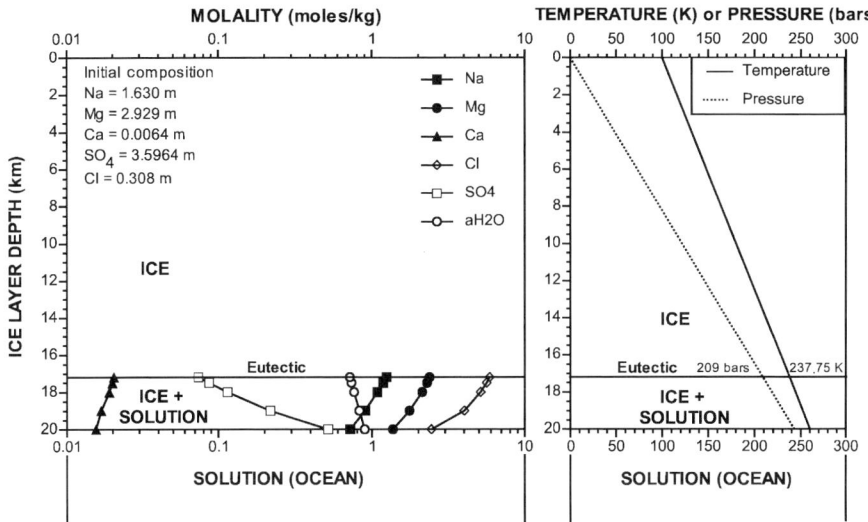

Fig. 5.25. The ionic composition of a hypothetical brine in a 20-km ice layer on Europa. Reprinted from Marion et al. (2005) with permission

and the physical and chemical environment of possible sea-floor biological communities (Kargel et al. 2000; Marion et al. 2003b)?

For the hypothetical pressure case, we used the same Na-Mg-Ca-Cl-SO_4-H_2O system that we used previously (Fig. 5.22a). How this system would respond to decreasing temperature down to the eutectic at 1 bar of pressure is shown in Fig. 5.22a; equilibria at 1460 bars of pressure is shown in Fig. 5.22b. The 1460 bars of pressure represents the pressure of a 20-km ice layer (243 bars = 0.929 × 130.9 × 20/10) (Eq. 5.2) plus 80 km of a brine layer (1216 bars = 1.161 × 130.9 × 80/10). The average density of the 20-km ice layer was 0.929 g cm^{-3} (calculated at 10 km and 180 K, Fig. 5.25); the average density of the brine (1.161 g cm^{-3}) is based on a model calculation of our assumed brine at 260 K at the midpoint of the brine layer (60 km beneath the surface). This pressure estimate is similar to that used by Zolotov and Shock (2001) of 1375 bars for a 100-km ocean on Europa.

There are several significant differences between equilibria at 1 bar and 1460 bars of pressure. High pressure favored the precipitation of $MgCl_2 \cdot 12H_2O$ at the eutectic (Fig. 5.22b), rather than $NaCl \cdot 2H_2O$ at low pressure (Fig. 5.22a). The temperature at which ice first forms decreases from 266 K to 254 K (Fig. 5.22). On the other hand, the temperature at which $MgSO_4 \cdot 12H_2O$ first starts to precipitate increases from 271 K to 282 K (Fig. 5.22). At 1 bar pressure, the original solution is supersaturated with respect to $CaSO_4 \cdot 2H_2O$ (gypsum) at 298 K; at 1460 bars of pressure, gypsum only begins precipitating at the eutectic (Fig. 5.22). Pressure favors chemical reactions that lower volume. Earlier in Chap. 4 we showed that the ΔV_r^0 for the reaction

$$H_2O(I,cr) \leftrightarrow H_2O(aq) \tag{5.6}$$

is $-1.63\,\text{cm}^3\,\text{mol}^{-1}$ at 1.01 bar and 0 °C. Similarly, ΔV_r^0 for the gypsum reaction

$$CaSO_4 \cdot 2H_2O \leftrightarrow Ca^{2-} + SO_4^{2-} + 2H_2O \tag{5.7}$$

is $-49.08\,\text{cm}^3\,\text{mol}^{-1}$. And ΔV_r^0 for the reaction

$$MgSO_4 \cdot 12H_2O \leftrightarrow MgSO_4 \cdot 7H_2O + 5H_2O(aq) \tag{5.8}$$

is $16.31\,\text{cm}^3\,\text{mol}^{-1}$. Bringing pressure to bear on these reactions will cause the reactions of Equations 5.6 and 5.7 to shift to the right, causing dissolution of the solid phase and a decline in the equilibrium temperature. On the other hand, the reaction of Eq. 5.8 will shift to the left, raising the equilibrium temperature (Fig. 5.22). This increase in the equilibrium temperature with increasing pressure is similar to what occurred with the NaCl-H_2O system (Eq. 3.83, Fig. 3.20).

A somewhat surprising result was that 1460 bars of pressure had little effect on the eutectic temperature (238.65 K vs. 237.45 K) (Fig. 5.22). A pressure of 1460 bars, per se, would decrease the freezing point of pure water by about -14.8 K (Fig. 3.3). Dropping the temperature at which ice first formed by 12 K had only a minor effect on the eutectic ($\Delta T = 1.2$ K) (Fig. 5.22). This is not, however, always the case. For example, for the simpler NaCl-H_2O system, the calculated eutectic temperature at 1 bar is -21.3 °C; at 1460 bars of pressure, the calculated eutectic temperature is -31.3 °C ($\Delta T = 10.0$ K). As we point out repeatedly, chemical systems and their response to temperature and pressure depend, ultimately, on thermal and volumetric properties of individual constituents, which makes every system response highly individualistic.

5.3.2 Ice Compositions

There is abundant geological and geophysical evidence that a liquid brine ocean exists on Europa beneath a floating, impure icy shell (Kargel and Consolmagno 1996; Khurana et al. 1998; Kivelson et al. 1999; Pappalardo et al. 1999). Cracking of the Europan ice layer may occasionally allow ocean brines to reach the surface (McKinnon 1997; Kargel et al. 2000; Greenberg and Geissler 2002). However, these brines will quickly freeze at the prevailing surface temperature of Europa (≈ 100 K; Spencer et al. 1999). Europa's surface has a complex heterogeneous distribution of hydrated impurities (thought to be mostly hydrated sulfates and/or sulfuric acid hydrates) and ice (Carlson et al. 1999b; Fanale et al. 1999; McCord et al. 1999). It is unknown how the surface distribution of materials relates to the subsurface distribution, but a definite relationship of impurities to geological features suggests that there is a complex crustal structure involving varying amounts and, perhaps, types of impurities. The physical arrangement of brines, salts, and ice within the floating shell depends on poorly known details of the roles of cryovolcanism; brine intrusion; sublimation, sputtering, and condensation at the surface; crust melt-through; ocean freezing and brine trapping; diapirism; impacts; fractional crystallization and fractional partial melting; brine drainage; and ocean processes and composition (Fig. 5.21).

How deep into the ice layer above the ocean can we expect to find thermodynamically stable unfrozen brine pockets? Liquid-filled brine inclusions or grain-boundary fluids may serve as habitats for life and may alter the rheological properties and geodynamics of the floating shell. Trapped brines also would affect the dielectric properties and radar penetration depth in and backscatter from the floating icy shell. These issues are thus directly relevant to the instrument design, exploration goals, and expected data interpretability of future Europa missions, such as the Jupiter Icy Moon orbiter.

In our hypothetical case, we initially assumed a 20-km ice layer overlying a deeper ocean. Later, we will examine the consequences of a thinner ice layer. We assumed that both temperature and pressure would decline linearly from the base of the 20-km ice layer where $T = 260$ K and $P = 243$ bars to the surface where $T = 100$ K and $P = 0$ bars (Fig. 5.25). The pressure in the ice layer was based on a calculation using an average density for ice of $0.929\,\text{cm}^3\,\text{g}^{-1}$ (calculated at 10 km and 180 K, Fig. 5.25), which leads to 243 bars at the 20-km base (see previous section for calculations). This model assumes a nonconvecting icy shell. In a convecting icy shell, temperatures of 240 K or higher can reach within 6 km of the surface of a 20-km ice layer (Zolotov and Shock 2004). We assumed the same hypothetical Na-Mg-Ca-SO_4-Cl-H_2O system for the briny ocean as was used previously (Fig. 5.22).

As temperature and pressure decline upward through the ice layer, ion concentrations of most brine constituents increase (Fig. 5.25). At these temperatures and pressures, the brine chemistry is dominated by Mg, Na, and Cl, in agreement with previous calculations (Fig. 5.22). Eventually the eutectic point is reached, below which only solid phases are thermodynamically stable. For our hypothetical case, this occurs at $T = 237.75$ K and $P = 209$ bars at a depth of 17.2 km (Fig. 5.25). This implies that potential aqueous habitats for life in the ice layer are limited to the basal zone of the ice. Over this 2.8-km interval, the activity of water (a_w) varies from 0.898 at 20 km to 0.718 at 17.2 km. Only halophilic organisms can survive such low a_w (Table 4.2). Solid phases precipitating in the ice/solution layers include ice, $NaCl \cdot 2H_2O$, $Na_2SO_4 \cdot 10H_2O$, $CaSO_4 \cdot 2H_2O$, and $MgSO_4 \cdot 12H_2O$.

In this particular case, the eutectic is largely set by temperature, with pressure playing only a minor role. For example, if we had run the above simulations entirely at 1.01 bar (1 atm) of pressure instead of making the equilibria pressure dependent, then the eutectic temperature would have been

238.65 K at a depth of 17.3 km. A pressure of 209 bars only decreased the eutectic temperature by 0.9 K and 0.1 km.

Had we run the above simulation with a thinner ice cover (e.g., 10 km) but with the same temperature boundary conditions at the surface (100 K) and base of the ice (260 K) (Fig. 5.25), then the temperature gradient would have been double (16 K km^{-1}) that used in our 20-km case (8K km^{-1}). Since the eutectic depends primarily on temperature (see previous discussion), the thickness of the ice/solution layer would be reduced from 2.8 km (20-km ice layer) to ca. 1.4 km (10-km ice layer). The only way to bring the ice/solution layer to within a few kilometers of the surface is to have an ice layer that is only a few kilometers thick. But even then the stable brine pockets within the ice layer will always be at the base of the ice. The cold surface of Europa precludes stable aqueous habitats for life anywhere near the surface, except in the case of a convecting icy shell (Zolotov and Shock 2004).

5.4 Application Limitations

In Chap. 3 (Sect. 3.6), we discussed limitations of the FREZCHEM model that were broadly grouped under Pitzer-equation parameterization and mathematical modeling. There exists another limitation related to equilibrium principles. The foundations of the FREZCHEM model rest on chemical thermodynamic equilibrium principles (Chap. 2). Thermodynamic equilibrium refers to a state of absolute rest from which a system has no tendency to depart. These stable states are what the FREZCHEM model predicts. But in the real world, unstable (also known as disequilibrium or metastable) states may persist indefinitely. Life depends on disequilibrium processes (Gaidos et al. 1999; Schulze-Makuch and Irwin 2004). As we point out in Chap. 6, if the Universe were ever to reach a state of chemical thermodynamic equilibrium, entropic death would terminate life. These nonequilibrium states are related to reaction kinetics that may be fast or slow or driven by either or both abiotic and biotic factors. Below are four examples of nonequilibrium thermodynamics and how we can cope, in some cases, with these unstable chemistries using existing equilibrium models.

A classic example of metastability is surface-seawater supersaturation with respect to calcite and other carbonate minerals (Morse and Mackenzie 1990; Millero and Sohl 1992). The degree of calcite supersaturation in surface seawater varies from 2.8- to 6.5-fold between 0 and 25 °C (Morse and Mackenzie 1990). In Fig. 3.18, experimental calcite solubility (metastable state) is approaching model calcite solubility (stable state) at subzero temperatures. In Table 5.1, the difference in seawater pH, assuring saturation or allowing supersaturation with respect to calcite, is 0.38 units. Moreover, in running these calculations, it was necessary to remove magnesite and dolomite from the minerals database (Table 3.1) because the latter minerals are more stable than calcite in seawater. But calcite is clearly the form that precipitates during either seawater evaporation or freezing (see Fig. 3.18 and 5.4 discussions). With calcite in the mineral database, the model will calculate the equilibrium composition, which assumes saturation with respect to calcite. To model seawater supersaturation with respect to calcite, you need to either remove calcite (plus magnesite, dolomite, aragonite, and vaterite) from the mineral database or adjust the calcite solubility product so that the model predicts the degree of calcite supersaturation (2.8- to 6.5-fold between 0 and 25 °C) found in natural seawater (Morse and Mackenzie 1990).

Some systems with sluggish kinetics can be ill suited for modeling by FREZCHEM because the occurrence during a reaction's progress of high abundances of both products and reactants, or the existence of intermediate products, is inconsistent with FREZCHEM's assumption that the minimum energy state is achieved at all times under all conditions within the range of the thermodynamic free energies. On the other hand, kinetics can be so sluggish and intermediate products so prevalent and seemingly stable that reactions between the intermediates and the original reactants can appear to be fully reversible in accordance with a revised notion of thermodynamic equilibrium. Those situations can be dealt with if the more stable materials are excluded from the computations. Examples of this are known from hydrocarbons in hydrothermal systems (Seewald 1994), in sulfate salt-water equilibria (Hogenboom et al. 1995), and many other chemical systems.

In the magnesium sulfate-water system, a commonly reported example of metastable equilibria involves the eutectic freezing and melting of epsomite and ice. For solutions of the appropriate compositions, this system only gradually – and only with significant patience and effort – transitions into a system of magnesium sulfate dodecahydrate and ice such that a lower eutectic freezing and melting temperature is observed (Kargel 1991; Hogenboom et al. 1995; Dalton et al. 2005). During thermal manipulations of these solutions and frozen mixtures, during the typical minutes to hours or days before magnesium sulfate dodecahydrate makes its appearance, the metastable freezing transitions involving epsomite can occur in close accord with thermodynamics if data for magnesium sulfate dodecahydrate is removed from the calculations. FREZCHEM – and the actual aqueous system – will only make ice and epsomite if that's all the system or the computer code "knows" to make. The trick is to know which stable phases to remove from FREZCHEM in order to model the metastable equilibria.

Microbial metabolic activity in general is known both to accelerate transitions to stable equilibria and to produce metastable intermediate dissolved species and mineral precipitates that otherwise would not exist or would not be abundant. In general, most metabolic schemes that intervene in the existence and abundance of one anionic species or complex will do so with others, too, and this also has a big effect on the evaporitic and freezing chemistry dealt with by FREZCHEM. For example, dolomite formation is linked to sulfate reduction in one biogeochemical scheme. Lacking microbial activity, the kinetics of dolomite precipitation from seawater are such that it simply does not happen, despite its precipitation being thermodynamically favored (Sanchez-Roman et al. 2005). Sulfate – even at low concentrations – frustrates the conversion of calcite to dolomite and the direct precipitation of dolomite; however, culture studies have shown that moderately halophilic sulfate-reducing bacteria mediate dolomite precipitation under a wide range of seawaterlike compositions. The origin of dolomite was long a classic mystery of the geosciences, as laboratory experimentation produced calcite, aragonite, hydromagnesite, siderite, and other carbonates, but not dolomite, yet dolomite is one of Earth's most abundant sedimentary rocks in the rock record and has been observed in certain shallow-water saline environments. It was recognized that abiogenic kinetics of dolomite precipitation from appropriate fluid compositions are just too sluggish to make synthetic dolomite rocks in a beaker and probably would be too slow in nature, too, if not for life. The critical role of microbial mediation in speeding up dolomite formation has been amply confirmed. For example, Van Lith et al. (2003) have observed dolomite formation in the laboratory when the key reactions are mediated by sulfate-reducing bacteria. Here we see a link between sulfate chemistry and carbonate chemistry that is not modeled by FREZCHEM; but the sulfate-precipitating conditions for which dolomite formation is promoted (for example, the freezing and evaporative chemistry of the anoxic sulfate brine) are handled by FREZCHEM.

Sulfate salts and related dissolved anionic complexes of sulfur are a key basis or product of dozens of well-documented microbial metabolic schemes, a few of which are mentioned here. On Earth, microbially driven biogeochemistry of sulfur is intimately interwoven into sulfur geochemistry. Although we may tend to treat low-temperature evaporitic processes in isolation from biogeochemistry, because thermodynamics can explain much of what happens to sulfur, we should at least recognize that microbial activity, on Earth at least, is both a chief source of soluble sulfate and the chief mediators of sulfur cycling amongst several oxidation states and innumerable minerals, compounds, and soluble complexes. What FREZCHEM effectively does is assume a state of thermodynamic equilibrium, or at least metastable equilibrium, and calculates the partitioning of sulfur when it remains stably in one oxidation state and no disequilibrated compounds are introduced into the system. An alternative to dealing exclusively with a single oxidation state is to use fixed percentages for system redox components (e.g., S^{2-} and SO_4^{2-} or Fe^{2+} and Fe^{3+}). In version 11 of the FREZCHEM model (currently under development), one can split total solution Fe into fixed percentages of Fe^{2+} and Fe^{3+}, which allows for ferrous, ferric, and mixed ferrous/ferric iron mineral precipitations. Kinetics ensures that even in a world devoid of life, many disequilibrated compounds involving sulfur will exist if that world is volcanically or hydrothermally dynamic; and any occurrence of life as varied and complex as Earth has would ensure that it will be really difficult to predict the molecular and phase state of sulfur compounds even if that world's geology and its rules of physical and kinetic disequilibria were understood. FREZCHEM effectively takes those sulfur-bearing components that are thermodynamically stable and then applies the Gibbs' Function to tell us what phase state those components will fall into for any set of cold aqueous conditions. This, in fact, is half the problem. The other half – kinetics and life (where life itself is a montage of kinetic and thermodynamic behaviors) – is to be recognized as both a perturbation on and determined by physical abiogenic thermodynamic behavior.

Despite these application limitations, it nevertheless is true that chemical equilibrium models can generally provide reasonable approximations of many, but not all, real-world chemical processes. The better we understand the biogeochemistry of natural systems, the better we can adapt our geochemical models for specific systems. One need only peruse standard geochemical textbooks (e.g., Nordstrom and Munoz 1994; Drever 1997; Millero 2001) to sense the importance of such models in geochemical applications.

6 The Search for and Future of Life in the Universe

"There are two possibilities. Maybe we're alone. Maybe we're not. Both are equally frightening."   Issac Asimov

"There are two possibilities. Maybe we're alone. Maybe we're not. Both are equally profound."   Giles Marion

In this chapter, we will first discuss a search strategy for life in the Universe and then discuss the future of life in the Universe.

6.1 A Search Strategy for Life in the Universe

Table 6.1 is a summary of environmental limitations for active "life as we know it." By this phrase we mean organic life existing under conditions where aqueous solutions and complex organic molecules (such as amino acids, proteins, and DNA) can exist; we consider here some environments that are extreme or exotic when gauged by conditions where life on Earth thrives. With this consideration, we do not exclude possibilities that unknown categories of life might thrive in some of the more "cosmic" environments mentioned below, but for which we have no means to evaluate their plausibility.

Although there are many definitions of life, none has general acceptance, as everybody seems to want to define it, but nobody quite knows how to put it into words that don't start a long and never-resolved discussion. Generally, we know life when we see it. Generally, life is capable of metabolism and Darwinian evolution. However we define it, energy, water, and nutrients are resources that all life forms (as we know them) require. The other seven factors in Table 6.1 are environmental constraints for active life as we know it (see Chap. 4 for details). Some of these factors are more easily sensed remotely than others. For example, detecting the presence or absence of surface liquid water is in principle easier than estimating whether or not all the essential nutrients are present in appropriate amounts.

In the search for extraterrestrial signatures of life, we need to distinguish among Solar System sites, extrasolar system sites, and the search for intelligent life. It is likely that within the 21st century most potential life-bearing

Table 6.1. Environmental requirements and limitations for "active" life on Earth Reprinted from Marion and Schulze-Makuch (2007) with permission

Limiting factor	Requirements/limitations*
Energy	Solar, geochemical, geothermal
Water	Presence, composition (see below)
Nutrients	C, O, H, N, K, Ca, P, Mg, S, Fe, Cl, Cu, Mn, Zn, Mo, B
Temperature	-20 to $121\,^\circ$C
Salinity	$a_w = 0.6$ to 1.0
Desiccation	RH = 60 to 100%
Acidity	pH ≈ 0 to > 12
Toxic elements	Pb, Hg, Cd, Zn, Cu, As, Al, B, Ni, Se, Ag, Mg, Mn, Mo
Radiation	Microbes up to 4000× more tolerant than humans
Pressure	≈ 0.1 to >1100 bars

* Energy, water, and nutrients are requirements (resources) for life. The remaining properties are environmental factors that can potentially limit life; ranges for the latter are known tolerances of "active" Earth life forms

bodies in our Solar System will be explored by lander or sample-return missions. Such missions can provide high technology assessments of whether life is present, or at least refine the potential for life on the planetary body. Exploration for life beyond the Solar System will have to rely entirely upon remote sensing for the foreseeable future. At present, the search for intelligent life in the Universe is largely restricted to radiowave monitoring by SETI (Search for Extraterrestrial Intelligence). The latter program is likely the best approach in the search for intelligent life in the Universe, no matter how rare such life might be (Ward and Brownlee 2000).

There are two biosignatures for life on Earth that could be especially important in the search for life beyond Earth, namely: the presence of complex organic chemicals and the disequilibrium concentrations of O_2 (or O_3) and methane in the atmosphere. Among the complex organic chemicals that are biosignatures are chlorophyll, proteins, polypeptides, and phospholipids. The presence of high concentrations of O_2 and O_3 in the atmosphere are powerful indicators of disequilibrium because these oxygen molecules are highly reactive (Leger et al. 1993; Akasofu 1999; Schidlowski 2002; Frey and Lummerzheim 2002). On Earth, atmospheric oxygen levels are maintained at a high level (21% by volume) because O_2 production through photosynthesis is in balance with O_2 uptake reactions (biological and nonbiological). The oxygen-rich atmosphere, per se, is a geosignature of a favorable environment for life, although the life so indicated is very peculiar and probably not a very common type of life. It required two to three billion years to evolve on Earth. For terrestrial animals, of course, an oxygen-rich atmosphere is just what is needed. More generally, atmospheres provide gases that are generally made of C-H-O-N-S compounds needed for biochemistry. Atmospheres also help moderate climate and reduce diurnal and seasonal variations in temperature, protect the surface from radiation and meteoritic strikes, and prevent liquids, especially the all-important water solvent, from dissipating into space (Schulze-Makuch and Irwin 2004).

A plausibility scale for life beyond Earth was developed by Irwin and Schulze-Makuch (2001) (Table 6.2). Earth-like planetary bodies with liquid water, readily available energy, and organic compounds were listed as Plausibility of Life (POL) Category I (high). At the other extreme are solar system bodies such as the Sun and Moon with conditions or compositions so harsh that life is unrealistic, which is Category V (remote verging on utterly impossible). The previously discussed planetary bodies, Mars and Europa, are Category II (favorable), which is why so much attention has been placed on these two Solar System bodies.

In the search for life in our Solar System, all of the above important resources for life, constraints for life, biosignatures, and geosignatures are appropriate criteria. In the search for life beyond our Solar System, attention should focus on (1) energy (especially solar) sources, (2) liquid water, (3)

Table 6.2. Astrobiology plausibility categories (Irwin and Schulze-Makuch 2001). Reprinted from Marion and Schulze-Makuch (2007) with permission

Category	Definition	Examples
I	Demonstrable presence of liquid water, readily available energy, and organic compounds	Earth
II	Evidence for the past or present existence of liquid water, availability of energy, and inference of organic compounds	Mars, Europa
III	Physically extreme conditions, but with evidence of energy sources and complex chemistry possibly suitable for life forms unknown on Earth	Venus*, Titan*, Triton, Enceladus
IV	Persistence of life very different from that on Earth conceivable in isolated habitats or reasonable inference of past conditions suitable for the origin of life prior to the development of conditions so harsh as to make its perseverance at present unlikely but conceivable in isolated habitats	Mercury, Jupiter
V	Conditions so unfavorable for life by any reasonable definition that its origin or persistence cannot be rated a realistic probability	Sun, Moon

* Venus and Titan have been upgraded to Category II in Schulze-Makuch and Irwin (2004) due to recent insights on their dynamic activity, planetary/lunar history, and presence of organic compounds.

complex organic chemistry, (4) the presence of an atmosphere, and (5) $O_2(O_3)$ disequilibrium. Three of these properties are those of Category I (Table 6.2). The other two properties are strong indicators of life for Terran-type life as it presently occurs on Earth. In a recent paper (Marion and Schulze-Makuch 2007), we proposed that planets or moons that possess these five attributes should be assigned a SupraEarth designation, Category Ia (very high, Earth-like). Other planetary bodies on which the plausibility of life is high, but which are not Earth-like were assigned a POL category of Ib.

This is, of course, a very Terra-centric categorization, but so far we only have the unitary example of Earth from which to draw examples. Furfaro et al. (2007) have developed a fuzzy-logic-based scheme that goes more toward the fundamentals of coupled physical/biological planetary systems, but still it takes a restricted view oriented toward Mars- and Earth-like planets. It is entirely possible that Earth has a thin biospheric veneer compared to potentially more massive biospheres on worlds with deep ice-covered oceans such as Europa; we just don't know.

Category Ia planets and satellites will become especially relevant in science when we are able to detect Earth-type planets in other solar systems. Our extrasolar-system search for life should focus on Earth-like planets where there are strong signatures of life. Such planets would also supplement the SETI search for intelligent life. Intelligent life may be rare in our Universe as suggested by Ward and Brownlee (2000), but our immediate search, especially the search in other solar systems, must continue and focus primarily on properties that are easily measured and are clear-cut necessities and strong indicators for life. Fortunately, there are several planned missions that will contribute to the search for extrasolar-system life including GAIA, COROT, EDDINGTON, KEPLER, and DARWIN (Foing 2002), though continuing budgetary crises threaten these missions. The first four missions are primarily designed to detect planets around stars. The DARWIN mission will determine the habitability of planets around nearby stars, including spectral signatures of gases such as CH_4 and O_3 that are potential biosignatures of life (Foing 2002). Coupled with ongoing and future solar system missions designed to look for chemical and physical conditions where life could exist (e.g., Cassini, MER, Mars Express, Europa or Enceladus orbiter, or Titan Balloon), we are in the midst of an unfolding scientific revolution regarding our search for extraterrestrial life or its potential habitats.

6.2 Entropic Death?

In Chap. 2 we introduced the first and second laws of thermodynamics, which can be represented as follows:

$$\text{First Law:} \quad \Delta U_{\text{universe}} = \Delta U_{\text{system}} + \Delta U_{\text{environment}} = 0.0; \quad (6.1)$$
$$\text{Second Law:} \quad \Delta S_{\text{universe}} = \Delta S_{\text{system}} + \Delta S_{\text{environment}} = \Delta S_{\text{irr}}, \quad (6.2)$$

where U is the internal energy and S the entropy. Assuming the validity of these laws for an isolated Universe (no transfer of energy/matter into or out of the Universe), then the total energy of the Universe is fixed, and processes, all of which are irreversible, lead to the degradation (or dispersal) of energy, which is measured by $T\Delta S_{\mathrm{irr}}$ (Chap. 2). So if total energy is fixed and some fraction is continuously being degraded, then the time must come when there is no longer any useful energy left, S_{universe} is maximized, and the Universe reaches a state of equilibrium. Life is based on disequilibrium processes (Gaidos et al. 1999; Schulze-Makuch and Irwin 2004). A state of Universal equilibrium implies entropic death and the end of life in the Universe, as well as an end to the chemistry, meteorology, and geology that is of such interest in planetary science.

In Chap. 1, we introduced the book with a quote from Albert Einstein (Schilpp 1949), which read in part that "classical thermodynamics... is the only physical theory of universal content concerning which I am convinced that, within the framework of the applicability of its basic concepts, it will never be overthrown." An important qualification to this statement is the phrase "within the framework of the applicability of its basic concepts." The laws of thermodynamics are based on laboratory-scale experiments. To assume that such laws are applicable to the Universe is a big assumption. However, we have no evidence yet that contradicts this assumption on the scales of problems relevant to life. Moreover, there remain vast cosmological questions with no answers and definitely no understanding of implications even if we knew the answers. For instance, does the proton have a very long but finite radioactive half-life? Does the neutrino have a very small but finite mass? Is the Universe "opened" or "closed" with respect to expansion and gravitational contraction? Also, the Universe may not be isolated with respect to matter/energy; or it could be isolated and cyclical.

The effects of dark energy, vacuum energy, and unconventional forms of dark matter (Albrecht and Skordis 2000; Huterer and Turner 2001; Huterer et al. 2002; Linder 2006) are known to be important in controlling the Universe's expansion (via the ill-defined cosmological constant); but they are, at best, poorly understood. The theoretical potential for these cosmological issues in quantum and relativistic gravitational models of the Universe extend surprisingly closely to our subject matter in this book, for example, the phase state of fluids (Linder 2006). Indeed, there is much interest in – but no firm, widely accepted understanding of – how nonclassical, highly quantum models of the Universe can produce ordinary physics, including thermodynamics (Husain and Winkler 2007). For outsiders to the field of cosmology, such as ourselves, it can seem that just about anything goes in this field, if the mathematical equations give solutions that match one or another set of observations; the potential for exceedingly bizarre phenomenology, maybe even mythic life forces, would seem almost unbounded by the variety of mathematical models of the Universe. The practical effect of various cosmological models on perturbing classical thermodynamics is unproven and presumably very small under the conditions of interest to us. Indeed, if anything, it would appear that concepts from classical thermodynamics, such as freezing and thawing potentials, are being adapted into and affecting cosmological theory (Linder 2006).

One fact is inescapable: our perceptions of the Universe around us, and insights that draw our minds onward, are impregnated by common phenomenology around us. Thus, our comfort zone is with classical thermodynamics and life as we know it (Lewis 1995; Lunine 1999). Thus, in this book we rely on "conventional" mass and energy and established thermodynamic principles, which seem to describe the flow and conservation of energy and mass with extraordinary precision in all measurements of common situations and in nearly all measurements under laboratory conditions (the exception being the measured negative value of vacuum energy). These concepts of classical, conventional thermodynamics seem to apply on geologic time scales and under all planetary conditions where life as we know it could possibly exist, a view consistent with Einstein's perspective that classical thermodynamics will not be overthrown.

We know very little about the crystallinity and phase states of matter in systems where the more far-future and far-out realms pertain or even in many ultraextreme extant (and definitively known) environments such as neutron stars (Glendenning and Pei 1995; Glendenning 2001), metallic hydrogen mantles of Jupiter-type planets, or ionic water in Uranus and Neptune. We are uncertain as well what will prevail in some of these ultraextreme environments in the distant future, such as when they eventually freeze into some sort of cryogenic lattice, and what the effects might be of a transition to a superconductive state. It seems likely, however, that the far future will entail either (1) a supercryogenic, highly expanded, and diffuse Universe and an inert state or (2) with Universal closure, a very hot and violent demise. We do not see much likelihood of life into the interminable future, unless it is a different or renewed Universe or involves some seemingly mystical or fictionlike transformation, maybe involving some exotic and unknown solvent (replacing water) and exotic polymer system (replacing DNA and proteins) that might involve one of these ultraextreme environments. Those are situations we cannot pretend to understand enough even to speculate on very well.

It is worth considering life as we know it in the future of the Solar System and the Universe. The future is a continuation of the past, and a last consideration here of the far future possibilities for life can take a lesson from the very distant past. The first entropic threat to life came before there was any life. In the first hundreds of millions of years after the Big Bang, before there was life, before there were stars, the Universe had already entered a period where a diffuse gas was under $10\,\mathrm{K}$ and a near vacuum pervaded what already had become a vast expanse of a still-expanding and chilling and attenuating Universe. There was virtually no chemistry, no nucleosynthesis, and seemingly no place for complexity of any type, much less a place for life; there was just H and He and traces of Li, Be, and B in a monotonous, dead, ethereal Universe. There were, however, slight variations in the density and temperature of the expanding gas, caused by factors still being explored by cosmologists. It was this slight heterogeneity that caused a hierarchy of structures to develop, and from these structures stars and galaxies and high-mass elements formed through operation of the carbon-nitrogen-oxygen nucleosynthetic cycle of main sequence stars, nucleosynthesis of heavier elements up to iron in massive dying stars and supernovae through silicon and oxygen burning, and elements heavier than iron through nucleosynthesis in supernovae during the r-process neutron capture reactions and other processes (Boesgaard and Steigman 1985; Lewis 1995; Wallerstein et al. 1999; Clayton 2003). Not only did life's main elements of organic chemistry form in these stars, but, just as critically, heavy radioactive elements were produced that ultimately would drive the heat engines of many potentially habitable planets (Lunine 1999). Lacking those radioisotopes, not even starlight-bathed planets like Earth would have life, as life not only needs warmth sufficient to maintain liquid water near the surface, but it requires geological activity to drive the system, including establishment of chemical disequilibria.

It was the formation of stars and then second-generation stars (such as our sun) with rocky planets that made life as we know it possible. Entropy still applied, yet the Universe became habitable in the period (which continues today) succeeding that first eon when it would have seemed entropic death was already gripping the Universe. Life was made possible by the nucleosynthesis of heavy elements and the condensation of solids and formation of planets where aqueous fluids could exist.

Among places where condensates accreted into significant solid bodies, such as planets, habitable realms have always been rarer than places that were either too cold or too hot for life to exist. Much of our Solar System's mass is still far too hot for life. Most of the deep interiors of the gas giants and rocky planets are too hot, as is, of course, the Sun itself. Most of the surface area of solid bodies in the Solar System are too cold – the icy satellites of the outer planets and the myriad comets and Kuiper Belt Objects on the far outer fringes of the Solar System. In this sense, places like the surfaces of Earth and Mars and Europa's subsurface ocean are indeed very rare places.

One can readily envisage a time, tens of billions of years from now, after the Sun has bloomed as a red giant and then died, after radiative cooling has permitted large fractions of Jupiter and Saturn to condense as liquid oceans, when most of the Solar System's planetary mass – or far more than now – finally has become habitable. These will be chemically active domains where peculiar forms of liquid water and no doubt some organic substances will be stable (Fig. 4.4); who can say what life may arise there, hundreds to thousands of kilometers below their cloud-capped atmospheric "surfaces"?

It would not exactly be life as we know it, as some of the key organic substances on which familiar life is based, probably including DNA and proteins, are not likely to be stable under prevailing temperature and pressure conditions, which tend to favor ionic forms of matter and simpler molecules. Some other organic chemistry might build high-mass polymers and perhaps living systems; although possibly organic based and water based, this would not be life as we know it.

Eventually, with no Sun to heat the deep interiors of these massive planets, the heat from the decay of radioactivities becoming nil, and the heat released by gravitational dissipation escaping, and with no ability to stem the radiative losses of stored energy, the gas giants will freeze solid and then plunge to sub-Kelvin temperatures perhaps trillions of years in the future (if the Universe is gravitationally "open" and lasts that long).

Any thought of life in a Universe Beyond is mystical and speculative at best and maybe nonsensical or immune from rational consideration; and any detailed consideration of future life even in Jupiter and Saturn billions of years from now is beyond our abilities, though modern experimental methods increasingly permit windows into the physics and odd chemistry of these environments. We give these situations no further consideration and instead move on to more familiar domains where ordinary water and ice exist, ordinary organic chemistry prevails, and life as we know it might exist.

6.3 Solar System Life

The nearer future of life as we know it on Earth, and life as we know it that might exist on Mars or Europa, is more readily predictable. We can say with certainty that mass extinction events will occur again and again, whether due to comet and asteroid impacts, fundamental epochal or transient chemical changes in the atmosphere, or human activities. Each of these has caused pronounced extinctions of a large percentage of Earth's life in the geologic past. Except for the continuing human-induced mass extinction (Wilson 2002), which has not been under way long enough for biological evolution to adapt to it, each of these types of mass extinctions was followed by reradiation of life (Kirchner and Weil 2000). Human-era global climate change (Stainforth et al. 2005; Strom 2007), though severe for humans and likely to cause extinctions of most species and many families of life, is not apt to be a fundamental limitation for most classes of life on Earth. We can assume that after some hundreds to thousands or a few million years of turmoil and extinction and other challenges to life on Earth wrought by humans, life will settle down to a vigorous state with class diversity as though humans never existed, but where many biological family-level lineages are terminated and their niches reoccupied by other families. The continuing mass extinctions caused by humans will heal, as it always has, with a reradiation of life's familial-, genus-, and species-level diversity within ten million years (Kirchner and Weil 2000).

Some ecological niches, especially those filled by predators, and some entire biomes, such as many coastal wetlands, already have been severely and permanently damaged by human beings. As global warming ensues, there may emerge a classic full-scale "hothouse" condition initiating sometime in this or the 22nd century (Strom 2007); alternatively, human impacts on climate may be no more than a major but brief blip on the climate evolution curve. In the former case, the cryosphere will disappear entirely, and in the latter case the cryosphere will constrict to a small fraction of its present area.

The balance of life is definitely undergoing change even now, and much greater changes are impending. Wherever the cryosphere disappears, so will disappear obligate pyschrophilic life – species, such as tundra vegetation and some microbial species in polar ice, that require cold conditions (Price 2000; Priscu et al. 1999; Breezee et al. 2004). Most of these species will become utterly extinct when the last permafrost and glaciers melt, if a full hothouse scenario emerges. Halophilic life, however, will survive. Some continental climates will be hot and very dry; thus, evaporative conditions will continue, and with those conditions, salt-tolerant and salt-requiring life will be sustained. Hyperthermophilic life, such as that in hot springs, fumaroles, and undersea hydrothermal "black smoker" vents, will continue, with some disruptions due to changes in the input of food from elsewhere in the biosphere. On the land, higher animals that are especially widespread and adaptable, especially omnivorous generalists and those that spend much of their time sheltered underground and which are small enough to evade human dietary interests, such as some small rodent species, will probably do well in whatever climate emerges from human disruptions of the global system.

Once Earth reestablishes a cooler climate equilibrium after humans become extinct or decivilized or mend their ways in the face of climate change and depletion of fossil fuels, permafrost and glaciers are apt to return. There is one school of thought that Earth will settle back to a state much as it is now or has been during most of the Holocene, with one millennium needed for the first major adjustment of climate and several more millennia required for regrowth and stabilization of the polar ice sheets. Another school of thought calls for a series of extreme oscillations of climate from extreme heat to extreme cold, with the oscillations taking many thousands of years and the period of unstable climate lasting millions of years; and a third school says that Earth will settle stably into a climate much colder than now as a new Ice Age takes place just as it might have even without human intervention. Although psychrophiles mainly will have become extinct if a full hothouse occurs, the genetic ability to readapt to cold conditions likely will be retained in halophiles because one common key to psychrophilic living is salt tolerance and salt utilization (Breezee et al. 2004; Wadham et al. 2004). Future psychrophiles thus will primarily also be halophiles, who will come to dominate the cold regions again.

Interestingly, many hyperthermophiles also are highly salt tolerant. This is an adaptation to life involving aqueous systems that evolve with high-pressure liquid/vapor and supercritical fluid-phase separation of hydrothermally heated seawater. Both psychrophiles and hyperthermophiles have large numbers of species that also require heavy-metal tolerance, due to the concentration of heavy metals by the thermodynamic phase-separation processes operative in both very cold and very hot aqueous systems (Breezee et al. 2004; Kaye and Baross 2002; Summit and Baross 1998).

Regardless of the human cause of the current/impending period of extreme climate change, such changes have occurred in the geologic past by natural causes. The future need of life to readapt to cooler conditions after the human era, and the inevitability of this adaptation, will not be fundamentally a different process than in periods in the geologic history of Earth. Permafrost and glaciers have been a periodic phenomenon on Earth since the planet began, but they have also periodically disappeared, probably at times completely, with not a crystal of ice anywhere on the surface or in the crust. There will come a time – perhaps a quarter billion years from now (though the timing depends on plate tectonics and orogenesis as much as on stellar physics) – when the brightening sun will forever render the Earth irrelevant to ice and FREZCHEM.

Hot deserts, torrid tropics, and warm-temperate regions will increasingly cover the Earth as hundreds of millions of years roll by. It may require another billion years for the last freezing nights to fade into the geologic past at all polar points and alpine heights on Earth, but that time will come. By then already most of Earth's surface will be minimally habitable or completely sterile, and atmospheric and oceanic oxygen abundances will have plummeted due to an extinction of animal life and heightened uptake of the free oxidizing agent. The planet will not be without its oxidized materials, with oxidizers produced by UV photolysis, volcanic exhalations, and geothermal dissociation of crustal salts. Microbial life, especially lineages descended from today's hyperthermophilic chemoautotrophs, may still thrive across most of Earth's surface and in the oceans.

Eventually, the oceans will evaporate and a humid warm atmosphere will emerge. The salts of the evaporating sea will eventually saturate progressively from one species to another and accumulate to a global layer roughly 30 m thick. This will add to salts already deposited over the eons, including over 800 m (global average) of carbonates. These last marine precipitates, however, will be deposited mainly in the drying dregs of the sea in quasiconcentrically zoned chemical deposits. By contrast, the carbonates will mainly rest on the continental uplands, or their eroded remnants – the Earth's former shallow marine platforms, in the future to be high and dry. In fact, the salts will likely be deposited in much thicker deposits across small parts of Earth's surface in those last hot seas. Earth eventually will transition to a Venus-like super-greenhouse planet. The last weak buffers against that transition will be the dehydration of hydrated salts (some originally laid down in today's era). If plate tectonics and crustal metamorphosis has not destroyed all remnants, dehydrated salt deposits will remain as testaments to aqueous activity of the past. Those salts will eventually dissociate, releasing their sulfur dioxide, carbon dioxide, and hydrochloric acid to the atmosphere. Indeed, we might look closely at Venus to see the future of Earth; or conversely, if Venus once had oceans, we should look to models of the future of Earth to see what may have happened long ago on Venus (Kargel et al. 1994).

While the Earth gradually, or in a sequence of steps, loses its habitability, other worlds in our Solar System may have improving conditions. Perhaps a few hundred million years after Earth becomes Venus-like, Mars may be just starting out as a world where surface life might exist and thrive and where FREZCHEM might find increasing applicability to more widespread, more frequent occurrences of cold, concentrated salt solutions. Throughout the geologic history of Mars, the upper crust has been subjected to pervasively cold conditions punctuated by brief periods of local or regional warmth probably driven by geothermal heating and some rare excursions of spin axis obliquity to orient the poles toward the Sun in summer. Whereas ice has been widespread on Mars, liquid water has been the exception (Kargel 2004). The future of Mars will be marked by a continuing decrease in geothermal heating but an increase in solar heating, such that episodes of surface melting will become more frequent, longer lasting, and more widespread, even though geothermal gradients are reduced and hydrothermal activity becomes less important.

If Mars is now a living world, its life is mostly underground, with variations possible from hyperthermophilic to psychrophilic. The predominant frozen state of Mars' water means that salts formed by 4 billion years of transient aqueous conditions and deep crustal aqueous geochemistry (Chap. 5) are concentrated in small amounts of hypersaline brines. Hence, besides being adapted for temperature extremes, any current Martian life probably is adapted for high salt tolerance, including high metal abundances; it probably depends on chemotrophy involving sulfur oxidation/reduction as well as using other redox systems, such as in the iron-, arsenic-, and selenium-bearing ecosystems on Earth (Neilands 1957; Stolz et al. 2006).

An increasing thaw state that will take place in Mars' future means that brines eventually may tend to freshen, particularly as the hydrologic cycle and salt precipitation sequesters salts in massive evaporitic deposits, and as ice melting tends to dilute salts that remain dissolved. A new diversity of life may emerge to take advantage of large variations and gradients in salinity and solute composition. The common present occurrence of chloride- or acid-dominated brines should be replaced by an increasing role of metal-sulfate-dominated and more chemically neutral aqueous systems and, perhaps for the first time, what we on Earth consider to be "fresh water." Photosynthetic life might evolve when and if surface conditions permit stable liquid water and the atmosphere can screen ultraviolet light.

There is, however, a conundrum facing any future Martian life. Because it is a small world, its radiogenic heat flux per unit area is about a third that of Earth; with the inexorable further decline of radionuclide abundances, hydrothermal and igneous activity is already much less important on Mars than on Earth and it is destined to become even less important (about a factor of two decline in radiogenic heat production every 1.3 billion years). This internal heat drives abiotic aqueous chemistry and biological chemotrophy, but it also drives crustal construction through volcanic and tectonic processes. As the surface warms and lakes and seas become more frequent, longer lived, and more widespread on Mars, water-driven erosion and flattening of the terrain also will become more efficient. Hence, just about the time the surface of Mars becomes pervasively wet, it may become flattened, with topographic highs eroded down and holes filled in; little tectonic and volcanic activity (aside from impacts) will occur to renew the relief. Mars will tend toward a global water ocean as crustal ice melts and water is expelled to the surface and the surface is flattened. With little or no land area, few nutrients formed by hydrolysis and other aqueous chemistry acting on the land, and few rivers sweeping dissolved and fine suspended matter to the sea, photosynthetic life will be almost exclusively marine, and those seas will be feeble in their primary photosynthetic productivity; meanwhile chemoautotrophic life will be increasingly starved at the sea floor. Hence, Mars will probably never resemble Earth in its highly productive and biologically diverse seas and land. The location and size of Mars are just such that it tends to miss out on strong solar heating at times when radiogenic heating is prodigious, and it misses out on radiogenic heating when solar heating eventually becomes right for surface life.

So long as life hangs on in the Martian sea, it will likely readapt to shallow marine, shore, and, perhaps, land conditions each time there is a large impact producing a mountainous rim and central peak. That relief may last for 10^6 to 10^8 years, depending on the size of crater and the climate. Such crater-based ecosystems may provide Mars with its only engine of biodiversity and primary productivity, with isolated craters serving similar ecological roles as do islands on Earth. However, the diversity of the gene pool that Martian craters have to work with will likely be much reduced from the one that Earth's islands inherit.

The situation in Europa's and Enceladus' ice-covered oceans or aquifers is very different from those on Earth and Mars. Energy availability to drive aqueous conditions seems not to be a problem, as tidal dissipative heating of both satellites appears to have maintained liquid layers for perhaps billions of years. The likelihood is that hydrothermal vent activity on the sea floors of both satellites is prodigious. However, we know very little about the chemistry and physical conditions in those realms, and because of that we can draw few conclusions about the biologically accessible energy – exploitable chemical disequilibria – that could produce and maintain life.

Kargel et al. (2000) suggested a bewildering variety of possible chemical models of Europa's ocean stemming from highly simplified scenarios. We have few clues and fewer hard constraints. We know that there are sulfates on the surface in deposits erupted from the interior or otherwise exposed by geologic activity; but these sulfates may be hydrates of sulfuric acid, magnesium sulfate, sodium sulfate, or many other materials. The reflectance spectroscopy and other observational data do not have features that uniquely identify specific substances, only broad classes of materials, such as sulfate hydrates and water ice. Each chemical interpretation has its own implications for the origin of the surface materials and the composition and physical conditions of an ocean or hydrosphere. The fact is, we do not know whether the ocean or hydrosphere is an ultra-acidic cryogenic brine, a caustic solution, a near-neutral concentrated salt solution, or a nearly freshwater environment. Life as we know it may be possible in just about whatever Europa throws at it (Kargel et al. 2000), but then again, life may be completely impossible in some of the more extreme possible conditions due either to the chemistry of the ocean or to its physical nature. For example, some of the more acidic or ultrahypersaline models may make organic polymer chemistry, and hence life, impossible.

Enceladus is much the same story as Europa, except that we know just a little more about the gases involved. They include carbon dioxide and methane, both important in terrestrial life and both possibly related to biological activity in Enceladus' interior. However, these materials may be purely abiogenic and not even related to organic chemistry at all. The observed eruptions of giant geysers, and the inferred deposition of geyser plume fallout, raises the intriguing possibility that we may go there and sample flash-frozen erupted oceanic brines directly, including any frozen life or biochemicals they may contain.

Titan presents one of the more fascinating worlds and clear prospect for subsurface life, as we know already that it has a rich organic chemistry cycle. It has a methane-rich, nitrogen-dominated atmosphere, and the methane may potentially be a biochemical. Photochemistry of the atmosphere produces a rich soup – mainly a frozen soup – of hydrocarbons and organic molecules. For the most part, these settle onto the surface in solid and liquid forms. According to Kargel et al. (2007), the interior may also be involved in processing these materials, which then undergo a fascinating array of familiar geologic processes in the near-surface and surface environment. What underlies the icy, organic/hydrocarbon-rich upper crust is uncertain. The low thermal conductivities of many organic and hydrocarbon solids and gas hydrates (such as methane clathrate) suggest that ordinary radiogenic heat is enough to make a warm crust. Aqueous solutions in the crust or beneath it are possible. Does Titan contain an ocean and maybe a rich variety of inorganic as well as organic solutes? Is Titan in fact a living world somewhere deep down? It has the biological precursors necessary for life, and it probably has thermal conditions right for life at some depth. Is the methane it belches into the atmosphere the product of methanogenic microbial life, or is it the belching from buried primordial cometlike ices? These are key questions and outstanding mysteries.

If not terminated by an impact of a giant asteroid or comet, Earth's life will be terminated by solar evolution, as will any life on Mars. Even Europa's ocean-preserving ice crust will sublimate in the days of a red giant Sun. Europa – in fact all the icy satellites – can never become what Arthur C. Clarke posited, as any degree of Earthly warmth removes the ice upon which subsurface hydrospheres depend. Enceladus and Titan, farther from the Sun, might also lose their ice or just might barely escape assaults by the red giant Sun. If its ice envelope survives, the tidal heat source driving Enceladus may or may not continue until then or even afterward. As Jupiter and Saturn slowly cool and contract, they will spin more rapidly, and tidal dissipation in these satellites might be ramped up, if the orbital resonance locking is maintained in these systems.

Titan's atmosphere will warm up dramatically during the red giant phase, and if it can avoid loss by hydrodynamic outflow, then Titan could become even more interesting geologically and biochemically than it is now. Any methanogenic life or abiogenic venting of methane into the atmosphere may yield a methane and water-vapor greenhouse atmosphere, which may turn the ice mantle into a free-surface liquid ocean at the peak of red giant solar heating. Methane, in fact, may by then have accumulated into a more abundant atmospheric species because the approaching red giant phase of the sun would terminate ultraviolet driven photolysis. Water vapor would add greatly to greenhouse conditions; water vapor condensation and rainfall and lack of UV photolysis would likely maintain water vapor in the atmosphere and prevent its escape. However, like Mars, the radiogenic heat engine of Titan also may be dying just as solar heating starts to make things interesting from that perspective. Without geologic activity, a maritime stage of Titan's climate might not do life any good. In any case, the red giant phase will last very briefly, the climax phase only about a million years; when the Sun suddenly collapses to a white dwarf, Titan's oceanic surface will freeze and its atmosphere will collapse in just weeks or months. Now with very little energy to heat the interior, and virtually nothing to heat the surface, any possible life in Titan will probably quickly (millions of years) become entombed in frozen ices, precipitated solutes, and hydrocarbons. It will be quite a nice, though depressing, problem for FREZCHEM!

Curiously, Enceladus, seemingly the least likely of the Solar System's water worlds, may be the very last refuge of life in this Solar System. This is because its energy is independent of the Sun: just pure tidal energy ultimately derived from the virtually inexhaustible rotational energy of Saturn. Even after the Sun becomes a white dwarf, so long as the tidal resonance with Dione is maintained, Enceladus can go on with a warm interior, even when the surface plunges to a few Kelvin. This requires, additionally, that it can hold onto its ice shell during the peak of the red giant phase; it will be a close call. Whereas Europa will probably lose its ice shell, it seems most probable that Enceladus will retain most of its ice. Geology will be affected as its surface first softens up during the red giant phase and then becomes steely hard, and very brittle, after the Sun begins its white dwarf phase. With such a cold surface, potentially tidal heat might not stave off solidification, but we think it is possible that the deep subsurface would remain liquid. Whether the resonance lock can be maintained as Saturn slowly cools and contracts and satellite orbits evolve, we are not sure. Whether Enceladus will by then have spewed and lost all its supervolatiles through geyser venting and surface sublimation – losing volatiles such as methane, carbon dioxide, and hydrocarbons, upon which its life might depend – we also do not know. But how ironic if Enceladus becomes the Solar System's last living refuge long after the lights have gone dim and cold! For sure it will not, however, be a place for human or similar life.

6.4 To the Stars or Bust?

Earlier, we examined the possibility that life could ultimately be curtailed throughout the Universe by the laws of thermodynamics. Whether this is true or not depends on the validity of certain assumptions (Sect. 6.2). In any case, such a termination would be tens or hundreds of billions of years into the future, if there is an "open," forever-expanding Universe. Of closer concern is our Solar System, where the time lines for life are shorter, but still billions of years into the future (Sect. 6.3). It seems a foregone conclusion that life in our Solar System will someday be terminated.

Human life may have a far shorter time span in the Solar System. Whether humans will eventually depart the Solar System in a positive quest for new places or depart as refugees from some horrific problem of natural or human creation, we cannot say. However, the icy satellites will never be a humanly habitable, terraformed abode. Mars may do well for humans for a very long time to come (Kargel 2004), and through technology even the asteroids might become places where humans set up outposts (Lewis 1997) even for possible economic/commercial benefit on Earth (Kargel 1994); but eventually nothing in this Solar System will be suitable for Earth's life. This leaves only the stars beyond the Solar System as potential refugia, which is one reason why the search for extrasolar system life or life-sustaining environments is important (Sect. 6.1).

Whereas panspermia arguably may have helped to sprinkle life and its ingredients around the Solar System, interstellar panspermia almost surely would require intelligent, engineered effort. Barring some completely unexpected surprise as to what may lurk in Europa's sea (as implied, in fun, by Arthur C. Clarke) or the rise of a successor intelligence on Earth in the time after humans (say, descendants of coyotes, dolphins, ants, or the ever-popular cockroaches), it would seem that it is humanity that will either push to the stars or keep Earth's life forever bound to our worlds near the Sun (or on Earth alone). Our optimism (perhaps betrayed by current events) is that we will succeed enough as a positive-minded civilization to push onward to other stars in a positive quest. Should we settle somewhere beyond our star system, it is likely that the place we choose, or where we will be successful, will be an icy, wet, or salty realm – or probably all three – of some rocky planet. The reasons are fourfold: (1) these materials have so much to offer those needing to "live off the land" during the initial development of an industrial infrastructure (Kargel, 2004); (2) these materials occur under conditions and as a product of planetary formation and geological evolution that would likely produce a rich diversity of other mineral and rock types needed to build and sustain an industrial infrastructure; (3) these materials and the conditions they imply also have much to offer possible endemic microbial life and, hence, complex ecosystems that may beckon for the sake of scientific interest and as a place where we may be better able to support ourselves; (4) finally, they are probably commonplace.

That Earth- and Mars- like rocky-salty-wet-and-icy worlds are common can be stated generally for six reasons. (1) Hydrogen and oxygen are both abundant in the Universe (hence, so is water). (2) Planets, and presumably many satellites, are common in the range of astrocentric distances also found in our Solar System (after allowing for selection effects in current planet search methods). (3) For most systems there will be a range of astrocentric distances and/or conditions of orbital resonances and tidal heating where water will be liquid or close to the triple point. (4) The conditions and planetary histories needed for having some water but not so much that there is no rocky land seem rather ordinary. (5) Chemical equilibria of water in contact with silicates and other rocky minerals (even in petrologic systems derived from nonsolar abundances of metals and oxygen) lead to salt formation under many cosmochemical conditions. And (6) the thermodynamic and radiative transfer buffer inherent in the ice/liquid water transition and water/rock reaction rates are so powerful that many of the aqueous rocky worlds will also be partially but incompletely icy. Mars and Earth may in fact be nearly end members of a very common type of planet, which would be within the Category I planets designated by Irwin and Schulze-Makuch (2001) and may include some Category Ia planets as designated by Marion and Schulze-Makuch (2007). In detail, of course, Solar System exploration has taught us to expect the unexpected; so in detail we would expect every new world and every new extrasolar planetary system encountered to be unique. Just because a world may be wet, icy, salty, and rocky might not be sufficient to identify it as a very hospitable world, as not all sets of environmental variables that allow these phases would necessarily allow humans.

FREZCHEM may have more relevance to extrasolar planetary exploration in the coming decades than some may recognize. In Planet Finder levels of planet detection, spectra will be returned indicating much about surface mineralogy. Planets that are covered in ices and salts and other hydrates – say, Europa or saltier versions of Mars or Death Valley – may produce hemisphere-integrated reflectance spectra indicative of particular salt assemblages. Though we would be at a loss to deconvolve the spectra into what likely are heterogeneous chemical mixtures, integrated spectra may, in some situations, provide much knowledge of aqueous chemical lines of descent of evaporitic phases or other comparably insight-rich lines of investigation; if so, FREZCHEM may find early applicability. However, systems well designed for chemical/mineralogical analysis of extrasolar planetary surfaces and spectral features such as the shapes of water absorption bands may require enormous arrays of large telescopic mirrors. These perhaps could be arrayed from Earth to Moon to Mars, and around the asteroid belt (perhaps spun and manufactured on asteroids). Such a super-large astronomical array of optical/near-infrared telescopes may be a century in coming. By that time, perhaps humans will have mastered fusion energy, and an interstellar wanderlust might be sparked.

Two fundamental questions will influence our drive to the stars: Are we alone? (Webb 2002). Should we reach for the stars? (Fig. 6.1).

If we are alone, do we have the psychological drive to seed the Universe with life? Is it part of our moral-neutral, but aggressive, DNA to spread ourselves out there? Some might ask, do we have a moral obligation to spread life

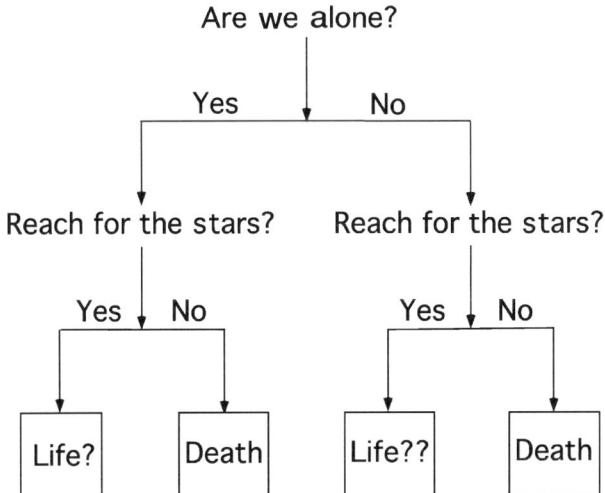

Fig. 6.1. The long-term consequences of: Are we alone? Do we reach for the stars?

if we are able? Or is that just a selfish question, where perhaps even the very question should be rejected? Will we discover that the effort to reach for the stars is just too great for comfort, or will we one day discover that the effort needed was easily within reach before resource depletion and other limits to life on Earth rendered the task impossible? If extending human life beyond our Solar System is impossible, might we simply extend life in the form of our most distant cousins, the microbes, by engineered effort or by accident of our exploration? Within our Solar System, accidental or engineered transfer of Earth's microbial life to sustainable colonies beyond Earth is possible today and may be an act in progress during our current exploration of Mars. Forward and back contamination of planetary bodies is a major concern today in planetary exploration (Horneck and Baumstark-Khan 2002; Wharton 2002; Kargel 2004).

But the question being asked here is on a different scale: not, what is the future of life on Mars? but rather, what is the future of life in the Universe? In the TV series *Star Trek*, the "prime directive" controlling interactions between humans and other life forms is **noninterference**. But if we do not reach for the stars and risk interference, then life as we know it will someday become extinct long before the thermodynamic death of the larger Universe (Fig. 6.1). If there are other life forms in the Universe, then our chances of success in propagating "life as we know it" are lessened if for no other reason than preventing interference, if absolute noninterference is the sole guiding ethic. In the Star Trek concept, noninterference was much more the ideal intent and far less the fictional reality of the crews' explorations. When considering these alternatives, always bear in mind that distant seeding from a dying solar system may be how life began on Earth (the Panspermia Hypothesis).

That TV series also brought to the fore the hypothetical possibility – perhaps an inevitability – that not all life out there (intelligent and otherwise) is entirely benign. What would be the consequences of finding a marvelous salty, icy world, seemingly perfect for us, only to find that it is already the home base for a life form (intelligent or some devastating pox) that, through our explorations, found our own planet inviting for its exclusive use? In Star Trek, most of the wonders and traumas of interstellar exploration were something that happened to space crews and planetary inhabitants light years away from Earth. But as we import materials from other planets in our own Solar System and beyond, or bring human explorers back to Earth, there are potentially devastating consequences to consider and protect against here at home. Thus, space exploration is something to do with great care, yet it is something very deep in the human spirit extending back tens of millennia but made possible in today's and tomorrow's technological eras.

As for the more sensationalistic possibilities – that we may be inviting to Earth very dreadful civilizations that will be intent on dominating or eating us (consider the TV series *V*), or engineering our planet to their own specifications and perhaps completely replacing Earth's biosphere, well, that calling card went out a century ago with our first telecommunications. Our space exploration will either be more sophisticated than any potential dreadful civilization's or probably far less sophisticated; either way, if this is our concern, we should shut down the telecom industry right away.

Our view is that as much as we cannot control the ballooning and nova of our future red giant Sun (a frightening thought), we cannot stop the human drive to explore and eventually settle worlds beyond Earth. We can, however, go about this eventuality reasonably safely or not, and we would choose to do so as safely as possible.

As pointed out in the epigraph for this chapter, either answer to the question Are we alone? is both frightening and profound. To reach for the stars, or not, is equally frightening and profound.

A FREZCHEM Program Guide

Program listings and FORTRAN codes for various version of the FREZCHEM model (versions 5.2 to 9.2) are available from the senior author (giles.marion@dri.edu; http://frezchem.dri.edu). In this appendix, we first describe the data input; then we examine four output files that deal with (1) seawater freezing, (2) strong acids, (3) gas hydrates, and (4) a pressure application. But before discussing these examples, there are some limitations that the user needs to be aware of.

A.1 Model Input Limitations

There are a few incompatibilities among model inputs. For example, the equations used to estimate pH for an alkaline system (Eq. 3.29) and an acidic system (Eq. 3.32) are incompatible. Therefore, when defining an alkaline system (pH > 4.5), be certain to assign "acidity" a value of 0.0, which bypasses the acidity algorithms. Similarly, when defining an acidic system (pH < 4.5), be certain to assign alkalinity a value of 0.0, which by-passes the alkalinity algorithms. The pH calculations (Eqs. 3.29 and 3.32) use a charge balance equation. If the initial input charges are unbalanced, then the model forces a charge balance by adjusting the pH and the ion concentrations that are a function of pH (e.g., HCO_3^-). If the initial charge is badly out of balance, then the calculated pH is also likely to be a poor estimate. As a **general rule**, it is best to adjust initial model inputs to provide a perfect charge balance.

An alkalinity specification can also conflict with CO_2 gas hydrates. Here the problem has to do with the magnitude of P_{CO_2}. The equations governing CO_2 solubility and carbonic acid dissociation are only defined for low P_{CO_2} values (0.0 to 1 bar) (Plummer and Busenberg 1982). At 10 °C, the equilibrium P_{CO_2} for $CO_2 \cdot 6H_2O$ in pure water is ca. 45 bars. Such high values are beyond the range of validity of the alkalinity relationships. Therefore caution is necessary in interpreting alkalinity/CO_2 gas hydrate applications. A way to get around this problem is illustrated with a gas hydrate example (Table A.4), where P_{CO_2} is assigned a constant value and P_{CH4} is allowed to change with pressure, which works for $CH_4 \cdot 6H_2O$. If $CO_2 \cdot 6H_2O$ is the desired gas hydrate, then an alternative is to remove alkalinity from the input database. For example, seawater alkalinity only represents 0.37% of the total anion charge (Table A.2); removing alkalinity and an equivalent amount of Ca (CaCO$_3$ is precipitating) solves the alkalinity/CO$_2$ gas hydrate problem.

The FREZCHEM model was designed to characterize aqueous electrolyte solutions. To work properly, there must always be ions in solution, even if only hypothetical. To simulate pure water, pure gas hydrate, pure ice, or other nonion equilibria, you need to add minor concentrations of ions (e.g., Na = Cl = 1×10^{-6} m). Such minor concentrations do not significantly affect the thermodynamic properties, but they do allow for proper model calculations.

A.2 Model Inputs

Table A.1 describes how data is input into the FREZCHEM model. This is specifically for version 9.2. Earlier versions have similar but fewer inputs. Pay particular attention to the units of gas concentrations. Versions of the model before 9.2 required gas concentrations in atm., while version 9.2 requires units of bars. Most of the program inputs are relatively self-explanatory (Table A.1). A few additional words are necessary to assure gas hydrates are handled correctly in the model.

There are two systems for inputting CH$_4$ and CO$_2$ gases into gas hydrate systems: an "open" carbon system, where the $P_{(g)}$ is fixed at a given total pressure, and a "closed" carbon system where the total carbon is fixed. The open carbon system would be appropriate for the base of a gas hydrate deposit where a large amount of free gas is present. In this open case, the gas content is present in excess and the water content is limiting. As a consequence, water may be converted entirely into hydrate ice. This is similar to what happens to water ice as the temperature decreases to the eutectic. On the other hand, the closed carbon system would be appropriate for minor amounts of CH$_4$ or CO$_2$ in, for example, an ice core. Here carbon is limiting and water is present in excess. This closed carbon case requires estimating the carbon present in the gas, aqueous, and solid phases. The existing model can input the amount of carbon in the aqueous and solid phases. The amount of carbon in the gas phase is arbitrarily estimated from the $P_{(g)}$ in 100 ml of gas, which is added to the carbon in the other phases. The total amount of carbon is such a system is small, and, as temperature or pressure changes, gas hydrates would rapidly precipitate at equilibrium. As a consequence, in the closed carbon system model, the assumption is made that all the carbon solidifies at the temperature or pressure where the system first reaches gas hydrate equilibrium. This closed carbon model is appropriate for estimating gas hydrate equilibrium temperatures and pressures but not the actual content of gas hydrates because of the arbitrary nature of estimating the carbon content of the gas phase.

To explore how pressure affects gas hydrate equilibria, you need to select the Pressure Pathway (Table A.1), plus you need to specify gas-phase mole fractions $[x_{(g)}]$ for $CO_{2(g)}$ and/or $CH_{4(g)}$ (Table A.1). Then the model calculates increasing $P_{(g)}$ using the equation

$$P_{(g)} = x_{(g)} P_T, \qquad (A.1)$$

where P_T is the specified total pressure.

If both $CH_{4(g)}$ and $CO_{2(g)}$ are present in the gas phase, there is an option to consider a mixed $(CH_4\text{-}CO_2) \cdot 6H_2O$ gas hydrate (MIX = 1, yes; MIX = 2, no). See Sect. 3.3.4 for a description of the mixture model. The option "MIX = 2, no" is a "nonsense" option that allows the program to treat the two gas systems as independent components. In reality, if both $CH_{4(g)}$ and $CO_{2(g)}$ are present, then a mixed structure I gas hydrate forms (Marion et al. 2006).

In addition of the instructions in Table A.1, there are also other, less frequently used, options that require changes in the program code. You might want to remove some solid phases from the mineral database for specific applications. For example, to simulate seawater, you might want to remove dolomite and magnesite from the mineral database because these minerals are not currently precipitating from seawater; instead, calcite is the normal seawater precipitate. Or you might want to remove all solid phases to simulate a pure solution-phase model. Instructions for removing selective minerals or all minerals are given at the end of the Parameter subroutine. If, for any reason, you want to assure that a solution is saturated with respect to a specific mineral (e.g., dolomite), then you can specify a mole amount that is sufficient to assure saturation (e.g., 0.1 moles). See comments in the middle of the main program. Also, you can refine a temperature change, evaporation, or pressure change step size as the model approaches equilibrium. See comments at the end of the main program.

A.3 Model Outputs

A.3.1 Seawater Freezing

Table A.2 is model output for seawater freezing at 253.15 K. Beneath the title, the output includes temperature, ionic strength, density of the solution (ρ), osmotic coefficient (ϕ), amount of unfrozen water, amount of ice, and pressure on the system. Beneath this line are the solution and gaseous species in the system. The seven columns include species identification, initial concentration, final (equilibrium) concentration, activity coefficient, activity, moles in the solution phase, and mass balance. The mass balance column only contains those components for which a mass balance is maintained. The number of these components minus 1 is generally the number of independent components in the system (in this case, $8-1=7$). The mass balances (col. 7) should equal the initial concentrations (col. 2). This mass balance comparison is a good check on the computational accuracy.

Solid phases that precipitate at $-20\,°C$ include ice, mirabilite, and calcite (Table A.2). In this particular case, we used the equilibrium crystallization option; as a consequence, the columns labeled "Moles" and "Accumulated moles" are identical. Had we used fractional crystallization, then the "Moles" column would have contained the moles of the solid phase that precipitated in the last temperature/evaporation/pressure step, and the "Accumulated moles" would include the total precipitation of that solid phase over all steps.

A.3.2 Strong Acid

The second output example (Table A.3) deals with strong acids. This is part of the simulation of the hypothetical acidic ocean for Europa described in Fig. 5.24.

At 263.15 K, the calculated pH is 0.02; this is just above the point where the pH dips into the negative range (Fig. 5.24). Because acidity (H^+) is specified as input (0.515 m), the system maintains a mass balance for H^+ (Table A.3); this is in contrast to the other examples (Tables A.2, A.4, and A.5) where the mass of H^+ is unconstrained. Note that a significant amount of solution "H" is present as H^+ and HSO_4^-. The number of independent components in this case (4) is again equal to the components under "Mass balance" (5) – 1.

Solids precipitating at 263.15 K include ice, mirabilite, and $MgSO_4 \cdot 12H_2O$ (Table A.3, Fig. 5.24). In this case, the bulk of the original water (1000 g) is actually present as hydrated salts (mirabilite = 132.2 g water; $MgSO_4 \cdot 12H_2O$ = 566.7 g water). In addition to using a comparison of "Initial conc." and "Mass balance" to check on the internal consistency of the model calculations (see above), solution activity products can be compared to equilibrium constants for solids that are precipitating. For example,

$$\left(a_{Mg^{2+}}\right)\left(a_{SO_4^{2-}}\right)(a_w)^{12} = K_{sp(MgSO_4 \cdot 12H_2O)}, \tag{A.2}$$

which gives

$$(0.13813)(0.043025)(0.90762)^{12} = 0.18570 \times 10^{-2}, \tag{A.3}$$

$$0.18572 \times 10^{-2} \approx 0.18570 \times 10^{-2}, \tag{A.4}$$

which is within convergence criteria of perfect agreement.

A.3.3 Gas Hydrates

The third example of model outputs deals with gas hydrates (Table A.4), which was part of our snowball Earth simulations (Fig. 5.6). The P_{CO_2}, in this case, was set equal to 0.12 bars and was independent of total pressure; this served to prevent P_{CO_2} from becoming too high and beyond the validity of the alkalinity relationships (Sect. A.1). The P_{CH_4}, on the other hand, was allowed to increase with pressure and served as a surrogate for all potential gas hydrates.

Ferrous iron [Fe(II)] was arbitrarily set for these simulations; note that virtually all the Fe(II) precipitates as siderite (FeCO$_3$) (Table A.4), as was pointed out earlier (Table 5.2). During the snowball Earth phase, the assumption was made that seawater would be saturated with dolomite. To assure dolomite saturation, we arbitrarily set dolomite content equal to 0.1 moles at the start of the simulations [X(61) = 0.1]; this was done by incorporating a new line of code at the beginning of the program where molal concentrations are inputted (see version 9.2 comments in main program). Note the especially high Ca, Mg, and alkalinity contents under "Mass balance." This is a case where mass balance does not equal initial concentration because of the addition of solid-phase dolomite. Adding the 0.1 moles of dolomite to the initial concentrations does equal the Ca, Mg, and alkalinity contents under "Mass balance."

In Fig. 5.7, we pointed out that the stability fields of ice and CH$_4$·6H$_2$O ($x = 1.0$) overlap. Table A.4 shows equilibria at 22.0 bars of pressure, where ice is stable, and at 23.0 bars of pressure, where CH$_4$·6H$_2$O is stable. The transition from ice to CH$_4$·6H$_2$O occurs precisely at 22.4 bars of pressure (0.220 km), which is what is plotted at $-3.37\,°C$ in Fig. 5.7.

The number of components under "Mass balance" is nine, which means eight independent components for this system. Had we specified a "closed" carbon system for both CO$_2$ and CH$_4$, then the total number of independent components would have been ten instead of eight because the masses associated with CO$_2$ and CH$_4$ would then be fixed.

A.3.4 Pressure Application

Earlier, we examined the movement of salts through a Martian regolith under a freezing regime (Fig. 5.18). Initially, each layer (0.5 km) was separately equilibrated; this initial step removed virtually all the Fe(II) as siderite, most of the Ca as calcite, and a lesser percent of Mg as hydromagnesite. In a second step, we froze the system from the top down, which led to salt exclusion, increasing salt concentrations with depth, and additional precipitation of calcite and hydromagnesite (Fig. 5.18).

Table A.5 is the output file for salts in the 4.5- to 5.0-km layer, where the system pressure is 484.5 bars (102 bars km^{-1} × 4.75 km). The temperature of 268.28 K is the freezing point depression for this particular composition and pressure; at 268.27 K, ice forms. The pH of this system is 8.02. The number of independent components is seven. This example deals with lithostatic pressures on solutions dispersed in a regolith, which is fundamentally different from the previous examples (Tables A.2–A.4) that dealt with seawaters.

Table A.1. Model Inputs (version 9.2) (hit return after every entry)

Title: Any alphanumeric character up to 50 characters.

Freeze (1) or evaporation (2) or pressure (3) pathway: Enter 1, 2, or 3 depending on whether you want to simulate a temperature change (1), an evaporation (2), or a pressure change (3). For evaluating a single point, enter 1.

Equilibrium (1) or fractional (2) crystallization: In equilibrium crystallization (1), precipitated solids are allowed to reequilibrate with the solution phase as environmental conditions change. In fractional crystallization (2), precipitated solids are removed and not allowed to reequilibrate with the solution phase as environmental conditions change.

Open (1) or closed (2) carbon system: If you want the gas partial pressure of CO_2 or CH_4 to be fixed at a given total pressure, enter 1. If you want the total carbon to be fixed, enter 2.

Sodium (m/kg): Enter sodium molality [moles/kg (water)]. Otherwise, enter 0.0.

Potassium (m/kg): Enter potassium molality [moles/kg (water)]. Otherwise, enter 0.0.

Calcium (m/kg): Enter calcium molality [moles/kg (water)]. Otherwise, enter 0.0.

Magnesium (m/kg): Enter magnesium molality [moles/kg (water)]. Otherwise, enter 0.0.

Iron (m/kg): Enter iron molality [moles/kg (water)]. Otherwise, enter 0.0.

Chloride (m/kg): Enter chloride molality [moles/kg (water)]. Otherwise, enter 0.0.

Sulfate (m/kg): Enter sulfate molality [moles/kg (water)]. Otherwise, enter 0.0.

Nitrate (m/kg): Enter nitrate molality [moles/kg (water)]. Otherwise, enter 0.0.

Carbonate alkalinity: Enter as equivalents/kg (water). If alkalinity = 0.0, then you **must** enter 0.0. The latter will cause the model to skip all bicarbonate-carbonate, pH chemistries in the model.

Initial pH: If alkalinity > 0.0, then the model will calculate pH, given an initial pH estimate that is specified here. If this estimate is far removed from the true pH, then the model may not converge.

Acidity: Enter as equivalents/kg (water). This is the total hydrogen concentration, if known initially. Generally this is only known for strong acid solutions. For example, for a 1 molal H_2SO_4 solution, enter 2.00. Otherwise, enter 0.0. The equations used to calculate pH for the alkalinity and acidity cases are incompatible. Thus a specification of either carbonate alkalinity or acidity requires that the other variable be assigned a value of 0.00. This will channel the calculations to the proper algorithm.

HCl (bars): If the HCl atmospheric concentration is known, then specify here. Otherwise, enter 0.0. If you specify 0.0, then the model will calculate HCl (bars). Note that if you specify HCl (bars) or the other acids below, then these properties override the total acidity specification (see above). That is, the solution is equilibrated with the atmospheric concentration. NB: you can, if desired, specify atmospheric concentrations for some acids (e.g., HCl and HNO_3) and leave other acid partial pressure unspecified (e.g., H_2SO_4 = 0.0).

HNO_3 (bars): If the HNO_3 atmospheric concentration is known, then specify here. Otherwise, enter 0.0.

H_2SO_4 (bars): If the H_2SO_4 atmospheric concentration is known, then specify here. Otherwise, enter 0.0.

Table A.1. (continued)

Initial total pressure (bars): Enter the initial total pressure of the system.

Initial CO_2 (bars): If alkalinity > 0.0 or CO_2 hydrates are simulated, then specify the initial concentration of $CO_2(g)$ in bars.

Mole fraction of CO_2: Enter the mole fraction of $CO_2(g)$ for the system [mole fraction = $CO_2(g)$/total pressure]. For pure CO_2, enter 1.0. If 0.0, then $CO_2(g)$ is fixed and independent of total pressure.

O_2 (bars): If the atmospheric concentration of oxygen is known, then specify here. Otherwise, enter 0.0. If you are interested in ferrous iron chemistry, then you may want to assign O_2 a value of 0.0. Otherwise, it is likely that the insolubility of ferric minerals in the presence of O_2 will cause all the iron to precipitate as a ferric mineral [see discussions in Marion et al. (2003a) iron paper].

Initial CH_4 (bars): If CH_4 hydrates are simulated, then specify the initial concentration of CH_4 (g) in bars.

Mole Fraction of CH_4: Enter the mole fraction of CH_4 (g) for the system (mole fraction = CH_4 (g)/total pressure). For pure CH_4, enter 1.0. If 0.0, then CH_4 (g) is fixed and independent of total pressure.

Mixed CH_4-CO_2 Gas Hydrate?: If both $CH_4(g)$ and $CO_2(g)$ are specified as inputs, then you can use these data to estimate the stability of a mixed CH_4-CO_2 gas hydrate (YES = 1) or treat the two gases as independent gas hydrates (NO = 2).

Initial temperature (K): Enter the temperature in absolute degrees (K) for start of simulation (e.g., 298.15).

For temperature change pathway (1):
  Final temperature (K): Enter final temperature of simulation (e.g., 273.15).
  Temperature decrement (K): The temperature interval between simulations (e.g., 5). For the above temperature designations, the model would calculate equilibrium starting at 298.15 K and ending at 273.15 K at 5-K intervals. If you want to change the decrement in a run (e.g., to reduce the step size near an equilibrium), see the comments near the end of the main program.

For evaporation pathway (2):
  Initial water (g): Normally enter "1000" at this point. The standard weight basis of the model is 1000 g water plus associated salts. If you enter 100 instead of 1000, the initial ion concentrations, specified above, will be multiplied by 10.0 (1000/100) as the starting compositions for calculations. This feature of the model is useful in precisely locating where minerals start to precipitate during the evaporation process without having to calculate every small change between 1000 g and 1 g.
  Final water (g): Enter the final amount of water that you want to remain in the system (e.g., 100).
  Water decrement (g): Enter the water decrement for simulations (e.g., 50 g). Specifying initial = 1000, final = 100, and decrement = 50 would result in calculations at 1000 g, 950 g,... 100 g. If you want to change the decrement in a run (e.g., to reduce the step size near an equilibrium), see the comments near the end of the main program.

Table A.1. (continued)

For pressure pathway (3):
Final pressure (bars): Enter the final pressure of the simulation [e.g., 101.01325 bars (100 atm)].
Pressure increment (bars): Enter the pressure increment. For example, if the initial pressure is 1.01 bars, the final pressure is 101.01 bars, and the pressure increment is 1.0 bars, then the simulation would calculate at 1.01, 2.01, 3.01,... 101.01325 bars. If you want to change the increment in a run, see the comments near the end of the main program.

Table A.2. Seawater Freezing

Temp(K)	Ion. str.	RHO	Phi	H2O (g)	Ice(g)	Press. (bars)
253.15	5.3658	1.1821	1.1949	119.15	876.08	1.0132

Solution SPECIES	Initial conc.	Final conc.	Act. coef.	Activity	Moles	Mass balance
NA	0.48610	3.6350	0.60965	2.2161	0.43311	0.48610
K	0.10580E-01	0.88795E-01	0.36930	0.32792E-01	0.10580E-01	0.10580E-01
CA	0.10650E-01	0.80088E-01	0.66744	0.53454E-01	0.95426E-02	0.10650E-01
MG	0.54750E-01	0.45949	0.83287	0.38269	0.54748E-01	0.54750E-01
H	0.43135E-07	0.74937E-08	3.7860	0.28371E-07	0.89288E-09	
MGOH	0.00000	0.30775E-06	0.54815	0.16870E-06	0.36669E-07	
CL	0.56664	4.7556	0.85911	4.0856	0.56664	0.56664
SO4	0.29270E-01	0.23293E-01	0.15949E-01	0.37150E-03	0.27754E-02	0.29270E-01
OH	0.80342E-07	0.14894E-06	0.23857E-01	0.35532E-08	0.17746E-07	
HCO3	0.23000E-02	0.67038E-03	0.35346	0.23695E-03	0.79877E-04	0.23000E-02
CO3	0.00000	0.72520E-05	0.11142E-01	0.80805E-07	0.86409E-06	
CO2	0.23919E-04	0.26151E-04	2.7363	0.71559E-04	0.31160E-05	
CACO3	0.00000	0.82036E-05	1.0000	0.82036E-05	0.97746E-06	
MGCO3	0.00000	0.14663E-04	1.0000	0.14663E-04	0.17471E-05	
CO2(BAR)	0.36447E-03	0.36447E-03	0.99108	0.36122E-03	0.00000	
H2O(BAR)	0.59398E-02			.10407E-02		
H2O(L)	55.50 8			.82311	6.6139	55.508

Solid SPECIES	Moles	Equil. constant	Accum. moles
ICE	48.630	0.82312	48.630
NACL.2H2O	0.00000	8.3964	0.00000
NACL	0.00000	23.777	0.00000
KCL	0.00000	1.6878	0.00000
CACL2.6H2O	0.00000	905.66	0.00000
MGCL2.6H2O	0.00000	50099.	0.00000
MGCL2.8H2O	0.00000	2765.6	0.00000
MGCL2.12H2O	0.00000	71.792	0.00000
KMGCL3.6H2O	0.00000	2328.8	0.00000
CACL2.2MGCL2.12H2O	0.00000	0.44358E+20	0.00000
NA2SO4.10H2O	0.26495E-01	0.26044E-03	0.26495E-01
NA2SO4	0.00000	0.47236	0.00000
MGSO4.6H2O	0.00000	0.17365E-01	0.00000
MGSO4.7H2O	0.00000	0.30124E-02	0.00000
K2SO4	0.00000	0.25490E-02	0.00000
MGSO4.K2SO4.6H2O	0.00000	0.14492E-05	0.00000
NA2SO4.MGSO4.4H2O	0.00000	0.29347E-02	0.00000
CASO4.2H2O	0.00000	0.14314E-04	0.00000
CASO4	0.00000	0.94256E-04	0.00000
MGSO4.12H2O	0.00000	0.76631E-03	0.00000
NA2SO4.3K2SO4	0.00000	0.67174E-10	0.00000

Table A.2. (continued)

CACO3(CALCITE)	0.11064E-02	0.43185E-08	0.11064E-02
MGCO3	0.00000	0.59380E-07	0.00000
MGCO3.3H2O	0.00000	0.29948E-04	0.00000
MGCO3.5H2O	0.00000	0.27452E-04	0.00000
CACO3.6H2O	0.00000	0.16425E-07	0.00000
NAHCO3	0.00000	0.12013	0.00000
NA2CO3.10H2O	0.00000	0.31966E-02	0.00000
NAHCO3.NA2CO3.2H2O	0.00000	0.30301E-01	0.00000
3MGCO3.MG(OH)2.3H2O	0.00000	0.36356E-33	0.00000
CAMG(CO3)2	0.00000	0.14022E-15	0.00000
NA2CO3.7H2O	0.00000	0.27789E-01	0.00000
KHCO3	0.00000	0.36768	0.00000
CACO3(ARAGONITE)	0.00000	0.65648E-08	0.00000
CACO3(VATERITE)	0.00000	0.21983E-07	0.00000
HNO3.3H2O	0.00000	171.86	0.00000
KNO3	0.00000	0.35949E-01	0.00000
NANO3	0.00000	2.9739	0.00000
HCL.3H2O	0.00000	10605.	0.00000
H2SO4.6.5H2O	0.00000	28.096	0.00000
H2SO4.4H2O	0.00000	999.90	0.00000
HCL.6H2O	0.00000	1000.0	0.00000
NANO3.NA2SO4.2H2O	0.00000	0.84122E-01	0.00000
NA3H(SO4)2	0.00000	0.55904E-01	0.00000
NAHSO4.H2O	0.00000	42.328	0.00000
K3H(SO4)2	0.00000	0.50489E-04	0.00000
K5H3(SO4)4	0.00000	0.88999E-08	0.00000
K8H6(SO4)7.H2O	0.00000	0.44537E-12	0.00000
KHSO4	0.00000	0.66187	0.00000
MGSO4.H2O	0.00000	47.746	0.00000
FESO4.7H2O	0.00000	0.11380E-02	0.00000
FESO4.H2O	0.00000	0.47599	0.00000
FECL2.6H2O	0.00000	8232.3	0.00000
FECL2.4H2O	0.00000	0.10000E+07	0.00000
FECO3	0.00000	0.18333E-10	0.00000
FE(OH)3	0.00000	0.14608E+15	0.00000
CO2.6H2O	0.00000	1.4185	0.00000
CH4.6H2O	0.00000	3.8708	0.00000

Iterations = 16

Table A.3. Strong acid

Temp(K)	Ion. str.	RHO	Phi	H2O(g)	Ice (g)	Press. (bars)
263.15	6.6373	1.1508	0.98802	235.25	65.800	1.0132

Solution SPECIES	Initial conc.	Final conc.	Act. coef.	Activity	moles	Mass Balance
NA	1.6200	0.64651	0.36435	0.23555	0.15209	1.6200
MG	2.8670	1.0436	0.13236	0.13813	0.24550	2.8670
H	0.51500	1.2939	0.74045	0.95807	0.30440	0.51500
SO4	3.9345	1.5662	0.27472E-01	0.43025E-01	0.36844	3.9345
OH	0.14587E-13	0.19128E-14	0.19846	0.37962E-15	0.45000E-15	
HSO4	0.00000	0.89523	1.7337	1.5521	0.21060	
H2SO4(BAR)	0.00000			0.33866E-22		
H2O(BAR)	0.49589E-02			0.25954E-02		
H2O(L)			55.508	0.90762	13.058	55.508

Solid SPECIES	Moles	Equil. constant	Accum. moles
ICE	3.6525	0.90762	3.6525
NACL.2H2O	0.00000	12.571	0.00000
NACL	0.00000	27.853	0.00000
KCL	0.00000	2.5902	0.00000
CACL2.6H2O	0.00000	1252.1	0.00000
MGCL2.6H2O	0.00000	55413.	0.00000
MGCL2.8H2O	0.00000	4682.9	0.00000
MGCL2.12H2O	0.00000	199.88	0.00000
KMGCL3.6H2O	0.00000	4526.6	0.00000
CACL2.2MGCL2.12H2O	0.00000	0.11070E+20	0.00000
NA2SO4.10H2O	0.73395	0.90537E-03	0.73395
NA2SO4	0.00000	0.48157	0.00000
MGSO4.6H2O	0.00000	0.18596E-01	0.00000
MGSO4.7H2O	0.00000	0.43652E-02	0.00000
K2SO4	0.00000	0.40945E-02	0.00000
MGSO4.K2SO4.6H2O	0.00000	0.39356E-05	0.00000
NA2SO4.MGSO4.4H2O	0.00000	0.31861E-02	0.00000
CASO4.2H2O	0.00000	0.18490E-04	0.00000
CASO4	0.00000	0.94050E-04	0.00000
MGSO4.12H2O	2.6215	0.18570E-02	2.6215
NA2SO4.3K2SO4	0.00000	0.30930E-09	0.00000
CACO3(CALCITE)	0.00000	0.43033E-08	0.00000
MGCO3	0.00000	999.90	0.00000
MGCO3.3H2O	0.00000	0.15229E-04	0.00000
MGCO3.5H2O	0.00000	0.96139E-05	0.00000
CACO3.6H2O	0.00000	0.32915E-07	0.00000
NAHCO3	0.00000	0.14567	0.00000
NA2CO3.10H2O	0.00000	0.69294E-02	0.00000
NAHCO3.NA2CO3.2H2O	0.00000	0.39156E-01	0.00000
3MGCO3.MG(OH)2.3H2O	0.00000	0.52738E-34	0.00000
CAMG(CO3)2	0.00000	999.90	0.00000
NA2CO3.7H2O	0.00000	0.49255E-01	0.00000
KHCO3	0.00000	0.50021	0.00000
CACO3(ARAGONITE)	0.00000	0.63986E-08	0.00000
CACO3(VATERITE)	0.00000	0.20194E-07	0.00000
HNO3.3H2O	0.00000	347.82	0.00000
KNO3	0.00000	0.80269E-01	0.00000
NANO3	0.00000	2.3218	0.00000
HCL.3H2O	0.00000	12483.	0.00000
H2SO4.6.5H2O	0.00000	22.486	0.00000
H2SO4.4H2O	0.00000	999.90	0.00000
HCL.6H2O	0.00000	1000.0	0.00000
NANO3.NA2SO4.2H2O	0.00000	0.84024E-01	0.00000
NA3H(SO4)2	0.00000	0.14086	0.00000
NAHSO4.H2O	0.00000	30.543	0.00000
K3H(SO4)2	0.00000	0.53417E-04	0.00000
K5H3(SO4)4	0.00000	0.17412E-07	0.00000

Table A.3. (continued)

K8H6(SO4)7.H2O	0.00000	0.41716E-12	0.00000
KHSO4	0.00000	0.97421	0.00000
MGSO4.H2O	0.00000	11.329	0.00000
FESO4.7H2O	0.00000	0.13913E-02	0.00000
FESO4.H2O	0.00000	0.36005	0.00000
FECL2.6H2O	0.00000	3983.3	0.00000
FECL2.4H2O	0.00000	17767.	0.00000
FECO3	0.00000	0.15200E-10	0.00000
FE(OH)3	0.00000	0.30291E+14	0.00000
CO2.6H2O	0.00000	3.8247	0.00000
CH4.6H2O	0.00000	9.8303	0.00000

Iterations = 12

Table A.4. Snowball Earth (P = 22 bars)

Temp (K)	Ion. str.	RHO	Phi	H2O (g)	Ice (g)	Press.(bars)
269.78	1.1948	1.0462	0.90033	872.39	127.61	22.013

Solution SPECIES	Initial conc.	Final conc.	Act. coef.	Activity	Moles	Mass balance
NA	0.65072	0.74590	0.58667	0.43760	0.65072	0.65072
K	0.14810E-01	0.16976E-01	0.54160	0.91943E-02	0.14810E-01	0.14810E-01
CA	0.14910E-01	0.28765E-01	0.18186	0.52313E-02	0.25094E-01	0.11491
MG	0.76650E-01	0.99533E-01	0.18345	0.18259E-01	0.86832E-01	0.17665
H	0.41971E-06	0.32686E-06	0.77747	0.25413E-06	0.28515E-06	
MGOH	0.00000	0.80770E-08	0.86688	0.70018E-08	0.70463E-08	
FE	0.14910E-01	0.18860E-04	1.0549	0.19896E-04	0.16454E-04	0.14910E-01
FEOH	0.00000	0.26680E-08	0.86469	0.23070E-08	0.23276E-08	
CL	0.79329	0.90933	0.67374	0.61265	0.79329	0.79329
SO4	0.40980E-01	0.46974E-01	0.72763E-01	0.34180E-02	0.40980E-01	0.40980E-01
OH	0.68471E-08	0.94432E-08	0.33906	0.32018E-08	0.82382E-08	
HCO3	0.32200E-02	0.16215E-01	0.52963	0.85877E-02	0.14146E-01	0.40322
CO3	0.00000	0.11352E-04	0.63513E-01	0.72099E-06	0.99033E-05	
CO2	0.85930E-02	0.73746E-02	1.2540	0.92476E-02	0.64336E-02	
FECO3	0.00000	0.21260E-06	1.0000	0.21260E-06	0.18547E-06	
CACO3	0.00000	0.44890E-05	1.0000	0.44890E-05	0.39161E-05	
MGCO3	0.00000	0.75996E-05	1.0000	0.75996E-05	0.66298E-05	
CH4	0.21503E-02	0.41284E-01	1.2636	0.52167E-01	0.36016E-01	
CH4(BAR)	1.0132	22.013	0.94879	20.886	0.00000	
CO2(BAR)	0.12000	0.12000	0.84531	0.10144	0.00000	
H2O(BAR)	0.46099E-02			.45102E-02		
H2O(L)	55.508			.96943	48.425	55.508

Solid SPECIES	Moles	Equil. constant	Accum. moles
ICE	7.0833	0.96943	7.0833
NACL.2H2O	0.00000	16.160	0.00000
NACL	0.00000	30.659	0.00000
KCL	0.00000	3.3876	0.00000
CACL2.6H2O	0.00000	1639.5	0.00000
MGCL2.6H2O	0.00000	56916.	0.00000
MGCL2.8H2O	0.00000	6256.8	0.00000
MGCL2.12H2O	0.00000	450.17	0.00000
KMGCL3.6H2O	0.00000	6785.2	0.00000
CACL2.2MGCL2.12H2O	0.00000	0.46835E+19	0.00000
NA2SO4.10H2O	0.00000	0.21216E-02	0.00000
NA2SO4	0.00000	0.51235	0.00000
MGSO4.6H2O	0.00000	0.20178E-01	0.00000
MGSO4.7H2O	0.00000	0.56803E-02	0.00000
K2SO4	0.00000	0.57253E-02	0.00000
MGSO4.K2SO4.6H2O	0.00000	0.75677E-05	0.00000
NA2SO4.MGSO4.4H2O	0.00000	0.36648E-02	0.00000
CASO4.2H2O	0.00000	0.21880E-04	0.00000
CASO4	0.00000	0.94803E-04	0.00000
MGSO4.12H2O	0.00000	0.33979E-02	0.00000
NA2SO4.3K2SO4	0.00000	0.10059E-08	0.00000
CACO3(CALCITE)	0.00000	0.44905E-08	0.00000
MGCO3	0.00000	0.29502E-07	0.00000
MGCO3.3H2O	0.00000	0.11135E-04	0.00000
MGCO3.5H2O	0.00000	0.72384E-05	0.00000
CACO3.6H2O	0.00000	0.52616E-07	0.00000
NAHCO3	0.00000	0.17664	0.00000
NA2CO3.10H2O	0.00000	0.12319E-01	0.00000
NAHCO3.NA2CO3.2H2O	0.00000	0.49625E-01	0.00000
3MGCO3.MG(OH)2.3H2O	0.00000	0.20069E-34	0.00000
CAMG(CO3)2	0.89812E-01	0.49654E-16	0.89812E-01
NA2CO3.7H2O	0.00000	0.75074E-01	0.00000
KHCO3	0.00000	0.62665	0.00000
CACO3(ARAGONITE)	0.00000	0.65686E-08	0.00000
CACO3(VATERITE)	0.00000	0.20047E-07	0.00000

Table A.4. (continued)

HNO3.3H2O	0.00000	552.06	0.00000
KNO3	0.00000	0.13366	0.00000
NANO3	0.00000	2.3780	0.00000
HCL.3H2O	0.00000	13809.	0.00000
H2SO4.6.5H2O	0.00000	16.942	0.00000
H2SO4.4H2O	0.00000	1031.8	0.00000
HCL.6H2O	0.00000	978.59	0.00000
NANO3.NA2SO4.2H2O	0.00000	0.11297	0.00000
NA3H(SO4)2	0.00000	0.15590	0.00000
NAHSO4.H2O	0.00000	31.454	0.00000
K3H(SO4)2	0.00000	0.60930E-04	0.00000
K5H3(SO4)4	0.00000	0.27170E-07	0.00000
K8H6(SO4)7.H2O	0.00000	0.48939E-12	0.00000
KHSO4	0.00000	1.2824	0.00000
MGSO4.H2O	0.00000	5.3255	0.00000
FESO4.7H2O	0.00000	0.17336E-02	0.00000
FESO4.H2O	0.00000	0.31455	0.00000
FECL2.6H2O	0.00000	2785.5	0.00000
FECL2.4H2O	0.00000	14020.	0.00000
FECO3	0.14893E-01	0.14345E-10	0.14893E-01
FE(OH)3	0.00000	0.11382E+14	0.00000
CO2.6H2O	0.00000	7.3200	0.00000
CH4.6H2O	0.00000	18.074	0.00000

Iterations = 3

Table A.4. Snowball Earth (P = 23 bars)

Temp(K)	Ion. str.	RHO	Phi	H2O (g)	Ice (g)	Press. (bars)
269.78	1.1985	1.0464	0.90052	869.57	0.00000	23.013

Solution SPECIES	Initial conc.	Final conc.	Act. coef.	Activity	Moles	Mass balance
NA	0.65072	0.74832	0.58658	0.43895	0.65072	0.65072
K	0.14810E-01	0.17031E-01	0.54140	0.92207E-02	0.14810E-01	0.14810E-01
CA	0.14910E-01	0.28833E-01	0.18184	0.52432E-02	0.25073E-01	0.11491
MG	0.76650E-01	0.99831E-01	0.18349	0.18318E-01	0.86810E-01	0.17665
H	0.41971E-06	0.32593E-06	0.77808	0.25360E-06	0.28342E-06	
MGOH	0.00000	0.81206E-08	0.86670	0.70382E-08	0.70615E-08	
FE	0.14910E-01	0.18880E-04	1.0567	0.19950E-04	0.16418E-04	0.14910E-01
FEOH	0.00000	0.26812E-08	0.86450	0.23179E-08	0.23315E-08	
CL	0.79329	0.91228	0.67378	0.61467	0.79329	0.79329
SO4	0.40980E-01	0.47127E-01	0.72612E-01	0.34219E-02	0.40980E-01	0.40980E-01
OH	0.68471E-08	0.94913E-08	0.33840	0.32118E-08	0.82534E-08	
HCO3	0.32200E-02	0.16167E-01	0.52940	0.85588E-02	0.14058E-01	0.40322
CO3	0.00000	0.11425E-04	0.63108E-01	0.72104E-06	0.99352E-05	
CO2	0.85930E-02	0.73213E-02	1.2546	0.91857E-02	0.63664E-02	
FECO3	0.00000	0.21320E-06	1.0000	0.21320E-06	0.18539E-06	
CACO3	0.00000	0.44964E-05	1.0000	0.44964E-05	0.39100E-05	
MGCO3	0.00000	0.76219E-05	1.0000	0.76219E-05	0.66278E-05	
CH4	0.21503E-02	0.42960E-01	1.2546	0.54329E-01	0.37357E-01	
CH4(BAR)	1.0132	23.013	0.94652	21.782	0.00000	
CO2(BAR)	0.12000	0.12000	0.83840	0.10061	0.00000	
H2O(BAR)	0.46099E-02			.45060E-02		
H2O(L)	55.508			.96930	48.269	55.508

Solid SPECIES	Moles	Equil. constant	Accum. moles
ICE	0.00000	0.96950	0.00000
NACL.2H2O	0.00000	16.167	0.00000
NACL	0.00000	30.679	0.00000
KCL	0.00000	3.3898	0.00000
CACL2.6H2O	0.00000	1640.1	0.00000
MGCL2.6H2O	0.00000	56946.	0.00000
MGCL2.8H2O	0.00000	6258.2	0.00000
MGCL2.12H2O	0.00000	450.02	0.00000

Table A.4. (continued)

KMGCL3.6H2O	0.00000	6794.8	0.00000
CACL2.2MGCL2.12H2O	0.00000	0.46969E+19	0.00000
NA2SO4.10H2O	0.00000	0.21253E-02	0.00000
NA2SO4	0.00000	0.51354	0.00000
MGSO4.6H2O	0.00000	0.20213E-01	0.00000
MGSO4.7H2O	0.00000	0.56891E-02	0.00000
K2SO4	0.00000	0.57362E-02	0.00000
MGSO4.K2SO4.6H2O	0.00000	0.75930E-05	0.00000
NA2SO4.MGSO4.4H2O	0.00000	0.36796E-02	0.00000
CASO4.2H2O	0.00000	0.21929E-04	0.00000
CASO4	0.00000	0.95046E-04	0.00000
MGSO4.12H2O	0.00000	0.34007E-02	0.00000
NA2SO4.3K2SO4	0.00000	0.10139E-08	0.00000
CACO3(CALCITE)	0.00000	0.45038E-08	0.00000
MGCO3	0.00000	0.29581E-07	0.00000
MGCO3.3H2O	0.00000	0.11161E-04	0.00000
MGCO3.5H2O	0.00000	0.72522E-05	0.00000
CACO3.6H2O	0.00000	0.52708E-07	0.00000
NAHCO3	0.00000	0.17682	0.00000
NA2CO3.10H2O	0.00000	0.12339E-01	0.00000
NAHCO3.NA2CO3.2H2O	0.00000	0.49783E-01	0.00000
3MGCO3.MG(OH)2.3H2O	0.00000	0.20294E-34	0.00000
CAMG(CO3)2	0.89833E-01	0.49933E-16	0.89833E-01
NA2CO3.7H2O	0.00000	0.75223E-01	0.00000
KHCO3	0.00000	0.62717	0.00000
CACO3(ARAGONITE)	0.00000	0.65872E-08	0.00000
CACO3(VATERITE)	0.00000	0.20107E-07	0.00000
HNO3.3H2O	0.00000	551.92	0.00000
KNO3	0.00000	0.13375	0.00000
NANO3	0.00000	2.3797	0.00000
HCL.3H2O	0.00000	13805.	0.00000
H2SO4.6.5H2O	0.00000	16.964	0.00000
H2SO4.4H2O	0.00000	1033.3	0.00000
HCL.6H2O	0.00000	977.59	0.00000
NANO3.NA2SO4.2H2O	0.00000	0.11328	0.00000
NA3H(SO4)2	0.00000	0.15590	0.00000
NAHSO4.H2O	0.00000	31.482	0.00000
K3H(SO4)2	0.00000	0.60930E-04	0.00000
K5H3(SO4)4	0.00000	0.27170E-07	0.00000
K8H6(SO4)7.H2O	0.00000	0.48939E-12	0.00000
KHSO4	0.00000	1.2836	0.00000
MGSO4.H2O	0.00000	5.3380	0.00000
FESO4.7H2O	0.00000	0.17363E-02	0.00000
FESO4.H2O	0.00000	0.31530	0.00000
FECL2.6H2O	0.00000	2787.5	0.00000
FECL2.4H2O	0.00000	14033.	0.00000
FECO3	0.14893E-01	0.14385E-10	0.14893E-01
FE(OH)3	0.00000	0.11382E+14	0.00000
CO2.6H2O	0.00000	7.3182	0.00000
CH4.6H2O	1.2066	18.066	1.2066

Iterations = 3

Table A.5. Mars Regolith

Temp(K)	Ion. str.	RHO	Phi	H2O (g)	Ice (g)	Press. (bars)
268.28	0.48032	1.0437	0.83074	999.08	0.00000	484.50

Solution SPECIES	Initial conc.	Final conc.	Act. coef.	Activity	Moles	Mass balance
NA	0.21510	0.21530	0.64014	0.13782	0.21510	0.21510
K	0.19200E-01	0.19218E-01	0.65199	0.12530E-01	0.19200E-01	0.19200E-01
CA	0.14070E-02	0.47339E-03	0.17557	0.83114E-04	0.47296E-03	0.14070E-02
MG	0.13340	0.64172E-01	0.20350	0.13059E-01	0.64113E-01	0.13340
H	0.15749E-07	0.14394E-07	0.66694	0.96001E-08	0.14381E-07	
MGOH	0.00000	0.16225E-06	0.71089	0.11535E-06	0.16211E-06	
CL	0.17680	0.17696	0.70657	0.12504	0.17680	0.17680
SO4	0.48000E-01	0.48044E-01	0.18075	0.86839E-02	0.48000E-01	0.48000E-01
OH	0.23143E-06	0.22936E-06	0.54025	0.12391E-06	0.22915E-06	
HCO3	0.23111	0.83117E-01	0.62800	0.52197E-01	0.83041E-01	0.23111
CO3	0.00000	0.38196E-02	0.52089E-01	0.19896E-03	0.38161E-02	
CO2	0.10524E-02	0.10685E-02	1.1046	0.11803E-02	0.10675E-02	
CACO3	0.00000	0.20076E-04	1.0000	0.20076E-04	0.20058E-04	
MGCO3	0.00000	0.14711E-02	1.0000	0.14711E-02	0.14698E-02	
CO2(BAR)	0.50000E-01	0.50000E-01	0.12096	0.60482E-02	0.00000	
H2O(BAR)	0.28776E-02			.28495E-02		
H2O(L)	55.508			.99073	55.458	55.508

Solid SPECIES	Moles	Equil. constant	Accum. moles
ICE	0.00000	0.99077	0.00000
NACL.2H2O	0.00000	18.212	0.00000
NACL	0.00000	39.472	0.00000
KCL	0.00000	4.1861	0.00000
CACL2.6H2O	0.00000	1764.9	0.00000
MGCL2.6H2O	0.00000	70328.	0.00000
MGCL2.8H2O	0.00000	6413.0	0.00000
MGCL2.12H2O	0.00000	315.35	0.00000
KMGCL3.6H2O	0.00000	11330.	0.00000
CACL2.2MGCL2.12H2O	0.00000	0.19267E+20	0.00000
NA2SO4.10H2O	0.00000	0.37515E-02	0.00000
NA2SO4	0.00000	1.3812	0.00000
MGSO4.6H2O	0.00000	0.42041E-01	0.00000
MGSO4.7H2O	0.00000	0.10526E-01	0.00000
K2SO4	0.00000	0.12019E-01	0.00000
MGSO4.K2SO4.6H2O	0.00000	0.27611E-04	0.00000
NA2SO4.MGSO4.4H2O	0.00000	0.20469E-01	0.00000
CASO4.2H2O	0.00000	0.56356E-04	0.00000
CASO4	0.00000	0.29233E-03	0.00000
MGSO4.12H2O	0.00000	0.42295E-02	0.00000
NA2SO4.3K2SO4	0.00000	0.22604E-07	0.00000
CACO3(CALCITE)	0.91398E-03	0.16536E-07	0.91398E-03
MGCO3	0.00000	3387.4	0.00000
MGCO3.3H2O	0.00000	0.33512E-04	0.00000
MGCO3.5H2O	0.00000	0.17425E-04	0.00000
CACO3.6H2O	0.00000	0.10157E-06	0.00000
NAHCO3	0.00000	0.25645	0.00000
NA2CO3.10H2O	0.00000	0.21644E-01	0.00000
NAHCO3.NA2CO3.2H2O	0.00000	0.18391	0.00000
3MGCO3.MG(OH)2.3H2O	0.16954E-01	0.34197E-32	0.16954E-01
CAMG(CO3)2	0.00000	13028.	0.00000
NA2CO3.7H2O	0.00000	0.16329	0.00000
KHCO3	0.00000	0.84309	0.00000

Table A.5. (continued)

CACO3(ARAGONITE)	0.00000	0.22903E-07	0.00000
CACO3(VATERITE)	0.00000	0.75891E-07	0.00000
HNO3.3H2O	0.00000	444.85	0.00000
KNO3	0.00000	0.16394	0.00000
NANO3	0.00000	3.2207	0.00000
HCL.3H2O	0.00000	11512.	0.00000
H2SO4.6.5H2O	0.00000	33.139	0.00000
H2SO4.4H2O	0.00000	2005.9	0.00000
HCL.6H2O	0.00000	614.08	0.00000
NANO3.NA2SO4.2H2O	0.00000	0.33811	0.00000
NA3H(SO4)2	0.00000	0.15607	0.00000
NAHSO4.H2O	0.00000	44.699	0.00000
K3H(SO4)2	0.00000	0.58754E-04	0.00000
K5H3(SO4)4	0.00000	0.24568E-07	0.00000
K8H6(SO4)7.H2O	0.00000	0.46538E-12	0.00000
KHSO4	0.00000	1.7528	0.00000
MGSO4.H2O	0.00000	17.449	0.00000
FESO4.7H2O	0.00000	0.32131E-02	0.00000
FESO4.H2O	0.00000	0.91684	0.00000
FECL2.6H2O	0.00000	4033.6	0.00000
FECL2.4H2O	0.00000	22219.	0.00000
FECO3	0.00000	0.49614E-10	0.00000
FE(OH)3	0.00000	0.14144E+14	0.00000
CO2.6H2O	0.00000	5.7776	0.00000
CH4.6H2O	0.00000	13.232	0.00000

Iterations = 3

B Parameter Tables

This appendix contains a complete listing of all parameters used in the FREZCHEM model (version 9.2). Tables B.1–B.6 deal primarily with model parameterizations as a function of temperature at 1.01 bar pressure. Tables B.7–B.11 list volumetric parameters used in developing a pressure dependence for the model. Table B.12 deals with gas hydrate equilibria.

Table B.1 lists all the chemical reactions and their temperature dependence. Table B.2 lists the Debye–Hückel constants (A_ϕ and A_v) as a function of temperature and pressure. Table B.3 lists the numerical arrays used for calculating unsymmetrical interactions (Equations 2.62 and 2.66). Table B.4 lists binary Pitzer-equation parameters for cations and anions as a function of temperature. Table B.5 lists ternary Pitzer-equation parameters for cations and anions as a function of temperature. Table B.6 lists binary and ternary Pitzer-equation parameters for soluble gases as a function of temperature. Table B.7 lists equations used to estimate the molar volume of liquid water and water ice as a function of temperature at 1.01 bar pressure and their compressibilities. Table B.8 lists equations for the molar volume and the compressibilities of soluble ions and gases as a function of temperature. Table B.9 lists the molar volumes of solid phases. Table B.10 lists volumetric Pitzer-equation parameters for ion interactions as a function of temperature. Table B.11 lists pressure-dependent coefficients for volumetric Pitzer-equation parameters. Table B.12 lists parameters used to estimate gas fugacities using the Duan et al. (1992b) model.

Several of these tables are fitted to the equation

$$P(T)_i = a_{1i} + a_{2i}T + a_{3i}T^2 + a_{4i}T^3 + \frac{a_{5i}}{T} + a_{6i}\ln(T) + \frac{a_{7i}}{T^2} + a_{8i}T^4. \quad (B.1)$$

In Table B.1, $P(T)_i = \text{Ln}(K)$, except for ice (see footnote); in Tables B.4–B.6, B.8, and B.10, $P(T)_i$ is a Pitzer-equation parameter.

Table B.1. Chemical reactions and their equation parameters used in the FREZCHEM model (version 9.2).[a] (Numbers are in computer scientific notation, where e ± xx stands for $10^{\pm xx}$)

Reaction Solution–solid-phase equilibria	Equation parameters							
	a_1	a_2	a_3	a_4	a_5	a_6	a_7	a_8
$H_2O(cr,I)$ ↔ $H_2O(l)$	1.906354e0	−1.880285e-2	6.603001e-5	−3.419967e-8				
NaCl(cr) ↔ Na^+(aq) + Cl^-(aq)	9.1483900le3	8.2234874 5e0	−8.1288759e-3	3.95552403e-6	−1.54040868e5	−1.83624247e3		
$NaCl·2H_2O$(cr) ↔ Na^+(aq) + Cl^-(aq) + $2H_2O$(l)	−1.2222551e4	−9.8806459e0	8.46685083e-3	−3.4459117e-6	2.09823965e5	2.42328528e3		
KCl(cr) ↔ K^+(aq) + Cl^-(aq)	−1.62917341e3	−1.51940390e0	1.45249679e-3	−6.9427505e-7	2.26012743e4	3.33075506e2		
$MgCl_2·6H_2O$(cr) ↔ Mg^{2+}(aq) + $2Cl^-$(aq) + $6H_2O$(l)	7.52225099e2	1.17584653e-1			−2.43223909e4	−1.21990076e2		
$MgCl_2·8H_2O$(cr) ↔ Mg^{2+}(aq) + $2Cl^-$(aq) + $8H_2O$(l)	2.2780197 6e3	6.49361616e-1			−6.23075123e4	−3.95438891e2		
$MgCl_2·12H_2O$(cr) ↔ Mg^{2+}(aq) + $2Cl^-$(aq) + $12H_2O$(l)	2.55008896e5	2.44532240e2	−2.48807876e-1	1.22425236e-4	−4.02983426e6	−5.18668604e4		
$CaCl_2·6H_2O$(cr) ↔ Ca^{2+}(aq) + $2Cl^-$(aq) + $6H_2O$(l)	1.42290062 6e5	1.61973105e2	−1.95332071e-1	1.17636119e-4	−2.04059847e6	−2.97464810e4		
$FeCl_2·4H_2O$(cr) ↔ Fe^{2+}(aq) + $2Cl^-$(aq)[c]	−4.594879e0	1.45731e-1	−3.461353e-4					
$FeCl_2·6H_2O$(cr) ↔ Fe^{2+}(aq) + $2Cl^-$(aq) + $6H_2O$(l)	−3.607762e2	4.61798e0	−1.886403e-2	2.525105e-5				
$KMgCl_3·6H_2O$(cr) ↔ K^+(aq) + Mg^{2+}(aq) + $3Cl^-$(aq) + $6H_2O$(l)	−4.45702171e1	2.32023790e-1	−7.14935692e-4	5.32658215e-7		−4.24817923e3	8.59110245e0	
$CaMgCl_2·2MgCl_2·12H_2O$(cr) ↔ Ca^{2+}(aq) + $2Mg^{2+}$(aq) + $6Cl^-$(aq) + $12H_2O$(l)	8.03777791e1	−1.388069e-1						

Table B.1. (continued)

Solution–solid-phase equilibria	a_1	a_2	a_3	a_4	a_5	a_6	a_7	a_8
$Na_2SO_4(cr)$ $\leftrightarrow 2Na^+(aq) + SO_4^{2-}(aq)$	-1.238537e0	1.929792e-3						
$Na_2SO_4 \cdot 10H_2O(cr)$ $\leftrightarrow 2Na^+(aq) + SO_4^{2-}(aq) + 10H_2O(l)$	-4.633773e1	1.753075e-1	-9.822103e-5					
$K_2SO_4(cr)$ $\leftrightarrow 2K^+(aq) + SO_4^{2-}(aq)$	6.500e0				-3.1573e3			
$MgSO_4 \cdot H_2O(cr)$ $\leftrightarrow Mg^{2+}(aq) + SO_4^{2-}(aq) + H_2O(l)$	1.306284e2	-8.44064e-1	1.356205e-3					
$MgSO_4 \cdot 6H_2O(cr)$ $\leftrightarrow Mg^{2+}(aq) + SO_4^{2-}(aq) + 6H_2O(l)$	-5.7876e0	6.8509e-3						
$MgSO_4 \cdot 7H_2O(cr)$ $\leftrightarrow Mg^{2+}(aq) + SO_4^{2-}(aq) + 7H_2O(l)$	3.956e0				-2.4710e3			
$MgSO_4 \cdot 12H_2O(cr)$ $\leftrightarrow Mg^{2+}(aq) + SO_4^{2-}(aq) + 12H_2O(l)$	-2.95818e1	8.851618e-2						
$CaSO_4(cr)$ $\leftrightarrow Ca^{2+}(aq) + SO_4^{2-}(aq)$	-7.822042e1	6.908174e-1	-2.246589e-3	2.344988e-6				
$CaSO_4 \cdot 2H_2O(cr)$ $\leftrightarrow Ca^{2+}(aq) + SO_4^{2-}(aq) + 2H_2O(l)$	-9.107165e1	7.584271e-1	-2.370863e-3	2.456876e-6				
$FeSO_4 \cdot H_2O(cr)$ $\leftrightarrow Fe^{2+}(aq) + SO_4^{2-}(aq) + H_2O(l)$	6.324332e0	-2.7915e-2						
$FeSO_4 \cdot 7H_2O(cr)$ $\leftrightarrow Fe^{2+}(aq) + SO_4^{2-}(aq) + 7H_2O(l)$	2.09618e1	-2.343349e-1	4.928070e-4					
$Na_2SO_4 \cdot 3K_2SO_4(cr)$ $\leftrightarrow 2Na^+(aq) + 6K^+(aq) + 4SO_4^{2-}(aq)$	-6.207986e1	1.527005e-1						

Table B.1. (continued)

Solution-solid-phase equilibria	a_1	a_2	a_3	a_4	a_5	a_6	a_7	a_8
$Na_2SO_4 \cdot MgSO_4 \cdot 4H_2O(cr)$ $\leftrightarrow 2Na^+(aq) + Mg^{2+}(aq)$ $+ 2SO_4^{2-}(aq) + 4H_2O(l)$	−7.9121e0	8.220223e-3						
$MgSO_4 \cdot K_2SO_4 \cdot 6H_2O(cr)$ $\leftrightarrow 2K^+(aq) + Mg^{2+}(aq)$ $+ 2SO_4^{2-}(aq) + 6H_2O(l)$	−8.661262e1	4.70966e-1	−7.186864e-4					1.144431e-7
$NaNO_3(cr)$ $\leftrightarrow Na^+(aq) + NO_3^-(aq)$	7.016265e2	−9.648812e0	5.093224e-2	−1.227837e-4				
$KNO_3(cr)$ $\leftrightarrow K^+(aq) + NO_3^-(aq)$	−4.652079e1	2.575014e-1	−3.431593e-4					
$NaNO_3 \cdot Na_2SO_4 \cdot 2H_2O(cr)$ $\leftrightarrow 3Na^+(aq) + NO_3^-(aq)$ $+ SO_4^{2-}(aq) + 2H_2O(l)$	1.419947e2	−1.11958e0	2.16824e-3					
$NaHCO_3(cr)$ $\leftrightarrow Na^+(aq) + HCO_3^-(aq)$	1.391669e2	−1.556298e0	5.625521e-3	−6.6461e-6				
$Na_2CO_3 \cdot 7H_2O(cr)$ $\leftrightarrow 2Na^+(aq) + CO_3^{2-}(aq)$ $+ 7H_2O(l)$	−1.807263e1	5.723688e-2						
$Na_2CO_3 \cdot 10H_2O(cr)$ $\leftrightarrow 2Na^+(aq) + CO_3^{2-}(aq)$ $+ 10H_2O(l)$	−7.861589e0	−5.802879e-2	2.622444e-4					
$NaHCO_3 \cdot Na_2CO_3 \cdot 2H_2O(cr)$ $\leftrightarrow 3Na^+(aq) + HCO_3^-(aq)$ $+ CO_3^{2-}(aq) + 2H_2O(l)$	−9.986743e0	2.563766e-2						
$KHCO_3(cr)$ $\leftrightarrow K^+(aq) + HCO_3^-(aq)$	−6.500201e0	1.301118e-2	3.442106e-5					
$MgCO_3(cr)$	−2.8902e1				3.1043e3			
$MgCO_3 \cdot 3H_2O(cr)$ $\leftrightarrow Mg^{2+}(aq) + CO_3^{2-}(aq)$ $+ 3H_2O(l)$	5.847459e1	−4.688674e-1	7.771429e-4					
$MgCO_3 \cdot 5H_2O(cr)$ $\leftrightarrow Mg^{2+}(aq) + CO_3^{2-}(aq)$ $+ 5H_2O(l)$	9.556371e2	−1.003715e1	3.464409e-2	−3.978274e-5				

Table B.1. (continued)

Solution–solid-phase equilibria	a_1	a_2	a_3	a_4	a_5	a_6	a_7	a_8
$MgCO_3 \cdot Mg(OH)_2 \cdot 3H_2O(cr)$ $\leftrightarrow 2Mg^{2+}(aq) + CO_3^{2-}(aq)$ $+ 2OH^-(aq) + 3H_2O(l)$	−1.27801e2				1.2861e4			
$CaCO_3(cr)$ $\leftrightarrow Ca^{2+}(aq) + CO_3^{2-}(aq)$ (calcite)	−3.958293e2	−1.79586e-1			6.537774e3	7.1595e1		
$CaCO_3(cr)$ $\leftrightarrow Ca^{2+}(aq) + CO_3^{2-}(aq)$ (aragonite)	−3.959924e2	−1.79586e-1			6.685079e3	7.1595e1		
$CaCO_3(cr)$ $\leftrightarrow Ca^{2+}(aq) + CO_3^{2-}(aq)$ (vaterite)	−3.963428e2	−1.79586e-1			7.079731e3	7.1595e1		
$CaCO_3 \cdot 6H_2O(cr)$ $\leftrightarrow Ca^{2+}(aq) + CO_3^{2-}(aq)$ $+ 6H_2O(l)$	3.6798e-1				−4.63073e3			
$CaMg(CO_3)_2(cr)$ $\leftrightarrow Ca^{2+}(aq) + Mg^{2+}(aq)$ $+ 2CO_3^{2-}(aq)$	−5.5261e1				4.7485e3			
$FeCO_3(cr)$ $\leftrightarrow Fe^{2+}(aq) + CO_3^{2-}(aq)$	−2.9654e1				1.24845e3			
$4Fe^{2+}(aq) + 10H_2O(l)$ $+ O_2(g)$ $\leftrightarrow 4Fe(OH)_3(cr)$ $+ 8H^+(aq)$	−8.786e0				1.04807e4			
$HCl \cdot 3H_2O(cr)$ $\leftrightarrow H^+(aq) + Cl^-(aq)$ $+ 3H_2O(l)$	5.142350e0	1.630170e-2						
$HCl \cdot 6H_2O(cr)$ $\leftrightarrow H^+(aq) + Cl^-(aq)$ $+ 6H_2O(l)$	4.367384e0	3.637139e-3						
$HNO_3 \cdot 3H_2O(cr)$ $\leftrightarrow H^+(aq) + NO_3^-(aq)$ $+ 3H_2O(l)$	−1.270087e1	7.050180e-2						

Table B.1. (continued)

Solution–solid-phase equilibria	a_1	a_2	a_3	a_4	a_5	a_6	a_7	a_8
$H_2SO_4 \cdot 4H_2O(cr)$ $\leftrightarrow 2H^+(aq) + SO_4^{2-}(aq) + 4H_2O(l)$	-7.103329e1	6.235489e-1	-1.279573e-3					
$H_2SO_4 \cdot 6.5H_2O(cr)$ $\leftrightarrow 2H^+(aq) + SO_4^{2-}(aq) + 6.5H_2O(l)$	-8.947832e1	7.407673e-1	-1.477903e-3					
$Na_3H(SO_4)_2(cr)$ $\leftrightarrow 3Na^+(aq) + H^+(aq) + 2SO_4^{2-}(aq)$	-2.122993e3	2.296880e1	-8.282936e-2	9.946746e-5				
$NaHSO_4 \cdot H_2O(cr)$ $\leftrightarrow Na^+(aq) + HSO_4^-(aq) + H_2O(l)$	6.581474e2	-6.926105e0	2.431396e-2	-2.830635e-5				
$K_3H(SO_4)_2(cr)$ $\leftrightarrow 3K^+(aq) + H^+(aq) + 2SO_4^{2-}(aq)$	4.561655e1	-4.356452e-1	8.546984e-4					
$K_5H_3(SO_4)_4(cr)$ $\leftrightarrow 5K^+(aq) + 3H^+(aq) + 4SO_4^{2-}(aq)$	-3.552640e1	6.711111e-2						
$K_8H_6(SO_4)_7 \cdot H_2O(cr)$ $\leftrightarrow 8K^+(aq) + 6H^+(aq) + 7SO_4^{2-}(aq) + H_2O(l)$	9.591803e1	-9.575202e-1	1.841905e-3					
$KHSO_4(cr)$ $\leftrightarrow K^+(aq) + HSO_4^-(aq)$	-1.019832e1	3.865551e-2						
$CO_2 \cdot 6H_2O(cr)$ $\leftrightarrow CO_2(g) + 6H_2O(l)$	-2.676925e1	1.147623e-1	-3.016638e-5					
$CH_4 \cdot 6H_2O(cr)$ $\leftrightarrow CH_4(g) + 6H_2O(l)$ $T \geq 285K$	1.649717e2	-1.214645e0	2.284327e-3					
$CH_4 \cdot 6H_2O(cr)$ $\leftrightarrow CH_4(g) + 6H_2O(l)$ $T < 285K$	-2.224001e1	9.319961e-2						

Table B.1. (continued)

Gas–solution-phase equilibria	a_1	a_2	a_3	a_4	a_5	a_6	a_7	a_8
$H_2O(g)$ $\leftrightarrow H_2O(l)^d$	9.053594e1	−7.215505e-1	2.112659e-3	−2.254724e-6				
$CO_2(g)$ $\leftrightarrow CO_2(aq)^d$	2.495691e2	4.570806e-2			−1.593281e4	−4.045154e1	1.541270e6	
$O_2(g)$ $\leftrightarrow O_2(aq)^d$	2.98399e-1				−5.59617e3		1.049668e6	
$CH_4(g)$ $\leftrightarrow CH_4(aq)^e$								
$HCl(g)$ $\leftrightarrow HCl(aq)^d$	7.812179e1	−3.696246e-1	6.735943e-4	−5.00638e-7				
$HNO_3(g)$ $\leftrightarrow HNO_3(aq)^d$	3.9400 5278e2	1.31442311e-1	−4.20928363e-5		−3.020 3522e3	−7.1001998e1		
$H_2SO_4(g)$ $\leftrightarrow H_2SO_4(aq)^d$	1.71812e2	−6.503309e-1	6.909049e-4					
$Mg^{2+}(aq) + H_2O(l)$ $\leftrightarrow MgOH^+(aq) + H^+(aq)$	−1.16680				−7.7593e3			
$Fe^{2+}(aq) + H_2O(l)$ $\leftrightarrow FeOH^+(aq) + H^+(aq)$	3.93e-1				−6.6392e3			
$CaCO_3^0(aq)$ $\leftrightarrow Ca^{2+}(aq) + CO_3^{2-}(aq)$	2.829430e3	6.89486e-1			−8.177113e4	−4.85818e2		
$MgCO_3^0(aq)$ $\leftrightarrow Mg^{2+}(aq) + CO_3^{2-}(aq)$	7.4201e1				−2.517845e3	−1.272433e1		
$FeCO_3^0(aq)$ $\leftrightarrow Fe^{2+}(aq) + CO_3^{2-}(aq)$	−1.46641e1				1.36517e3			
$H_2O(l)$ $\leftrightarrow H^+(aq) + OH^-(aq)$	−1.4816780e2	8.933802e-1	−2.332199e-3	2.146860e-6				
$CO_2(aq) + H_2O(l)$ $\leftrightarrow H^+(aq) + HCO_3^-(aq)$	−8.204327e2	−1.4027266e-1			5.027549e4	1.268339e2	−3.879660e6	

200 B Parameter Tables

Table B.1. (continued)

Gas–solution-phase equilibria	a_1	a_2	a_3	a_4	a_5	a_6	a_7	a_8
HCO_3^- (aq) $\leftrightarrow H^+$(aq) $+ CO_3^{2-}$(aq)	-2.84192e2	-7.48996e-2			1.186243e4	3.892561e1	-1.297999e6	
HSO_4^{-1}(aq) $\leftrightarrow H^+$(aq) $+ SO_4^{2-}$(aq)	1.2956527e3	5.70474e-1	-2.57206e-4		-3.056394e4	-2.360504e2		
$2H_2O(l) \leftrightarrow 2H_2(g) + O_2(g)^d$	3.92869e1				-6.87562e4			

[a] See Tables 3.2 and 3.3 for the primary source of these parameters.
[b] Parameters are for the equation $\ln(K) = a_1 + a_2 T + a_3 T^2 + a_4 T^3 + a_5/T + a_6 \text{Ln}(T) + a_7/T^2 + a_8 T^{-4}$, where T is temperature (K), except for ice [H_2O(cr,I)], which is given as $K = a_1 + a_2 T \ldots$.
[c] This equation does not extrapolate well to temperatures below ≈ 263 K. Removed from model mineral database below 263 K.
[d] The gas equations in Table B.1 are written in units of "atm." These are converted to bars within the FREZCHEM model.
[e] $\ln(K) = -4.3021034$5e1 $+ 6.8327721$e-2$T + 5.6871873$0e3$/T - 3.5663628$1e-5$T^2 + 5.7913379$1e1$/(680 - T) - 6.1161666$2e-3$P + 7.8552810$3e-4$P\text{Ln}(T)$ + 9.4254075$9e-2$P/T - 1.9213204$0e-2$P/(680 - T) + 9.1718689$9e-6$P^2/T$, where T is temperature (K) and P is pressure (bars).

Table B.2. The Debye–Hückel constants used in the FRZCHEM model. T is temperature (K), and P is pressure (bars). (Numbers are in computer scientific notation, where e$\pm xx$ stands for $10^{\pm xx}$)

Equation	Reference
$A_\phi = 8.66836498\text{e}1 + 8.48795942\text{e-}2T - 8.88785150\text{e-}5T^2$ $+4.88096393\text{e-}8T^3 - 1.32731477\text{e}3/T$ $-1.76460172\text{e}1\text{Ln}(T)$	Spencer et al. 1990
$A_v = 3.73387 - 2.89662\text{e-}2T + 1.29461\text{e-}3P - 5.62291\text{e-}6TP$ $+7.62143\text{e-}5T^2 + 4.09944\text{e-}8P^2$	Marion et al. 2005

Table B.3. Numerical arrays for calculating $J(x)$ and $J'(x)$ (Eqs. 2.62 and 2.66) (Pitzer, 1991)

k	a_k^I	a_k^{II}
0	1.925154014814667	0.628023320520852
1	−0.060076477753119	0.462762985338493
2	−0.029779077456514	0.150044637187895
3	−0.007299499690937	−0.028796057604906
4	0.000388260636404	−0.036552745910311
5	0.000636874599598	−0.001668087945272
6	0.000036583601823	0.006519840398744
7	−0.000045036975204	0.001130378079086
8	−0.000004537895710	−0.000887171310131
9	0.000002937706971	−0.000242107641309
10	0.000000396566462	0.000087294451594
11	−0.000000202099617	0.000034682122751
12	−0.000000025267769	−0.000004583768938
13	0.000000013522610	−0.000003548684306
14	0.000000001229405	−0.000000250453880
15	−0.000000000821969	0.000000216991779
16	−0.000000000050847	0.000000080779570
17	0.000000000046333	0.000000004558555
18	0.000000000001943	−0.000000006944757
19	−0.000000000002563	−0.000000002849257
20	−0.000000000010991	0.000000000237816

Table B.4. Binary Pitzer-equation parameters for cations and anions used in the FREZCHEM model.[a] (Numbers are in computer scientific notation, where e±xx stands for $10^{\pm xx}$)

Pitzer Parameter	a_1	a_2	a_3	a_4	a_5	a_6	a_7	a_8
$C^\phi_{Na,Cl}$	1.70761824e0	2.32970177e-3	−2.46665619e-6	1.21543380e-9.	−1.35583596e0	−3.87767714e-1		
$B^{(0)}_{Na,Cl}$	7.87239712e0	−8.3864096e-3	1.44137774e-5	−8.780301e-9.	−4.96920671e2	−8.20972560e-1		
$B^{(1)}_{Na,Cl}$	8.66915291e2	6.06166931e-1	−4.8048921e-4	1.88503857e-7	−1.70460145e4	−1.67171296e2		
$C^\phi_{K,Cl}$	−3.27571680e0	−1.27222054e-3	4.71374283e-7	1.1162507e-11	9.0747766e1	5.80513562e-1		
$B^{(0)}_{K,Cl}$	2.65718766e1	9.92715099e-3	−3.62323330e-6	−6.28427180e-11	−7.55707220e2	−4.67300770e0		
$B^{(1)}_{K,Cl}$	1.69742977e3	1.22270943e0	−9.99044490e-4	4.04786721e-7	−3.28684422e4	−3.28813848e2		
$C^\phi_{Mg,Cl}$	5.9532e-2	−2.49949e-4	2.41831e-7					
$B^{(0)}_{Mg,Cl}$	3.13852913e2	2.61769099e-1	−2.46268460e-4	1.15764787e-7	−5.53133381e3	−6.21616862e1		
$B^{(1)}_{Mg,Cl}$	−3.18432525e4	−2.86710358e1	2.78892838e-2	−1.3279705e-5	5.24032958e5	6.40770396e3		
B^ϕ_{MgCl}	2.64231655e1	2.46922993e-2	−2.48298510e-5	1.22421864e-8	−4.18098427e2	−5.35350322e0		
$C^\phi_{Ca,Cl}$	−5.62764702e1	−3.00771997e-2	1.05630400e-5	3.3331626e-9.	1.11730349e3	1.06664743e1		
$B^{(0)}_{Ca,Cl}$	3.4787e0	−1.5417e-2	3.1791e-5					
$B^{(1)}_{Ca,Cl}$	−8.61e-3							
$B^{(0)}_{Fe,Cl}$	3.359e-1							
$B^{(1)}_{Fe,Cl}$	3.83836e1	−1.236e-1						
$B^\phi_{Fe,Cl}$	1.470571e1	−2.293478e-1	1.350491e-3	−3.537708e-6			3.466102e-9.	
C^ϕ_{Na,SO_4}	−3.447035e1	5.312162e-1	−3.119749e-3	8.181394e-6			−8.022815e-9.	
$B^{(0)}_{Na,SO_4}$	5.161188e1	−8.245243e-1	4.902233e-3	−1.280795e-5			1.252258e-8	
$B^{(1)}_{Na,SO_4}$	7.0e-4	4.8e-5	9.0e-9.	3.26e-10	−7.68e0	2.835e-3		
C^ϕ_{K,SO_4}	−7.568e-1	2.529e-3	3.65e-8	5.31e-10	−1.08e0	−1.25e-3		
$B^{(0)}_{K,SO_4}$	1.953e0	−3.996e-3	3.55e-7	1.669e-8	2.67e1	−4.785e-2		
$B^{(1)}_{K,SO_4}$	2.230e-1	−6.101e-4	−1.0e-9.	−1.096e-9.	4.265e1	−1.792e-2		
C^ϕ_{Mg,SO_4}	1.678e0	−5.514e-3	5.97e-7	1.5651e-8	−2.2392e2	6.594e-2		
$B^{(0)}_{Mg,SO_4}$	1.484e0	6.274e-3	5.41e-6	8.84e-8	−1.3210e3	3.0605e-1		
$B^{(1)}_{Mg,SO_4}$	1.8829e2	−1.03999e0	1.2242e-3	3.4974e-6	8.975e4	−6.79235e1		
$B^{(2)}_{Mg,SO_4}$	3.3e-2	−1.529e-4	8.97e-7	1.569e-9.	1.1e0	−1.2755e-2		
C^ϕ_{Ca,SO_4}								

Table B.4. (continued)

Pitzer Parameter	a_1	a_2	a_3	a_4	a_5	a_6	a_7	a_8
$B^{(0)}_{Ca,SO_4}$	7.95e-2	−1.22e-4	5.001e-6	6.704e-9.	−1.5228e2	−6.885e-3		
$B^{(1)}_{Ca,SO_4}$	2.8945e0	7.434e-3	5.287e-6	−1.01513e-7	−2.0850e3	1.345e0		
$B^{(2)}_{Ca,SO_4}$	−5.704e1	−1.028e-2	−2.235e-4	3.526e-7	5.788e3	−1.8378e0		
C^ϕ_{Fe,SO_4}	2.09e-2							
$B^{(0)}_{Fe,SO_4}$	1.29506e1	−3.86e-2	3.91e-5					
$B^{(1)}_{Fe,SO_4}$	8.758343e1	−6.383194e-1	1.190408e-3		−1.38964e3			
$B^{(2)}_{Fe,SO_4}$	−4.20e1							
C^ϕ_{Na,NO_3}	0.00							
$B^{(0)}_{Na,NO_3}$	7.088889e-1	−5.599812e-3	1.085e-5					
$B^{(1)}_{Na,NO_3}$	8.312979e1	−6.161264e-1	1.1334e-3					
C^ϕ_{K,NO_3}	6.669e-2	−2.0505e-4						
$B^{(0)}_{K,NO_3}$	−2.948e-1	7.326e-4						
$B^{(1)}_{K,NO_3}$	−6.2735e0	2.1149e-2						
C^ϕ_{Mg,NO_3}	−2.062e-2							
$B^{(0)}_{Mg,NO_3}$	5.207e-1	−5.1525e-4						
$B^{(1)}_{Mg,NO_3}$	2.9242e0	−4.4925e-3						
C^ϕ_{Ca,NO_3}	−2.014e-2							
$B^{(0)}_{Ca,NO_3}$	3.687e-1	−5.295e-4						
$B^{(1)}_{Ca,NO_3}$	4.1485e0	−9.1875e-3						
C^ϕ_{Fe,NO_3} [c]	−2.062e-2							
$B^{(0)}_{Fe,NO_3}$ [c]	5.207e-1	−5.1525e-4						
$B^{(1)}_{Fe,NO_3}$ [c]	2.9242e0	−4.4925e-3						
C^ϕ_{Na,HCO_3}	0.00							
$B^{(0)}_{Na,HCO_3}$	−3.7262419e1	−1.445932e-2			6.82885977e2	6.899557e0		

Table B.4. (continued)

Pitzer Parameter	a_1	a_2	a_3	a_4	a_5	a_6	a_7	a_8
$B^{(1)}_{\mathrm{Na,HCO_3}}$	-6.14635193e1	-2.44673e-2			1.129389146e3	1.14108589e1		
$C^{\phi}_{\mathrm{Na,CO_3}}$	5.2e-3							
$B^{(0)}_{\mathrm{Na,CO_3}}$	-6.053877702e1	-2.301655e-2			1.10837601518e3	1.119855531e1		
$B^{(1)}_{\mathrm{Na,CO_3}}$	-2.375156616e2	-9.989121e-2			4.412511973e3	4.4582070301		
$C^{\phi}_{\mathrm{K,HCO_3}}$	0.00							
$B^{(0)}_{\mathrm{K,HCO_3}}$	-3.088232e-1	1.00e-3			-6.9869e-4	-4.70148e-6		
$B^{(1)}_{\mathrm{K,HCO_3}}$	-2.802e-1	1.09999e-3			9.36932e-4	6.15660566e-6		
$C^{\phi}_{\mathrm{K,CO_3}}$	5.0e-4							
$B^{(0)}_{\mathrm{K,CO_3}}$	-1.991649e-1	1.10e-3			1.8063362e-5			
$B^{(1)}_{\mathrm{K,CO_3}}$	1.330579e-1	4.36e-3			1.1899e-3			
$C^{\phi}_{\mathrm{Mg,HCO_3}}$	0.00							
$B^{(0)}_{\mathrm{Mg,HCO_3}}$	1.369710e4	8.250840e0	-4.34e-3		-2.734061716e5	-2.6071152202e3		
$B^{(1)}_{\mathrm{Mg,HCO_3}}$	-1.578398351e5	-9.27779354e1	4.77642e-2		3.20320969486e6	2.992715150e4		
$C^{\phi}_{\mathrm{Ca,HCO_3}}$	0.00							
$B^{(0)}_{\mathrm{Ca,HCO_3}}$	2.957653405e4	1.844730e1	-9.989e-3		-5.765205185e5	-5.6611237e3		
$B^{(1)}_{\mathrm{Ca,HCO_3}}$	-1.028851052e3	-3.72587671e-1	8.9691e-5		2.649224030e4	1.83131556722		
$C^{\phi}_{\mathrm{Fe,HCO_3}}$ c	0.00							
$B^{(0)}_{\mathrm{Fe,HCO_3}}$ c	1.369710e4	8.250840e0	-4.34e-3		-2.734061716e5	-2.6071152202e3		
$B^{(1)}_{\mathrm{Fe,HCO_3}}x^c$	-1.578398351e5	-9.27779354e1	4.77642e-2		3.20320969486e6	2.992715150e4		
$C^{\phi}_{\mathrm{Na,OH}}$	-2.011610e-1	2.492005e-3	-9.411507e-6	1.127798e-8				
$B^{(0)}_{\mathrm{Na,OH}}$	-1.290052e1	1.108975e-1	-3.208817e-4	3.124540e-7				
$B^{(1)}_{\mathrm{Na,OH}}$	-7.698673e0	7.152585e-2	-2.187257e-4	2.290117e-7				
$C^{\phi}_{\mathrm{K,OH}}$	4.1e-3							
$B^{(0)}_{\mathrm{K,OH}}$	1.298e-1							
$B^{(1)}_{\mathrm{K,OH}}$	3.20e-1							
$C^{\phi}_{\mathrm{Ca,OH}}$	0.00							

Table B.4. (continued)

Pitzer Parameter	a_1	a_2	a_3	a_4	a_5	a_6	a_7	a_8
$B^{(0)}_{Ca,OH}$	$-1.747e\text{-}1$							
$B^{(1)}_{Ca,OH}$	$-2.303e\text{-}1$							
$B^{(2)}_{Ca,OH}$	$-5.72e0$							
$C^{\phi}_{MgOH,Cl}$	0.00							
$B^{(0)}_{MgOH,Cl}$	$-1.0e\text{-}1$							
$B^{(1)}_{MgOH,Cl}$	$1.658e0$							
$C^{\phi}_{FeOH,Cl}$ ^{c}	0.00							
$B^{(0)}_{FeOH,Cl}$ ^{c}	$-1.0e\text{-}1$							
$B^{(1)}_{FeOH,Cl}$ ^{c}	$1.658e0$							
$C^{\phi}_{H,Cl}$	$-2.78619e\text{-}3$	$1.735035e\text{-}5$	$-5.937631e\text{-}8$					
$B^{(0)}_{H,Cl}$	$3.56122e\text{-}1$	$-5.305245e\text{-}4$						
$B^{(1)}_{H,Cl}$	$-7.420199e0$	$1.158740e\text{-}1$	$-6.666408e\text{-}4$	$1.685445e\text{-}6$				$-1.564321e\text{-}9$
C^{ϕ}_{H,NO_3}	$1.264791e\text{-}1$	$-8.808517e\text{-}4$	$1.469263e\text{-}6$					
$B^{(0)}_{H,NO_3}$	$-1.019608e0$	$8.052811e\text{-}3$	$-1.414706e\text{-}5$					
$B^{(1)}_{H,NO_3}$	$-6.829838e0$	$7.588276e\text{-}2$	$-2.843828e\text{-}4$	$3.695995e\text{-}7$				
$C^{(0)}_{H,SO_4}$	$-5.315201e\text{-}1$	$8.619821e\text{-}3$	$-5.14571e\text{-}5$	$1.350347e\text{-}7$				$-1.314709e\text{-}10$
$C^{(1)}_{H,SO_4}$	$-7.766817e2$	$1.164255e1$	$-6.537128e\text{-}2$	$1.626702e\text{-}4$				$-1.513403e\text{-}7$
$B^{(0)}_{H,SO_4}$	$1.388618e1$	$-2.200263e\text{-}1$	$1.314763e\text{-}3$	$-3.475869e\text{-}6$				$3.420786e\text{-}9$
$B^{(1)}_{H,SO_4}$	$-1.679737e2$	$2.609701e0$	$-1.526685e\text{-}2$	$3.984122e\text{-}5$				$-3.908818e\text{-}8$
$C^{(0)}_{H,HSO_4}$	$9.773929e\text{-}2$	$-1.148641e\text{-}3$	$4.422169e\text{-}6$	$-5.765707e\text{-}9$				
$C^{(1)}_{H,HSO_4}$	$1.099940e2$	$-1.969903e0$	$1.289248e\text{-}2$	$-3.672285e\text{-}5$				$3.850406e\text{-}8$
$B^{(0)}_{H,HSO_4}$	$-6.888679e\text{-}1$	$1.195453e\text{-}2$	$-4.667571e\text{-}5$	$5.878913e\text{-}8$				
$B^{(1)}_{H,HSO_4}$	$-3.815780e2$	$6.261241e0$	$-3.789588e\text{-}2$	$1.006085e\text{-}4$				$-9.906089e\text{-}8$
C^{ϕ}_{Na,HSO_4}	0.00							

Table B.4. (continued)

Pitzer Parameter	a_1	a_2	a_3	a_4	a_5	a_6	a_7	a_8
$B^{(0)}_{Na,HSO_4}$	4.54e-2							
$B^{(1)}_{Na,HSO_4}$	-6.7010e0	2.5352e-2						
C^{ϕ}_{K,HSO_4}	0.00							
$B^{(0)}_{K,HSO_4}$	-3.0e-4							
$B^{(1)d}_{K,HSO_4}$	1.004759e2	-7.127881e-1	1.262349e-3					
$C^{\phi d}_{Mg,HSO_4}$	3.075223e0	-2.094837e-2	3.566667e-5					
$B^{(0)d}_{Mg,HSO_4}$	1.175741e0	-2.374627e-3						
$B^{(1)d}_{Mg,HSO_4}$	-4.091076e2	2.838782e0	-4.902359e-3					
C^{ϕ}_{Ca,HSO_4}	0.00							
$B^{(0)}_{Ca,HSO_4}$	6.19e-2							
$B^{(1)}_{Ca,HSO_4}$	2.602e0							
C^{ϕ}_{Fe,HSO_4}	0.00							
$B^{(0)}_{Fe,HSO_4}$	6.758464e1	-7.649696e-1	2.894494e-3	-3.636364e-6				
$B^{(1)}_{Fe,HSO_4}$	3.48e0							

[a] See Tables 3.2 and 3.3 for the primary source of these parameters.
[b] Parameters are for the equation: Pitzer parameter = $a_1 + a_2T + a_3T^2 + a_4T^3 + a_5/T + a_6\text{Ln}(T) + a_7/T^2 + a_8T^4$, where T is temperature (K).
[c] Assumed the same as the Mg analogs.
[d] These equations are only valid between 0 and 25°C. At subzero temperatures, use $B^{(1)}_{K,HSO_4} = -0.0371$, $\Psi_{K,H,SO_4} = -0.0064$, $C^{\phi}_{Mg,HSO_4} = 0.0143$, $B^{(0)}_{Mg,HSO_4} = 0.5315$, $B^{(1)}_{Mg,HSO_4} = 0.5361$, $\Psi_{Mg,H,HSO_4} = -0.0234$, and $\Psi_{SO_4,HSO_4,Mg} = -0.0884$.

Table B.5. Ternary Pitzer-equation parameters for cations and anions used in the FREZCHEM model.[a] (Numbers are in computer scientific notation, where e±xx stands for $10^{\pm xx}$)

Pitzer parameter	a_1	a_2	a_3	a_4	a_5	a_6	a_7	a_8
$\Theta_{Na,K}$	−1.8226741e1	−3.6903847e-3			6.1241501e2	3.0299498e0		
$\Psi_{Na,K,Cl}$	6.4810812e0	1.4680346e-3			−2.0435401e2	−1.0944304e0		
Ψ_{Na,K,SO_4}	−5.63e-2	1.4146e-3			−2.566e2	1.8538e-1		
Ψ_{Na,K,NO_3}	−0.0012		2.3e-8	−2.1088e-8				
Ψ_{Na,K,HCO_3}	−0.0079							
Ψ_{Na,K,CO_3}	0.003							
$\Theta_{Na,Mg}$	0.070							
$\Psi_{Na,Mg,Cl}$	−3.109870e-2	5.4464780e-5			1.9940421e0			
Ψ_{Na,Mg,SO_4}	−1.207e-1	5.235e-4	−5.39e-7	−4.39e-10	−1.723e1	1.2645e-2		
Ψ_{Na,Mg,NO_3}	−0.0099							
$\Theta_{Na,Ca}$	3.0e-2	−1.9e-5		9.5e-10	−2.50e0	1.3e-3		
$\Psi_{Na,Ca,Cl}$	−7.63980e0	−1.2990e-2	1.1060e-5			1.8475e0		
Ψ_{Na,Ca,SO_4}	−8.08e-2	4.6565e-3	5.546e-6	−1.4107e-7	−1.0915e3	9.6985e-1		
Ψ_{Na,Ca,NO_3}	−3.481452e-2	1.285714e-4						
$\Theta_{Na,Fe}$	0.080							
$\Psi_{Na,Fe,Cl}$	−0.014							
Ψ^c_{Na,Fe,SO_4}	−1.207e-1	5.235e-4	−5.39e-7	−4.39e-10	−1.723e1	1.2645e-2		
$\Theta_{Na,H}$	0.036							
$\Psi_{Na,H,Cl}$	−0.0037							
Ψ_{Na,H,NO_3}	−7.707985e-2	2.482609e-4						
Ψ_{Na,H,HSO_4}	−0.0129							
$\Theta_{K,Mg}$	0.1167							
$\Psi_{K,Mg,Cl}$	5.0362230e-2	−8.750820e-6	−3.27e-7	−9.37e-10	−2.899090e1	−8.84e-3		
Ψ_{K,Mg,SO_4}	−1.18e-1	−4.78e-5			3.344e1			
Ψ_{K,Mg,NO_3}	−6.494e-1	2.00e-3						
$\Theta_{K,Ca}$	2.365710e0	−4.540e-3			−2.84940e2			
$\Psi_{K,Ca,Cl}$	−5.930e-2	2.54280e-4			−1.34390e1			
Ψ_{K,Ca,NO_3}	−9.011042e-1	2.944255e-3						
$\Theta^c_{K,Fe}$	0.1167							
$\Psi^c_{K,Fe,Cl}$	5.0362230e-2	−8.750820e-6	−3.27e-7	−9.37e-10	−2.899090e1	−8.84e-3		
Ψ^c_{K,Fe,SO_4}	−1.18e-1	−4.78e-5			3.344e1			
$\Theta_{K,H}$	0.005							
$\Psi_{K,H,Cl}$	−0.0114							

208 B Parameter Tables

Table B.5. (continued)

Pitzer parameter	a_1	a_2	a_3	a_4	a_5	a_6	a_7	a_8
Ψ_{K,H,NO_3}	-2.089075e0	1.399737e-2	-2.360221e-5					
Ψ_{K,H,SO_4}	-1.4967e0	5.456e-3						
Ψ_{K,H,HSO_4}	-0.0303							
$\Theta_{Mg,Ca}$	5.31274136e0	-6.342248e-3			-9.83113847e2	-7.4061986e0		
$\Psi_{Mg,Ca,Cl}$	4.15790220e1	1.30377312e-2			-9.8165826e2			
Ψ_{Mg,Ca,SO_4}	0.024							
$\Theta_{Mg,Fe}$	0.0							
$\Psi_{Mg,Fe,Cl}$	0.0							
Ψ_{Mg,Fe,SO_4}	0.0							
$\Theta_{Mg,H}$	0.10							
$\Psi_{Mg,H,Cl}$	-0.0077							
Ψ_{Mg,H,HSO_4}	-4.525314e-2	7.995736e-5						
$\Psi_{Mg,MgOH,Cl}$	0.028							
$\Theta_{Ca,Fe}^c$	5.3127 4136e0	-6.342248e-3			-9.83113847e2	-7.4061986e0		
$\Psi_{Ca,Fe,Cl}^c$	4.15790220e1	1.30377312e-2			-9.8165826e2			
Ψ_{Ca,Fe,SO_4}^c	0.024							
$\Theta_{Ca,H}$	0.092							
$\Psi_{Ca,H,Cl}$	-0.0142							
$\Theta_{Fe,H}$	0.0							
$\Psi_{Fe,H,Cl}$	-1.4157e-1	5.15e-4						
Ψ_{Fe,H,SO_4}^c	0.0							
Ψ_{Fe,H,HSO_4}	5.75716e1	-5.7767e-1	1.924796e-3	-2.129138e-6				
$\Psi_{Fe,FeOH,Cl}^c$	0.028							
Θ_{Cl,SO_4}	7.0e-2				-1.00e0			
$\Psi_{Cl,SO_4,Na}$	2.554e-2	-6.138e-5	-9.0e-9	-7.8e-10	-8.9e-1	-2.275e-3		
$\Psi_{Cl,SO_4,K}$	6.08e-2	-1.824e-4	-2.15e-8	3.04e-10	5.22e0	-3.01e-3		
$\Psi_{Cl,SO_4,Mg}$	5.869e-2	-8.97e-5	4.7e-8	-3.28e-10	-2.413e1	4.345e-3		
$\Psi_{Cl,SO_4,Ca}$	-2.63e-2	-9.46e-5	-3.125e-7	6.5e-11	2.944e1	-6.49e-3		
$\Psi_{Cl,SO_4,Fe}^c$	5.869e-2	-8.97e-5	4.7e-8	-1.28e-9	-2.413e1	4.345e-3		
$\Psi_{Cl,SO_4,H}$	0.0			6.5e-11				
Θ_{Cl,NO_3}	0.016							
$\Psi_{Cl,NO_3,Na}$	-5.773862e-1	3.94084e-3	-6.80e-6					
$\Psi_{Cl,NO_3,K}$	-1.065548e0	6.957931e-3	-1.142531e-5					
$\Psi_{Cl,NO_3,Mg}$	0.0046							

Table B.5. (continued)

Pitzer parameter	a_1	a_2	a_3	a_4	a_5	a_6	a_7	a_8
$\Psi_{Cl,NO_3,Ca}$	−0.0409							
$\Psi_{Cl,NO_3,H}$	0.0							
Θ_{Cl,HCO_3}	0.03							
$\Psi_{Cl,HCO_3,Na}$	−3.9e-3	−3.25e-5	−6.6e-8	−2.74e-10	5.83e0	−9.85e-4		
$\Psi_{Cl,HCO_3,Mg}$	−0.096							
$\Psi^c_{Cl,HCO_3,Fe}$	−0.096							
Θ_{Cl,CO_3}	−0.02							
$\Psi_{Cl,CO_3,Na}$	9.6e-3	4.7e-6	−3.75e-8	−2.23e-10	2.94e0	−5.2e-4		
$\Psi_{Cl,CO_3,K}$	0.004							
Θ_{Cl,HSO_4}	−0.006							
$\Psi_{Cl,HSO_4,Na}$	−0.006							
$\Psi_{Cl,HSO_4,H}$	0.013							
Θ_{NO_3,SO_4}	2.559033e0	−1.679067e-2	2.892045e-5					
$\Psi_{NO_3,SO_4,Na}$	−2.222887e-1	1.0313e-3	−1.0e-6					
$\Psi_{NO_3,SO_4,K}$	−4.715997e0	3.156712e-2	−5.304224e-5					
$\Psi_{NO_3,SO_4,Mg}$	−2.9392e-1	9.60e-4						
$\Psi_{NO_3,SO_4,Ca}$	−0.2281							
$\Psi_{NO_3,SO_4,H}$	−3.75677e-1	3.241125e-3	−8.920938e-6	7.36791e-9				
$\Psi_{NO_3,HCO_3,Na}$	0.0061							
$\Psi_{NO_3,CO_3,Na}$	0.0132							
$\Psi_{NO_3,CO_3,K}$	0.0075							
Θ_{HCO_3,SO_4}	0.01							
$\Psi_{HCO_3,SO_4,Na}$	−0.005							
$\Psi_{HCO_3,SO_4,Mg}$	−0.161							
$\Psi^c_{HCO_3,SO_4,Fe}$	−0.161							
Θ_{HCO_3,CO_3}	−0.04							
$\Psi_{HCO_3,CO_3,Na}$	0.002							
$\Psi_{HCO_3,CO_3,K}$	0.012							
Θ_{CO_3,SO_4}	0.02							
$\Psi_{CO_3,SO_4,Na}$	3.9e-3	3.73e-5	1.64e-7	6.16e-10	−1.822e1	3.455e-3		
$\Psi_{CO_3,SO_4,K}$	−0.009							
Θ_{NO_3,HSO_4}	4.151797e0	−4.445727e-2	1.57303e-4	−1.876457e-7				
$\Psi_{NO_3,HSO_4,H}$	−6.187135e0	9.307292e-2	−5.252546e-4	1.316116e-6				
$\Psi_{SO_4,HSO_4,Na}$	−0.0094							

Table B.5. (continued)

Pitzer parameter	Equation parameter[b]							
	a_1	a_2	a_3	a_4	a_5	a_6	a_7	a_8
$\Psi_{SO_4,HSO_4,K}$	−0.0677							
$\Psi_{SO_4,HSO_4,Mg}$	−7.656214e-1	2.4871e-3						
$\Psi_{SO_4,HSO_4,Fe}$	0.0							
$\Theta_{OH,Cl}$	−0.050							
$\Psi_{OH,Cl,Na}$	−0.006							
$\Psi_{OH,Cl,K}$	−0.006							
$\Psi_{OH,Cl,Ca}$	−0.025							
Θ_{OH,SO_4}	−0.013							
$\Psi_{OH,SO_4,Na}$	−0.009							
$\Psi_{OH,SO_4,K}$	−0.050							
Θ_{OH,CO_3}	0.10							
$\Psi_{OH,CO_3,Na}$	−0.017							
$\Psi_{OH,CO_3,K}$	−0.01							

[a] See Tables 3.2 and 3.3 for the primary source of these parameters.
[b] Parameters are for the equation: Pitzer parameter $= a_1 + a_2 T + a_3 T^2 + a_4 T^3 + a_5/T + a_6 \text{Ln}(T) + a_7/T^2 + a_8 T^4$, where T is temperature (K).
[c] Assumed the same as the Mg analogs.

Table B.6. Binary and ternary soluble gas Pitzer-equation parameters used in the FREZCHEM model.[a] (Numbers are in computer scientific notation where e±xx stands for $10^{\pm xx}$)

Pitzer parameter	a_1	a_2	a_3	a_4	a_5	a_6	a_7	a_8
$\lambda_{CO_2,H}$	0.0							
$\lambda_{CO_2,Na}$	−5.49638465e3	−3.326566e0		1.7532e-3	1.09399341e5	1.047021567e3		
$\lambda_{CO_2,K}$	2.856528099e3	1.7670079e0	−9.487e-4		−5.59541929e4	−5.46074467e2		
$\lambda_{CO_2,Mg}$	−4.79362533e2	−5.41843e-1	3.8812e-4		3.589474052e3	1.043452732e2		
$\lambda_{CO_2,Ca}$	−1.27746472e4	−8.101555e0	4.42472e-3		2.455415435e5	2.452509720e3		
$\lambda^c_{CO_2,Fe}$	−4.79362533e2	−5.41843e-1	3.8812e-4		3.589474052e3	1.043452732e2		
$\lambda_{CO_2,Cl}$	1.659449942e3	9.964326e-1	−5.2122e-4		−3.31591774e4	−3.15827883e2		
λ_{CO_2,SO_4}	2.274656591e3	1.8270948e0	−1.14272e-3		−3.39277625e4	−4.57015738e2		
$\zeta_{CO_2,H,Cl}$	−8.04121738e2	−4.70474e-1	2.40526e-4		1.633438917e4	1.523838752e2		
$\zeta_{CO_2,Na,Cl}$	−3.794591785e2	−2.58005e-1	1.47823e-4		6.879030871e3	7.374511574e1		
$\zeta_{CO_2,K,Cl}$	−3.79686097e2	−2.57891e-1	1.47333e-4		6.853264129e3	7.37977116e1		
$\zeta_{CO_2,Mg,Cl}$	−1.34260256e3	−7.72286e-1	3.91603e-4		2.772680974e4	2.536231940 6e2		
$\zeta_{CO_2,Ca,Cl}$	−1.66065290e2	−1.8002e-2	−2.47349e-5		5.256844332e3	2.737745241 5e1		
$\zeta^c_{CO_2,Fe,Cl}$	−1.34260256e3	−7.72286e-1	3.91603e-4		2.772680974e4	2.536231940 6e2		
ζ_{CO_2,H,SO_4}	0.0							
ζ_{CO_2,Na,SO_4}	6.70300 2482e4	3.7930519e1	−1.894730e-2		−1.39908237e6	−1.263027457e4		
ζ_{CO_2,K,SO_4}	−2.90703326e3	−2.860763e0	1.951086e-3		3.075686749e4	6.1137560512e2		
ζ_{CO_2,Mg,SO_4}	−7.37424392e3	−4.608331e0	2.489207e-3		1.431626076e5	1.412302898e3		
$\zeta^c_{CO_2,Ca,SO_4}$	0.0							
$\zeta^c_{CO_2,Fe,SO_4}$	−7.37424392e3	−4.608331e0	2.489207e-3		1.431626076e5	1.412302898e3		
$\lambda_{O_2,H}$	−2.379e-1				8.1450e1			
$\lambda_{O_2,Na}$	−3.9548e-1		9.19882e-7		1.41307e2			
$\lambda_{O_2,K}$	−5.1698e-1				1.99431e2			
$\lambda_{O_2,Mg}$	−7.9489e-1				3.05513e2			
$\lambda_{O_2,Ca}$	2.497e-1							
$\lambda^c_{O_2,Fe}$	−7.9489e-1				3.05513e2			
$\lambda_{O_2,Cl}$	0.0							
λ_{O_2,SO_4}	1.00706e0				−2.74085e2			
$\lambda_{O_2,OH}$	9.3318e-1				−4.30552e2		4.98608e4	
λ_{O_2,NO_3}	−3.77e-2							

Table B.6. (continued)

Pitzer parameter	a_1	a_2	a_3	a_4	a_5	a_6	a_7	a_8
λ_{O_2,HCO_3}	8.54e-2							
λ_{O_2,CO_3}	1.0258e0				-2.77074e2			
$\zeta_{O_2,H,Cl}$	-7.7e-3				-2.739e0			
$\zeta_{O_2,Na,Cl}$	-2.11e-2							
$\zeta_{O_2,K,Cl}$	-5.65e-3							
$\zeta_{O_2,Mg,Cl}$	-1.69e-2							
$\zeta_{O_2,Ca,Cl}$	-5.65e-3							
$\zeta_{O_2,Fe,Cl}^c$	-4.60e-2							
ζ_{O_2,Na,SO_4}	0.0							
ζ_{O_2,K,SO_4}	0.0							
ζ_{O_2,Mg,SO_4}	0.0							
ζ_{O_2,Fe,SO_4}^c	0.0							
ζ_{O_2,Na,NO_3}	-1.20e-2							
ζ_{O_2,K,NO_3}	-2.81e-2							
ζ_{O_2,Ca,NO_3}	0.0							
$\zeta_{O_2,Na,OH}$	-1.25e-2							
$\zeta_{O_2,K,OH}$	2.342e-3						-8.3615e2	
ζ_{O_2,Na,HCO_3}	0.0							
ζ_{O_2,Na,CO_3}	-1.81e-2							
$\lambda_{CH_4,Na}^d$	9.92230792e-2	2.57906811e-5						
$\lambda_{CH_4,K}$	0.13909							
$\lambda_{CH_4,Mg}$	0.24678							
$\lambda_{CH_4,Ca}^e$	-5.64278808e0	8.51392725e-3			1.00057752e3			
$\lambda_{CH_4,Fe}^c$	0.24678							
$\lambda_{CH_4,Cl}$	0.00							
λ_{CH_4,SO_4}	0.03041							
λ_{CH_4,HCO_3}	0.00669							
λ_{CH_4,CO_3}	0.16596							

Table B.6. (continued)

Pitzer parameter	Equation parameter[b]							
	a_1	a_2	a_3	a_4	a_5	a_6	a_7	a_8
$\zeta_{CH_4,Na,Cl}$	−0.00624							
$\zeta_{CH_4,K,Cl}$	−0.00382							
$\zeta_{CH_4,Mg,Cl}$	−0.01323							
$\zeta_{CH_4,Ca,Cl}$	−0.00468							
$\zeta_{CH_4,Fe,Cl}$	−0.01323							

[a] See Tables 3.2 and 3.3 for the primary source of these parameters.
[b] Parameters are for the equation: Pitzer parameter $= a_1 + a_2T + a_3T^2 + a_4T^3 + a_5/T + a_6\mathrm{Ln}(T) + a_7/T^2 + a_8T^4$, where T is temperature (K).
[c] Assumed the same as the Mg analogs.
[d] This equation also contains the terms $+1.834514 02\mathrm{e}{-2} \times xP/T - 8.07196716\mathrm{e}{-6} \times P^2/T$, where P is pressure (bars) and T is temperature (K).
[e] This equation also contains the term $+5.27816886\mathrm{e}{-5} \times P$, where P is pressure (bars).

Table B.7. Equations for the molar volumes (cm^3/mole) of liquid water and water ice at 1.01 bar pressure and their compressibilities [cm^3/(mol·bar)] (Marion et al. 2005), where T is temperature (K) and P is pressure (bars)

$1000\rho_w^0 = (999.83952 + 16.94518 \cdot t - 7.98704 \times 10^{-3} \cdot t^2$
$ - 4.617046 \times 10^{-5} \cdot t^3 + 1.05563 \times 10^{-7} \cdot t^4 \quad\quad (t = °C)$
$ - 2.805425 \times 10^{-10} \cdot t^5)/(1.0 + 1.687985 \times 10^{-2} \cdot t)$

$\overline{V}_l^0 = \dfrac{18.01528}{\rho_w^0} \quad\quad T \geq 273.15 \text{ K}$

$\overline{V}_l^0 = 18.0182 - 1.407964 \times 10^{-3}(T - 273.15) + 1.461418 \times 10^{-4}(T - 273.15)^2$
$\phantom{\overline{V}_l^0 =} T < 273.15 \text{ K}$

$\overline{K}_l^0 = 8.62420 \times 10^{-3} - 5.06616 \times 10^{-5}T - 3.78615 \times 10^{-6}P + 2.27679 \times 10^{-8}PT$
$\phantom{\overline{K}_l^0 =} + 1.18481 \times 10^{-10}P^2 + 8.20762 \times 10^{-8}T^2$
$\phantom{\overline{K}_l^0 =} - 3.63261 \times 10^{-11}PT^2 - 2.87099 \times 10^{-13}TP^2$

$\overline{V}_{\text{I,cr}}^0 = 19.30447 - 7.988471 \times 10^{-4}T + 7.563261 \times 10^{-6}T^2$

$\overline{K}_{\text{I,cr}}^0 = 2.790102 \times 10^{-2} - 2.235440 \times 10^{-4} \cdot T + 4.497731 \times 10^{-7} \cdot T^2$

Table B.8. Equations for the molar volumes (cm^3/mole) and the isothermal compressibilities [cm^3/(mole.bar)] of soluble ions and gases at infinite dilution. Derived from the database of Millero (1983) over a temperature range of 273 to 298 K. (Numbers are in computer scientific notation, where e±xx stands for $10^{\pm xx}$). Reprinted from Marion et al. (2005) with permission

	Equation parameter[a]		
Molar volume	a_1	a_2	a_3
H$^+$	0.0	0.0	0.0
Na$^+$	−9.0589e1	5.2876e−1	−7.68e−4
K$^+$	−7.2019e1	4.9128e−1	−7.36e−4
Mg^{2+}	−1.65407e2	9.6088e−1	−1.600e−3
Fe^{2+}[b]	−1.65407e2	9.6088e−1	−1.600e−3
Ca^{2+}	−1.42301e2	7.9195e−1	−1.256e−3
Cl$^-$	−1.02412e2	7.8012e−1	−1.264e−3
OH$^-$	−3.34884e2	2.16412e0	−3.536e−3
HSO$_4^-$	−1.85898e2	1.41308e0	−2.248e−3
NO$_3^-$	−1.74089e2	1.27266e0	−1.984e−3
HCO$_3^-$	−1.97188e2	1.41308e0	−2.248e−3
CO$_3^{2-}$	−3.93904e2	2.52016e0	−4.064e−3
SO$_4^{2-}$	−3.52433e2	2.36431e0	−3.808e−3
CO$_2$	33.6		
O$_2$	32.0		
CH$_4$	37.4		
Isothermal compressibility			
H$^+$	0.0	0.0	0.0
Na$^+$	−3.0891e−2	1.36923e−4	−1.56e−7
K$^+$	−1.0616e−2	7.613e−6	5.6e−8
Mg^{2+}	6.2899e−2	−4.8948e−4	8.44e−7
Fe^{2+}[b]	6.2899e−2	−4.8948e−4	8.44e−7
Ca^{2+}	6.3299e−2	−4.8948e−4	8.44e−7
Cl$^-$	−1.5576e−1	1.0101e−3	−1.644e−6
OH$^-$	−1.2874e−1	7.6471e−4	−1.160e−6
HSO$_4^-$	−1.2737e−1	7.6471e−4	−1.160e−6
NO$_3^-$	−7.1475e−2	4.2295e−4	−6.08e−7
HCO$_3^-$	−1.2763e−1	7.6471e−4	−1.160e−6
CO$_3^{2-}$	−3.0024e−1	1.8935e−3	−3.056e−6
SO$_4^{2-}$	−2.9963e−1	1.8935e−3	−3.056e−6

[a] $Y = a_1 + a_2 T + a_3 T^2$, where T is temperature (K).
[b] Assumed to be the same as Mg^{2+}.

Table B.9. Molar volumes (cm^3/mole) of solids at 298 K. These molar volumes were derived from (atomic weight/density), except for footnoted entries (Marion et al. 2005)

Solid phase	Molar volume	Solid phase	Molar volume
NaCl(cr) (halite)	27.02	NaHCO$_3$(cr) (nahcolite)	38.91
NaCl·2H$_2$O(cr) (hydrohalite)	57.96	Na$_2$CO$_3$·7H$_2$O(cr)	153.71
KCl(cr) (sylvite)	37.52	Na$_2$CO$_3$·10H$_2$O(cr) (natron)	198.71
MgCl$_2$·6H$_2$O(cr) (bischofite)	129.57	NaHCO$_3$·Na$_2$CO$_3$·2H$_2$O (trona)	107.02
MgCl$_2$·8H$_2$O(cr)[a]	159.08	KHCO$_3$(cr) (kalicinite)	46.14
MgCl$_2$·12H$_2$O(cr)[a]	218.10	MgCO$_3$(cr) (magnesite)	28.02
FeCl$_2$·4H$_2$O(cr)	103.01	MgCO$_3$·3H$_2$O(cr) (nesquehonite)	74.79
FeCl$_2$·6H$_2$O(cr)[a]	133.65	MgCO$_3$·5H$_2$O(cr) (lansfordite)	100.80
CaCl$_2$·6H$_2$O(cr) (antarcticite)	128.12	3MgCO$_3$·Mg(OH)$_2$·3H$_2$O(cr) (hydromagnesite)	169.13
KMgCl$_3$·6H$_2$O(cr) (carnallite)	172.58	FeCO$_3$(cr) (siderite)	30.49
CaCl$_2$2MgCl$_2$·12H$_2$O(cr) (tachyhydrite)	311.81	CaCO$_3$(cr) (calcite)	36.93
Na$_2$SO$_4$(cr) (thenardite)	53.33	CaCO$_3$(cr) (aragonite)	34.15
Na$_2$SO$_4$·10H$_2$O(cr) (mirabilite)	219.80	CaCO$_3$(cr) (vaterite)	37.72
K$_2$SO$_4$(cr) (arcanite)	65.50	CaCO$_3$·6H$_2$O(cr) (ikaite)	117.54
MgSO$_4$·H$_2$O(cr) (kieserite)	56.60	CaMg(CO$_3$)$_2$(cr) (dolomite)	64.34
MgSO$_4$·6H$_2$O(cr) (hexahydrite)	132.58	HCl·3H$_2$O(cr)[a]	62.51
MgSO$_4$·7H$_2$O(cr) (epsomite)	146.71	HCl·6H$_2$O(cr)[a]	101.06
MgSO$_4$·12H$_2$O(cr)[a]	220.50	HNO$_3$·3H$_2$O(cr)[b]	73.09
FeSO$_4$·H$_2$O(cr) (szomolnokite)	57.21	H$_2$SO$_4$·4H$_2$O(cr)[a]	113.98
FeSO$_4$·7H$_2$O(cr) (melanterite)	146.48	H$_2$SO$_4$·6.5H$_2$O(cr)[a]	154.86
CaSO$_4$(cr) (anhydrite)	45.94	NaHSO$_4$·H$_2$O(cr)	65.65
CaSO$_4$·2H$_2$O(cr) (gypsum)	74.69	KHSO$_4$(cr) (mercallite)	58.64
Na$_2$SO$_4$3K$_2$SO$_4$(cr) (aphthitalite)	246.24	Fe(OH)$_3$(cr) (ferrihydrite)	28.12
Na$_2$SO$_4$·MgSO$_4$·4H$_2$O(cr) (bloedite)	149.98	CH$_4$·6H$_2$O(cr)	135.75
MgSO$_4$·K$_2$SO$_4$·6H$_2$O(cr) (picromerite)	191.78	CO$_2$·6H$_2$O(cr)	135.75
NaNO$_3$(cr) (nitratine)	37.60		
KNO$_3$(cr) (niter)	48.04		
NaNO$_3$·Na$_2$SO$_4$·2H$_2$O(cr) (darapskite)	119.57		

[a] Calculated from linear fits to degree of hydration.
[b] HNO$_3$·3H$_2$O = HCl·3H$_2$O + NaNO$_3$ − NaCl.

Table B.10. Volumetric Pitzer-equation parameters for ion interactions (Marion et al. 2005). (Numbers are in computer scientific notation, where e±xx stands for $10^{\pm xx}$). Reprinted from Marion et al. (2005) with permission

Pitzer parameter	Equation parameter[a]			Temperature range (K)
	a_1	a_2	a_3	
$B^{(0)V}_{Na,Cl}$	1.088468e-4	−3.2412e-7		
$B^{(1)V}_{Na,Cl}$	6.193806e-3	−4.135834e-5	6.911453e-8	252.6–298.15
$C^{V}_{Na,Cl}$	−5.174534e-6	1.514940e-8		
$B^{(0)V}_{K,Cl}$	6.593033e-5	−1.782e-7		
$B^{(1)V}_{K,Cl}$	4.987530e-3	−3.224932e-5	5.215803e-8	273.15–298.15
$C^{V}_{K,Cl}$	−7.112e-7			
$B^{(0)V}_{Mg,Cl}$	6.446574e-4	−4.168596e-6	6.920e-9	
$B^{(1)V}_{Mg,Cl}$	1.624497e-3	−5.618320e-6		273.15–298.15
$C^{V}_{Mg,Cl}$	−5.567e-7			
$B^{(0)V}_{Ca,Cl}$	1.409126e-4	−4.293e-7		
$B^{(1)V\,b}_{Ca,Cl}$	7.702445e-2	−7.785779e-4	2.612103e-6	236.9–298.15
$C^{V}_{Ca,Cl}$	−6.723184e-6	2.216e-8		
$B^{(0)V}_{Fe,Cl}$	1.19712e-4	−2.42e-7		
$B^{(1)V}_{Fe,Cl}$	−2.40397e-3	1.80e-6		288.15–298.15
$C^{V}_{Fe,Cl}$	0.00			
$B^{(0)V}_{Na,SO_4}$	5.589854e-4	−1.698e-6		
$B^{(1)V}_{Na,SO_4}$	3.323357e-2	−2.218483e-4	3.717169e-7	273.15–298.15
C^{V}_{Na,SO_4}	−6.561572e-5	2.106e-7		
$B^{(0)V}_{K,SO_4}$	−3.303430e-3	1.1125e-5		
$B^{(1)V}_{K,SO_4}$	1.122179e-2	−3.6788e-5		273.15–298.15
C^{V}_{K,SO_4}	1.741518e-3	−5.791e-6		
$B^{(0)V}_{Mg,SO_4}$	1.789039e-4	−4.330e-7		
$B^{(1)V}_{Mg,SO_4}$	3.157137e-2	−2.055400e-4	3.358883e-7	273.15–298.15
$B^{(2)V}_{Mg,SO_4}$	9.096288e-2	−2.570e-4		
C^{V}_{Mg,SO_4}	8.232980e-6	−2.629e-8		
$B^{(0)V}_{Fe,SO_4}$	5.246895e-4	−1.196667e-6		
$B^{(1)V}_{Fe,SO_4}$	−3.368357e-3	5.846667e-6		288.15–291.15
$B^{(2)V}_{Fe,SO_4}$	0.00			
C^{V}_{Fe,SO_4}	0.00			
$B^{(0)V}_{Na,HCO_3}$	2.630581e-2	−1.820515e-4	3.141869e-7	
$B^{(1)V}_{Na,HCO_3}$	−9.485264e-2	6.639475e-4	−1.157075e-6	273.15–298.15
C^{V}_{Na,HCO_3}	1.374e-5			
$B^{(0)V}_{Na,CO_3}$	2.567300e-4	−6.597813e-7		
$B^{(1)V}_{Na,CO_3}$	1.488352e-2	−9.170975e-5	1.410818e-7	273.15–298.15
C^{V}_{Na,CO_3}	−6.326112e-6	1.030e-8		

Table B.10. (continued)

Equation parameter[a]

Pitzer parameter	a_1	a_2	a_3	Temperature range (K)
$B^{(0)V}_{K,HCO_3}$	−7.241e-5			
$B^{(1)V}_{K,HCO_3}$	7.527977e-2	−5.051817e-4	8.503591e-7	273.15–298.15
C^{V}_{K,HCO_3}	3.9791e-5			
$B^{(0)V}_{K,CO_3}$	1.061377e-4	−2.348475e-7		
$B^{(1)V}_{K,CO_3}$	2.307677e-2	−1.421728e-4	2.179648e-7	273.15–298.15
C^{V}_{K,CO_3}	−8.468e-7			
$B^{(0)V}_{Na,NO_3}$	6.772663e-4	−4.469461e-6	7.432663e-9	
$B^{(1)V}_{Na,NO_3}$	6.112177e-3	−3.714953e-5	5.670603e-8	273.15–298.15
C^{V}_{Na,NO_3}	0.00			
$B^{(0)V}_{K,NO_3}$	4.017245e-3	−2.727479e-5	4.639794e-8	
$B^{(1)V}_{K,NO_3}$	−9.970759e-3	7.222127e-5	−1.295960e-7	273.15–298.15
C^{V}_{K,NO_3}	0.00			
$B^{(0)V}_{Mg,NO_3}$	1.921787e-3	−1.276510e-5	2.135779e-8	
$B^{(1)V}_{Mg,NO_3}$	2.306674e-3	−8.051390e-6		273.15–298.15
C^{V}_{Mg,NO_3}	0.00			
$B^{(0)V}_{Ca,NO_3}$	2.034795e-4	−6.169134e-7		
$B^{(1)V}_{Ca,NO_3}$	2.100958e-3	−7.119747e-6		279.15–298.15
C^{V}_{Ca,NO_3}	0.00			
$B^{(0)V}_{Fe,NO_3}$	1.05448e-3			
$B^{(1)V}_{Fe,NO_3}$	−5.97153e-3			293.15
C^{V}_{Fe,NO_3}	0.0			
$B^{(0)V}_{H,Cl}$	1.710768e-4	−1.080235e-6	1.693469e-9	
$B^{(1)V}_{H,Cl}$	0.00			268.15–298.15
$C^{V}_{H,Cl}$	−5.584006e-6	3.505464e-8	−5.380952e-11	
$B^{(0)V}_{H,NO_3}$	2.908819e-4	−1.909668e-6	3.163571e-9	
$B^{(1)V}_{H,NO_3}$	2.191849e-3	−1.421625e-5	2.255071e-8	273.15–298.15
C^{V}_{H,NO_3}	0.00			
$B^{(0)V}_{H,HSO_4}$	9.447669e-4	−6.127516e-6	9.939028e-9	
$B^{(1)V}_{H,HSO_4}$	−2.229583e-2	1.474382e-4	−2.418969e-7	
C^{V}_{H,HSO_4}	0.00			273.15–298.15
$B^{(0)V}_{H,SO_4}$	5.374354e-4	−3.519066e-6	5.951988e-9	
$B^{(1)V}_{H,SO_4}$	1.326979e-2	−7.288212e-5	1.010339e-7	
C^{V}_{H,SO_4}	0.00			

Table B.10. (continued)

Pitzer parameter	Equation parameter[a]			Temperature range (K)
	a_1	a_2	a_3	
$B^{(0)V}_{\text{Na,HSO}_4}$	7.848698e-3	−5.199439e-5	8.721005e-8	
$B^{(1)V}_{\text{Na,HSO}_4}$	−3.960782e-2	2.627059e-4		273.15–298.15
$C^{V}_{\text{Na,HSO}_4}$	0.00			
$B^{(0)V}_{\text{K,HSO}_4}$	1.199133e-2	−7.845685e-5	1.287364e-7	
$B^{(1)V}_{\text{K,HSO}_4}$	−4.699762e-2	3.063952e-4	−5.001161e-7	273.15–298.15
$C^{V}_{\text{K,HSO}_4}$	0.00			

[a] Fitted to $Y = a_1 + a_2 T + a_3 T^2$, where Y = Pitzer-equation volumetric parameter and T = temperature (K).
[b] This equation contains a fourth term: $-2.909328\text{e-}9 T^3$.

Table B.11. Pressure-dependent parameters for Eqs. 3.23 and 3.24

Chemistry	T range (K)	P range (bars)	$K_0^{B^{(1)\nu}}$	$K_0^{C\nu}$	References
NaCl[a]	273.15 to 298.15	1 to 1002	= 1.644134e-6 − 5.245763e-9 T	= −5.658952e-8 + 1.896576e-10 T	Chen et al. (1980); Rogers and Pitzer (1982); Gates and Wood (1985)
KCl	298.15	1 to 406	= 8.78e-8	= −1.969e-9	Chen et al. (1980); Gates and Wood (1985)
MgCl$_2$	298.15	1 to 406	= 1.568e-7	= 1.062e-9	Chen et al. (1980); Gates and Wood (1985)
CaCl$_2$	298.15	1 to 407	= 1.447e-7	= 5.18e-10	Gates and Wood (1985)
Na$_2$SO$_4$	273.15 to 298.15	1 to 1002	= 5.585892e-6 − 1.7572e-8 T	= −5.890e-7 + 1.78424e-9 T	Chen et al. (1980)
MgSO$_4$	273.15 to 303.15	1 to 1014	= 3.315e-7	= 6.57e-9	Chen et al. (1980); Hogenboom et al. (1995); Hogenboom 2005 (pers. comm.)

[a] If Na $\geq$ 3.4 m, then $K_0^{B^{(1)\nu}} = 8.11$e-8, $K_0^{C\nu} = 2.41$e-10.

Table B.12. Parameters for Duan et al. (1992b) gas fugacity model (Eqs. 3.37–3.48). (Numbers are in computer scientific notation where e±xx stands for $10^{\pm xx}$)

	Gases		
Parameter	$CO_2(g)$	$CH_4(g)$	$H_2O(g)$[a]
T_c	304.20	190.60	647.25
P_c	73.825	46.41	221.19
a_1	8.99288497e-2	8.72553928e-2	8.64449220e-2
a_2	−4.94783127e-1	−7.52599476e-1	−3.96918955e-1
a_3	4.77922245e-2	3.75419887e-1	−5.73334886e-2
a_4	1.03808883e-2	1.07291342e-2	−2.93893e-4
a_5	−2.82516861e-2	5.49626360e-3	−4.15775512e-3
a_6	9.49887563e-2	−1.84772802e-2	1.99496791e-2
a_7	5.20600880e-4	3.18993183e-4	1.18901426e-4
a_8	−2.93540971e-4	2.11079375e-4	1.55212063e-4
a_9	−1.77265112e-3	2.01682801e-5	−1.06855859e-4
a_{10}	−2.51101973e-5	−1.65606189e-5	−4.93197687e-6
a_{11}	8.93353441e-5	1.19614546e-4	−2.73739155e-6
a_{12}	7.88998563e-5	−1.08087289e-4	2.65571238e-6
α	−1.66727022e-2	4.48262295e-2	8.96079018e-3
B	1.398e0	7.5397e-1	4.02e0
γ	2.96e-2	7.7167e-2	2.57e-2

[a] These parameters for $H_2O(g)$ are not currently being used in the FREZCHEM model.

References

Abe F, Kato C, Horikoshi K (1999) Pressure-regulated metabolism in microorganisms. Trends Microbiol 7:447–453

Adams LH, Gibson RE (1930) The melting curve of sodium chloride dihydrate. An experimental study of an incongruent melting at pressures up to twelve thousand atmospheres. J Am Chem Soc 52:4252–4264

Adisamito S, Frank RJ, III, Sloan Jr ED (1991) Hydrates of carbon dioxide and methane mixtures. J Chem Eng Data 36:68–71

Akasofu S (1999) Auroral spectra as a tool for detecting extraterrestrial life. EOS 35:397

Albrecht A, Skordis C (2000) Phenomenology of a realistic accelerating Universe using only Planck-scale physics. Phys Rev Lett 84:2076–2079

Ananthaswamy J, Atkinson G (1984) Thermodynamics of concentrated electrolyte mixtures. 4. Pitzer-Debye-Hückel limiting slopes for water from 0 to $100\,°C$ and from $1\,atm$ to $1\,kbar$. J Chem Eng Data 29:81–87

Anderson JD, Schubert G, Jacobson RA, Lau EL, Moore WB, Sjogren WL (1998) Europa's differentiated internal structure: inferences from four Galileo encounters. Science 281:2019–2022

Archer DG, Rard JA (1998) Isopiestic investigation of the osmotic and activity coefficients of aqueous $MgSO_4$ and the solubility of $MgSO_4 \cdot 7H_2O(cr)$ at $298.15\,K$: thermodynamic properties of the $MgSO_4 + H_2O$ system to $440\,K$. J Chem Eng Data 43:791–806

Archer DG, Wang P (1990) The dielectric constant of water and Debye-Hückel limiting law slopes. J Phys Chem Ref Data 19:371–411

Arrhenius G, Lepland A (2000) Accretion of Moon and Earth and the emergence of life. Chem Geol 169:69–82

Assur A (1958) Composition of sea ice and its tensile strength. In: Arctic Sea Ice, Publication 598, National Academy of Sciences-National Research Council, pp 106–138

Bachofen R (1986) Microorganisms in extreme environments: introduction. Experientia 42:1179–1182

Bakker RJ, Dubessy J, Cathelineau M (1996) Improvements in clathrate modelling: I. The H_2O-CO_2 system with various salts. Geochim Cosmochim Acta 60:1657–1681

Banin A, Han FX, Kan I, Cicelsky A (1997) Acidic volatiles and the Mars soil. J Geophys Res 102:13,341–13,356
Baird AK, Clark BC (1981) On the original igneous source of Martian fines. Icarus 45:113–123
Barr AC, Pappalardo RT, Nimmo F (2002) Shear heating and strike-slip motion on Europa: implications for a radiation-driven ecosystem. In: Greeley R (ed) Europa focus group workshop 3, pp 1–2. Available at: http://astrobiology.asu.edu/focus/europa/intro.html
Baum SK, Crowley TJ (2003) The snow/ice instability as a mechanism for rapid climate change: a Neoproterozoic Snowball Earth model example. Geophys Res Lett 30, doi:10.1029/2003GL017333, 2003
Baumstark-Khan C, Facius R (2002) Life under conditions of ionizing radiation. In: Horneck G, Baumstark-Khan C (eds) Astrobiology: the Quest for the Conditions of Life. Springer, Berlin Heidelberg New York, pp 261–284
Bernard A, Symonds RB (1989) The significance of siderite in the sediments from Lake Nyos, Cameroon. J Volcanol Geotherm Res 39:187–194
Bischoff JL, Fitzpatrick JA, Rosenbauer RJ (1993) The solubility and stabilization of ikaite ($CaCO_3 \cdot 6H_2O$) from 0 to 25 °C: environmental and paleoclimatic implications for thinolite tufa. J Geol 101:21–33
Blaney DL, McCord TB (1989) An observational search for carbonates on Mars. J Geophys Res 94:10,159–10,166
Blöchl E, Rachel R, Burggraf S, Hafenbradl D, Jannasch HW, Stetter KO (1997) *Pyrolobus fumarii*, gen. and sp. nov., represents a novel group of archaea, extending the upper temperature limit for life to 113 °C. Extremophiles 1:14–21
Blunier T (2000) "Frozen" methane escapes from the sea floor. Science 288:68–69
Bodiselitsch B, Koeberl C, Master S, Reimold WU (2005) Estimating duration and intensity of Neoproterozoic snowball glaciations from Ir anomalies. Science 308:239–242
Boesgaard AM, Steigman G (1985) Big-Bang nucleosynthesis – theories and observations. Ann Rev Astron Astrophys 23:319–378
Brantner B, Fierlinger H, Puxbaum H, Berner A (1994) Cloudwater chemistry in the subcooled droplet regime at Mount Sonnblick (3106 m a.s.l., Salzburg, Austria). Water Air Soil Pollut 74:363–384
Breezee J, Cady N, Staley JT (2004) Subfreezing growth of the sea ice bacterium, Psychromonas ingramii. Microb Ecol 47:300–304
Bridges JC, Catling DC, Saxton JM, Swindle TD, Lyon IC, Grady MM (2001) Alteration assemblages in Martian meteorites: Implications for near-surface processes. Space Sci Rev 96:365–392
Bridges JC, Grady MM (2000) Evaporite mineral assemblages in the nakhlite (Martian) meteorites. Earth Planet Sci Lett 176:267–280
Burns RG (1993) Rates and mechanisms of chemical weathering of ferromagnesian silicate minerals on Mars. Geochim Cosmochim Acta 57:4555–4574

Buseck PR, Schwartz SE (2004) Tropospheric aerosols. In: Holland HD, Turekian KK (eds) Treatise on geochemistry, Vol. 4, The atmosphere [Keeling RF (ed)] Elsevier, Amsterdam, pp 91–142

Cabrol NA, Wynn-Williams DD, Crawford DA, Grin EA (2001) Recent aqueous environments in Martian impact craters: an astrobiological perspective. Icarus 154:98–112

Calvin WM, King TV, Clark RN (1994) Hydrous carbonates on Mars? Evidence from Mariner infrared spectrometer and ground-based telescopic spectra. J Geophys Res 99:14,659–14,675

Canfield DE, Teske A (1996) Late Proterozoic rise in atmospheric oxygen concentration inferred from phylogenetic and sulphur-isotope studies. Nature 382:127–132

Cano RJ, Borucki MK (1995) Revival and identification of bacterial spores in 25- to 40-million-year-old Dominican amber. Science 268:1060–1064

Carlson RW, Anderson MS, Johnson RE, Smythe WD, Hendrix AR, Barth CA, Soderblom LA, Hansen GB, McCord TB, Dalton JB, Clark RN, Shirley JH, Ocampo AC, Matson DL (1999a) Hydrogen peroxide on the surface of Europa. Science 283:2062–2064

Carlson RW, Johnson RE, Anderson MS (1999b) Sulfuric acid on Europa and the radiolytic sulfuric cycle. Science 26:97–99

Carney RS (1994) Consideration of the oasis analogy for chemosynthetic communities at Gulf of Mexico hydrocarbon vents. Geo-Marine Lett 14:149–159

Carr MH (1996) Water on Mars. Oxford University Press, New York

Carslaw KS, Clegg SL, Brimblecombe P (1995) A thermodynamic model of the system $HCl-HNO_3-H_2SO_4-H_2O$, including solubilities of HBr, from < 200 to 328 K. J Phys Chem 99:11,557–11,574

Carslaw KS, Peter T, Clegg SL (1997) Modeling the composition of liquid stratospheric aerosols. Rev Geophys 35:125–154

Catling DC (1999) A chemical model for evaporites on early Mars: possible sedimentary tracers of the early climate and implications for exploration. J Geophys Res 104:16,453–16,469

Catling DC, Moore J (2000) Iron oxide deposition from aqueous solution and iron formations on Mars. Lunar Planet Sci Conf XXXI, Houston, TX. Abstract 1517

Catling DC, Moore J (2003) The nature of coarse-grained crystalline hematite and its implications for the early environment of Mars. Icarus 165:277–300

Chen C-TA, Chen JH, Millero FJ (1980) Densities of NaCl, $MgCl_2$, Na_2SO_4, and $MgSO_4$ aqueous solutions at 1 atm from 0 to 50 °C and from 0.001 to 1.5 m. J Chem Eng Data 25:307–310

Chen C-T, Fine RA, Millero FJ (1977) The equation of state of pure water determined from sound speeds. J Chem Phys 66:2142–2144

Christensen PR, Anderson DL, Chase SC, Clancy RT, Clark RN, Conrath, BJ, Kieffer HH, Kuzmin RO, Malin MC, Pearl JC, Roush TL, Smith

MD (1998) Results from the Mars Global Surveyor Thermal Emission Spectrometer. Science 279:1692–1698

Christensen PR, Bandfield JL, Clark RN, Edgett KS, Hamilton VE, Hoefen T, Kieffer HH, Kuzmin RO, Lane MD, Malin MC, Morris RV, Pearl JC, Pearson R, Roush TL, Ruff SW, Smith MD (2000a) Detection of crystalline hematite mineralization on Mars by the Thermal Emission Spectrometer: evidence for near-surface water. J Geophys Res 105:9623–9642

Christensen PR, Bandfield JL, Smith MD, Hamilton VE, Clark RN (2000b) Identification of a basaltic component on the Martian surface from Thermal Emission Spectrometer data. J Geophys Res 105:9609–9621

Christensen PR, Morris RV, Lane MD, Bandfield JL, Malin MC (2001) Global mapping of Martian hematite mineral deposits: Remnants of water-driven processes on early Mars. J Geophys Res 106:23,873–23,885

Christner BC, Mosley-Thompson E, Thompson LG, Zagorodnov V, Sandman K, Reeve JN (2000) Recovery and identification of viable bacteria immured in glacial ice. Icarus 144:479–485

Chyba CF (2000) Energy for microbial life on Europa. Nature 403:381–382

Chyba CF, Hand KP (2001) Life without photosynthesis. Science 292:2026–2027

Chyba CF, Phillips CB (2001) Possible ecosystems and the search for life on Europa. Proc Natl Acad Sci USA 98:801–804

Clark BC (1993) Geochemical components in Martian soil. Geochim Cosmochim Acta 57:4575–4581

Clark BC, Van Hart DC (1981) The salts of Mars. Icarus 45:370–378

Clayton DD (2003) Handbook of isotopes in the cosmos. Cambridge University Press, Cambridge, UK

Clegg SL, Brimblecombe P (1990a) Equilibrium partial pressures and mean activity and osmotic coefficients of 0–100% nitric acid as a function of temperature. J Phys Chem 94:5369–5380

Clegg SL, Brimblecombe P (1990b) The solubility and activity coefficient of oxygen in salt solutions and brines. Geochim Cosmochim Acta 54:3315–3328

Clegg SL, Brimblecombe P (1995) Application of a multicomponent thermodynamic model to activities and thermal properties of $0–40\,\mathrm{mol\,kg^{-1}}$ aqueous sulfuric acid from <200 to $328\,\mathrm{K}$. J Chem Eng Data 40:43–64

Clegg SL, Rard JA, Pitzer KS (1994) Thermodynamic properties of $0–6\,\mathrm{mol\,kg^{-1}}$ aqueous sulfuric acid from 273.15 to 328.15 K. Chem Soc Faraday Trans 90:1875–1894

Cockell CS (1999) Life on Venus. Plant Space Sci 47:1487–1501

Cometta S, Sonnleitner B, Sidler W, Fiechter A (1982) Population distribution of aerobic extremely thermophilic microorganisms in an Icelandic natural hot spring. Eur J Appl Microbiol Biotechnol 16:151–156

Condon DJ, Prave AR, Benn DI (2002) Neoproterozoic glacial-rainout intervals: Observations and implications. Geology 30:35–38

Cooper JF (2001) Jovian magnetospheric irradiation effects on Europa surface composition. In: Greeley R (ed) Europa focus group workshop 1, pp 6–7. Available at: http://astrobiology.asu.edu/focus/europa/intro.html

Cooper JF, Phillips CB, Green JR, Wu X, Carlson RW, Tamppari LK, Terrile RJ, Johnson RE, Eraker JH, Makris NC (2002) Europa exploration: science and mission priorities for 2003–2013 and beyond. In Sykes MV (ed) ASP Conference Series: the future of solar system exploration, 2003–2013, ASP, San Francisco, pp 1–36

Couzin J (2002) Weight of the world on microbes' shoulders. Science 295:1444–1445

Crowley TJ, Hyde WT, Peltier WR (2001) CO_2 levels required for deglaciation of a 'near-snowball' Earth. Geophys Res Lett 28:283–286

Dalton JB, Prieto-Ballesteros O, Kargel JS, Jamieson CS, Jolivet J, Quinn R (2005) Spectral comparison of heavily hydrated salts with disrupted terrains on Europa. Icarus 177:472–490

Debye P, Hückel E (1923) Zür Theorie der Electrolyte: I. Gefrierpunkterniedindung und verwandte Ersheinungen. Physik Z 24:185–206

Delsemme AH (2001) An argument for the cometary origin of the biosphere. Am Sci 89:432–442

Dholabhai PD, Bishnoi PR (1994) Hydrate equilibrium conditions in aqueous-electrolyte solutions – mixtures of methane and carbon dioxide. J Chem Eng Data 39:191–194

Dholabhai PD, Kalogerakis N, Bishnoi PR (1993) Equilibrium conditions for carbon dioxide hydrate formation in aqueous electroyte solutions. J Chem Eng Data 38:650–654

Dholabhai PD, Parent JS, Bishnoi PR (1997) Equilibrium conditions for hydrate formation from binary mixtures of methane and carbon dioxide in the presence of electrolytes, methanol and ethylene glycol. Fluid Phase Equilibria 141:235–246

Donnadieu Y, Godderis Y, Ramstein G, Nedelec A, Meert J (2004) A 'snowball Earth' climate triggered by continental break-up through changes in runoff. Nature 428:303–306

Dose K, Bieger-Dose A, Ernst B, Feister U, Gomez-Silva B, Klein A, Risi S, Stridde C (2001) Survival of microorganisms under the extreme conditions of the Atacama Desert. Orig Life Evol Biosph 31:287–303

Dougherty AJ, Hogenboom DL, Kargel JS (2007) Volumetric and optical studies of high-pressure phases of $MgSO_4 nH_2O$ with applications to Europa. Lunar Planet Sci Conf XXXVIII, Houston, TX. Abstract 2275

Drever JI (1997) The geochemistry of natural waters. Surface and groundwater environments, 3rd edn. Prentice-Hall, Upper Saddle River, NJ

Duan Z, Møller N, Greenberg J, Weare JH (1992a) The prediction of methane solubility in natural waters to high ionic strength from 0 to 250 °C and from 0 to 1600 bar. Geochim Cosmochim Acta 56:1451–1460

Duan Z, Møller N, Weare JH (1992b) An equation of state for the CH_4-CO_2-H_2O: I. Pure systems from 0 to 1000°C and 0 to 8000 bar. Geochim Cosmochim Acta 56:2605–2617

Duckworth AW, Grant WD, Jones BE, van Steenbergen R (1996) Phylogenetic diversity of soda lake alkaliphiles. FEMS Microbiol Ecol 19:181–191

Edwards KJ, Gihring TM, Banfield JF (1999) Seasonal variations in microbial populations and environmental conditions in an extreme acid mine drainage environment. Appl Environ Microbiol 65:3627–3632

Ehrenfreund P, Menten KM (2002) From molecular clouds to the origin of life. In: Horneck G, Baumstark-Khan (eds) Astrobiology: the quest for the conditions of life, Springer, Berlin Heidelberg New York, pp 7–23

Elberling B (2001) Environmental controls of the seasonal variation in oxygen uptake in sulfidic tailings deposited in a permafrost-affected area. Water Resour Res 37:99–107

Englezos P, Hall S (1994) Phase equilibrium data on carbon dioxide hydrate in the presence of electrolytes, water soluble polymers and montmorillonite. Can J Chem Eng 72:887–893

Eugster HP, Hardie LA (1978) Saline lakes. In: Lerman A (ed) Lakes: chemistry, geology, physics, Springer, Berlin Heidelberg New York, pp 237–293

Fanale FP, Li YH, Decarlo E, Domergue-Schmidt N, Sharma SK, Horton K, Granahan JC, Galileo NIMS Team (1998) Laboratory simulation of the chemical evolution of Europa's aqueous phase. Lunar and Planetary Science Conference XXIX, Houston, TX. Abstract 1248

Farmer JD (1998) Thermophiles, early biosphere evolution, and the origin of life on Earth: implications for the exobiological exploration of Mars. J Geophys Res 103:28,457–28,461

Farmer JD (2000) Hydrothermal systems: Doorways to early biosphere evolution. GSA Today 10:1–9

Feistel R (2003) A new extended Gibbs thermodynamic potential of seawater. Prog Ocean 58:43–114

Fernandez-Remolar DC, Rodriquez N, Gomez F (2003) Geological record of an acidic environment driven by iron hydrochemistry: The Tinto River system. J Geophys Res 108(E7), 5080, doi:10.1029/2002JE001918, 2003

Fischman J (1995) Have 25-million-year-old bacteria returned to life? Science 268:977

Fisher CR, MacDonald IR, Sassen R, Young CM, Macko SA, Hourdez S, Carney RS, Joye S, McMullin E (2000) Methane ice worms: *Hesiocaeca methanicola* colonizing fossil fuel reserves. Naturwissenschaften 87:184–187

Fisk MR, Giovannoni SJ (1999) Sources of nutrients and energy for a deep biosphere on Mars. J Geophys Res 104:11,805–11,815

Foing BH (2002) Space activities in exo-astrobiology. In G Horneck and C Baumstark-Khan (ed) Astrobiology: the Quest for the Conditions of Life. Springer, Berlin Heidelberg New York, pp 389–398

Forget F, Pierrehumbert RT (1997) Warming early Mars with carbon dioxide clouds that scatter infrared radiation. Science 278:1273–1276

Frey HU, Lummerzheim D (2002) Can conditions for life be inferred from optical emissions of extra-solar-system planets, In Atmospheres in the solar system: comparative aeronomy. Geophysical Monograph 130, American Geophysical Union, Washington, DC, pp 381–388

Friedmann EI, Ocampo R (1976) Endolithic blue-green algae in the Dry Valleys: Primary producers in the Antarctic desert ecosystem. Science 193:1247–1249

Friedmann EI, Ocampo-Friedmann R (1984) The Antarctic crytoendolithic ecosystem: relevance to exobiology. Orig Life 14:771–776

Fritsen CH, Priscu JC (1998) Cyanobacterial assemblages in permanent ice covers on Antarctic lakes: distribution, growth rate, and temperature response of photosynthesis. J Phytol 34:587–597

Furfaro R et al (12) (2007) The search for life beyond Earth through fuzzy expert systems. Planet Space Sci in press.

Fyfe WS (1996) The biosphere is going deep. Science 273:448

Gaidos EJ, Marion GM (2003) Geological and geochemical legacy of a cold early Mars. J Geophys Res 108(E6), 5055, doi:10.1029/2002JE002000, 2003

Gaidos EJ, Nealson KH, Kirschvink JL (1999) Life in ice-covered oceans. Science 284:1631–1633

Garrels RM, Christ CL (1965) Solutions, minerals, and equilibria. Harper & Row, New York

Gates JA, Wood RH (1985) Densities of aqueous solutions of NaCl, $MgCl_2$, KCl, NaBr, LiCl, and $CaCl_2$ from 0.05 to $5.0\,\text{mol}\,\text{kg}^{-1}$ and 0.1013 to 40 MPa at 298.15 K. J Chem Eng Data 30:44–49

Gibbs JW (1948) The collected works of J. Willard Gibbs. Yale University Press, New Haven, CT

Gibson Jr EK, McKay DS, Thomas-Keprta K, Romanek CS (1997) The case for relic life on Mars. Sci Am, December 1997, pp 58–65

Gilichinsky DA (2002) Permafrost model of extraterrestrial habitat. In: Horneck G, Baumstark-Khan C (eds) Astrobiology: the quest for the conditions of life, Springer, Berlin Heidelberg New York, pp 125–142

Gitterman KE (1937) Thermal analysis of sea water. CRREL TL 287. US-ACRREL, Hanover, NH

Glendenning NK (2001) Phase transitions and crystalline structures in neutron star cores. Phys Rep 342:393–447

Glendenning NK, Pei S (1995) Crystalline structure of the mixed confined-deconfined phase in neutron stars. Phys Rev C 52:2250–2253

Godderis Y, Donnadieu Y, Nedelec A, Dupre B, Dessert C, Grard A, Ramstein G, Francois LM (2003) The Sturtian 'snowball' glaciation: fire and ice. Earth Planet Sci Lett 211:1–12

Gooding JL (1992) Soil mineralogy and chemistry on Mars: Possible clues from salts and clays in SNC meteorites. Icarus 99:28–41

Gow AJ (1971) Relaxation of ice in deep drill cores from Antarctica. J Geophys Res 76:2533–2541

Gow AJ, Ueda HT, Garfield DE (1968) Antarctic ice sheet: preliminary results of first core hole to bedrock. Science 161:1011–1013

Greenberg R (2002) Tides and the biosphere of Europa. American Scientist 90:48–55

Greenberg R, Geissler P (2002) Europa's dynamic icy crust. Meteor Planet Sci 37:1685–1710

Greenberg R, Tufts BR, Geissler P, Hoppa GV (2002) Europa's crust and ocean: How tides create a potentially habitable physical setting. In: Horneck G, Baumstark-Khan C (eds) Astrobiology: the Quest for the Conditions of Life, Springer, Berlin Heidelberg New York, pp 111–124

Griffith LL, Shock EL (1995) A geochemical model for the formation of hydrothermal carbonates on Mars. Nature 377:406–408

Grotzinger JP, Knoll AH (1995) Anomalous carbonate precipitates: is the Precambrian the key to the Permian? Paloios 10:578–596

Hall DL, Sterner SM, Bodnar RJ (1988) Freezing point depression of aqueous sodium chloride solutions. Econ Geol 83:197–202

Harvie CE, Eugster HP, Weare JH (1982) Mineral equilibria in the six-component seawater system, Na-K-Mg-Ca-SO_4-Cl-H_2O at 25 °C: II. Compositions of the saturated solutions. Geochim Cosmochim Acta 46:1603–1618

Harvie CE, Greenberg JP, Weare JH (1987) A chemical equilibrium algorithm for highly non-ideal multiphase systems: free energy minimization. Geochim Cosmochim Acta 51:1045–1057

Harvie CE, Møller, N, Weare, JH (1984) The prediction of mineral solubilities in natural waters: the Na-K-Mg-Ca-H-Cl-SO_4-OH-HCO_3-CO_3-CO_2-H_2O system to high ionic strengths at 25 °C. Geochim Cosmochim Acta 48:723–751

Hazen RM, Roedder E (2001) How old are bacteria from the Permian age? Nature 411:155

He S, Morse JW (1993) The carbonic acid system and calcite solubility in aqueous Na-K-Ca-Mg-Cl-SO_4 solutions from 0 to 90 °C. Geochim Cosmochim Acta 57:3533–3555

Herut B, Starinsky A, Katz A, Bein A (1990) The role of seawater freezing in the formation of subsurface brines. Geochim Cosmochim Acta 54:13–21

Hoffman PF, Kaufman AJ, Halverson GP, Schrag DP (1998) A Neoproterozoic snowball Earth. Science 281:1342–1346

Hoffman PF, Schrag DP (2000) Snowball Earth. Sci Am, January 2000, pp 68–75

Hoffman PF, Schrag DP (2002) The snowball Earth hypothesis: testing the limits of global change. Terra Nova 14:129–155

Hogenboom DL, Kargel JS, Ganasan JP, Lee L (1995) Magnesium sulfate-water to 400 MPa using a novel piezometer: densities, phase equilibria, and planetological implications. Icarus 115:258–277

Hogenboom DL, Kargel JS, Pahalawatta PV (1999) Densities and phase relationships at high pressures of the sodium sulfate-water system. Lunar and Planetary Science Conference XXX, Houston, TX. Abstract 1793

Holland HD (2004) The geologic history of seawater. In: Elderfield H (ed) Treatise on Geochemistry, Vol. 6, The Oceans and Marine Geochemistry. Elsevier, Amsterdam, pp 583–625

Holland TJB, Powell R (1998) An internally consistent thermodynamic data set for phases of petrological interest. J Metamorph Geol 16:309–343

Horneck G, Baumstark-Khan C (eds) (2002) Astrobiology: the Quest for the Conditions of Life. Springer, Berlin Heidelberg New York

Hubbard WB, Podolak M, Stevenson DJ (1995) The interior of Neptune. In: Cruikshank DP (ed) Neptune and Triton. University of Arizona Press, Tucson, pp 109–140

Huber C, Wächtershäuser G (2006) α-hydroxy and α-amino acids under possible Hadean, volcanic origin-of-life conditions. Science 314:630–632

Huber H, Stetter KO (1998) Hyperthermophiles and their possible potential in biotechnology. J Biotechnol 64:39–52

Hurtgen MT, Arthur MA, Halverson GP (2005) Neoproterozoic sulfur isotopes, the evolution of microbial sulfur species, and the burial efficiency of sulfide as sedimentary pyrite. Geology 33:41–44

Husain V, Winkler O (2007) Semiclassical states for quantum cosmology. Phys Rev D 75:024014

Huterer D, Turner MS (2001) Probing dark energy: methods and strategies. Phys Rev D 64:123527

Huterer D, Starkman GD, Trodden M (2002) Is the universe inflating? Dark energy and the future of the universe. Phys Rev D 66:043511

Hyde WT, Crowley TJ, Baum SK, Peltier WR (2000) Neoproterozoic 'snowball Earth' simulations with a coupled climate/ice-sheet model. Nature 405:425–429

Irwin LN, Schulze-Makuch D (2001) Assessing the plausibility of life on other worlds. Astrobiology 1:143–160

Jacobsen SB (2001) Gas hydrates and deglaciations. Nature 412:691–693

Jakosky BM, Shock EL (1998) The biological potential of Mars, the early Earth, and Europa. J Geophys Res 103:19,359–19,364

Jawad A, Snelling AM, Heritage J, Hawkey PM (1998) Exceptional desiccation tolerance of *Acinetobacter radioresistens*. J Hosp Infect 39:235–240

Jenkins GS (2000) The "snowball Earth" and Precambrian climate. Science 288:975–976

Jiang G, Kennedy MJ, Christie-Blick N (2003) Stable isotopic evidence for methane seeps in Neoproterozoic postglacial cap carbonates. Nature 426:822–826

Johnson DB (1998) Biodiversity and ecology of acidophilic microorganisms. FEMS Microbiol Ecol 27:307–317

Jones BF, Eugster HP, Rettig SL (1977) Hydrochemistry of the Lake Magadi basin, Kenya. Geochim Cosmochim Acta 41:53–72

Jouzel J, Barkov NI, Barnola JM, Bender M, Chappellaz J, Genthon C, Kotlyakov VM, Lipenkov V, Lorius C, Petit JR, Raynaud D, Raisbeck G, Ritz C, Sowers T, Stievenard M, Yiou F, Yiou P (1993) Extending the Vostok ice-core record of palaeoclimate to the penultimate glacial period. Nature 364:407–412

Junge K (2002) Bacterial abundance, activity, and diversity at extremely cold temperatures in Arctic sea ice, Ph.D. dissertation, University of Washington, Seattle

Junge K, Eicken H, and Deming JW (2004) Bacterial activity at -2 to $-20\,°C$ in Arctic wintertime sea ice. Appl Environ Microbiol 70:550–557

Junge K, Krembs C, Deming JW, Stierle A, Eicken H (2001) A microscopic approach to investigate bacteria under in-situ conditions in sea-ice samples. Ann Glaciol 33:304–310

Kargel JS (1991) Brine volcanism and the interior structures of asteroids and icy satellites. Icarus 94:368–390

Kargel JS (1994) Metalliferous asteroids as potential sources of precious metals. J Geophys Res 99:21,129–21,141

Kargel JS, Kirk RL, Fegley B Jr, Treiman A (1994) Carbonate-sulfate volcanism on Venus? Icarus 112:219–252

Kargel JS (2001) Roles of Europa's stratified crust and ocean in diapirism and melt-through. In: Greeley R (ed) Europa focus group workshop 2, pp 19–20. Available at: http://astrobiology.asu.edu/focus/europa/intro.html

Kargel JS (2004) Mars: a warmer, wetter planet. Springer-Praxis, Berlin-Chichester, UK

Kargel JS (2006) Enceladus: cosmic gymnast, volatile miniworld. Science 311:1389–1391

Kargel JS, Consolmagno GJ (1996) Magnetic fields and the detectability of brine oceans in Jupiter's icy satellites. Lunar Planet Sci 27:643–644

Kargel J, Kaye J, Head JW III, Marion GM, Sassen R, Crowley J, Prieto O, Grant SA, Hogenboom D (2000) Europa's salty crust and ocean: origin, composition, and the prospects for life. Icarus 148:226–265

Kargel JS, Head JW III, Hogenboom DL, Khurana KK, Marion GM (2001) The system sulfuric acid-magnesium sulfate-water: Europa's ocean properties related to thermal state. Lunar and Planetary Science Conference XXXII, Houston TX. Abstract 2138

Kargel, JS, Furfaro R, Prieto-Ballesteros O, Rodriguez JAP, Montgomery DR, Gillespie AR, Marion GM, Wood SE (in press) Martian hydrogeology sustained by thermally insulating gas and salt hydrates. Geology.

Kargel JS, Furfaro R, Hays CC, Lopes RMC, Lunine JI, Mitchell KL, Wall SD, Cassini Radar Team (2007) Titan's GOO-sphere: glacial, permafrost, evaporite, and other familiar processes involving exotic materials. Lunar Planet Sci Conf XXXVIII, Houston, TX. Abstract 1992.

Kashefi K, Lovley DR (2003) Extending the upper temperature limit for life. Science 301:934

Kato C, Li L, Nogi Y, Nakamura Y, Tamaoka J, and Horikoshi K (1998) Extremely barophilic bacteria isolated from the Mariana Trench, Challenger Deep, at a depth of 11,000 meters. Appl Environ Microbiol 64:1510–1513

Kaye JZ, Baross JA (2000) High incidence of halotolerant bacteria in Pacific hydrothermal-vent and pelagic environments. FEMS Microbiology Ecol 32:249–260

Kaye JZ, Baross JA (2002) Salinity, pressure, and heavy-metal stress response of moderately halophilic bacteria isolated from hydrothermal-vent environments. EOS 83:F1450

Kell GS (1975) The density, thermal expansivity and compressibility of liquid water from 0 to 150°C and 0 to 1 kilobar. J Chem Eng Data 20:97–105

Kelley DS, Baross JA, Delaney JR (2002) Volcanoes, fluids, and life at mid-ocean ridge spreading centers. Ann Rev Earth Planet Sci 30:385–491

Kempe S, Kazmierczak J (1997) A terrestrial model for an alkaline Martian hydrosphere. Planet Space Sci 45:1493–1499

Kempe S, Kazmierczak J (2002) Biogenesis and early life on Earth and Europa: favored by an alkaline ocean? Astrobiology 2:123–130

Kennedy MJ, Christie-Blick N, Prave AR (2001a) Carbon isotopic composition of Neoproterozoic glacial carbonates as a test of paleoceanographic models for snowball Earth phenomena. Geology 29:1135–1138

Kennedy MJ, Christie-Blick N, Sohl LE (2001b) Are Proterozoic cap carbonates and isotopic excursions a record of gas hydrate destabilization following Earth's coldest intervals? Geology 29:443–446

Kerr RA (1997) Life goes to extremes in the deep Earth—and elsewhere? Science 276:703–704

Kerr RA (2000) An appealing snowball Earth that's still hard to swallow. Science 287:1734–1736

Khurana KK, Kivelson MG, Stevenson DJ, Schubert G, Russell CT, Walker RJ, Polanskey C (1998) Induced magnetic fields as evidence for subsurface oceans in Europa and Callisto. Nature 395:777–780

Kieffer HH, Jakosky BM, Snyder CW (1992) The planet Mars: from antiquity to the present. In: Kieffer HH, Snyder CW, Matthews MS (eds) Mars. The University of Arizona Press, Tucson, pp 1–33

Kirchner JW, Weil A (2000) Delayed biological recovery from extinctions throughout the fossil record. Nature 404:177–180

Kirschvink JL (1992) Late Proterozoic low-latitude global glaciation: The snowball Earth. In: Schopf JW, Klein C (eds) The Proterozoic Biosphere. Cambridge University Press, Cambridge, UK, pp 51–52

Kirschvink JL, Gaidos EJ, Bertani LE, Beukes NJ Gutzmer J, Maepa LN, Steinberger RE (2000) Paleoproterozoic snowball Earth: Extreme climatic and geochemical global change and its biological consequences. Proc Natl Acad Sci 97:1400–1405

Kivelson MG, Khurana KK, Stevenson DJ, Bennett L, Joy S, Russell CT, Walker RJ, Zimmer C, Polanskey C (1999) Europa and Callisto: Induced or intrinsic magnetic fields in a periodically varying plasma environment. J Geophys Res 104:4609–4625

Klein HP, Horowitz NH, Biemann K (1992) The search for extant life on Mars. In: Kieffer HH, Snyder CW, Mathews MS (eds) Mars. University of Arizona Press, Tucson, pp 1221–1233

Krishnaswamy R, Hanger RA (1998) Phycomicrobial ecology of acid mine drainage in the Piedmont of Virginia. In: Proceedings of the 15th Annual National Meeting of the American Society for Surface Mining and Reclamation, Princeton, WV, pp 299–308

Krumgalz BS, Pogorelsky R, Pitzer KS (1995) Ion interaction approach to calculations of volumetric properties of aqueous multiple-solute electrolyte solutions. J Soln Chem 24:1025–1038

Krumgalz BS, Pogorelsky R, Pitzer KS (1996) Volumetric properties of single aqueous electrolytes from zero to saturation concentration at $298.15\,°K$ represented by Pitzer's ion-interaction equations. J Phys Chem Ref Data 25:663–689

Krumgalz BS, Starinsky A, Pitzer KS (1999) Ion-interaction approach: pressure effect on the solubility of some minerals in submarine brines and seawater. J Soln Chem 28:667–692

Krumgalz BS, Hecht A, Starinsky A, Katz A (2000) Thermodynamic constraints on Dead Sea evaporation: can the Dead Sea dry up? Chem Geol 165:1–11

Kushner D (1981) Extreme environments: are there any limits to life? In: Ponnamperuma C (ed) Comets and the Origin of Life. D. Reidel, Dordrecht, pp 241–248

Kvenvolden KA (1993) Gas hydrates—geological perspective and global change. Rev Geophys 31:173–187

Lane M (2004) Thermal emission spectroscopy of sulfates: Possible hydrous iron-sulfate in the soil at the MER-A Gusev Crater Landing site. Lunar and Planetary Science Conference XXXV. Houston, TX. Abstract 1858

Larson SD (1955) Phase studies of the two-component carbon dioxide-water system involving the carbon dioxide hydrate. PhD dissertation, University of Illinois

Leach DL, Marsh E, Emsbo P, Rombach CR, Kelley KD, Anthony M (2004) Conditions of hydrothermal fluids at the shale-hosted Red Dog Zn-Pb-Ag deposits, Brooks Range, Alaska. Econ Geol 99:1449–1480

Leger A, Pirre M, Marceau FJ (1993) Search for primitive life on a distant planet: relevance of O_2 and O_3 detections. Astron Astrophys 277:309–313

Lewis JS (1995) Physics and chemistry of the solar system. Academic, New York

Lewis JS (1997) Mining the sky: untold riches from the asteroids, comets, and planets. Helix, New York

L'Haridon S, Reysenbach AL, Glénat P, Prieur D, Jeanthon C (1995) Hot subterranean biosphere in a continental oil reservoir. Nature 377:223–224

Lide DR (ed) (1994) CRC handbook of chemistry and physics, 75th edn. CRC Press, Boca Raton, FL

Likens GE, Bormann FH, Pierce RS, Eaton JS, Johnson NM (1977) Biogeochemistry of a forested ecosystem. Springer, Berlin Heidelberg New York

Linder EV (2006) Theory challenges of the accelerating Universe. eprint arXiv:astroph/0610173, Publ Date: 10/2006

Linke WF (1958) Solubilities of inorganic and metal organic compounds, Vol. I, 4th edn. American Chemical Society, Washington, DC

Linke WF (1965) Solubilities of inorganic and metal organic compounds, Vol. II, 4th edn. American Chemical Society, Washington, DC

List RJ (1951) Smithsonian meteorological tables, 6th edn. Smithsonian Institution, Washington, DC

Lockwood JP, Rubin M (1989) Origin and age of the Lake Nyos maar, Cameroon. J Volcanol Geotherm Res 39:117–124

Lopez-Archilla AI, Marin I, Amils R (2001) Microbial community composition and ecology of an acidic aquatic environment: the Tinto River, Spain. Microb Ecol 41:20–35

Lorentz NJ, Corsetti FA, Link PK (2004) Seafloor precipitates and C-isotope stratigraphy from the Neoproterozoic Scout Mountain Member of the Pocatello Formation, southeast Idaho: implications for neoproterozoic earth system behavior. Precambrian Res 130:57–70

Lo Surdo A, Alzona EM, Millero FJ (1982) The (p, V, T) properties of concentrated aqueous electrolytes. I. Densities and apparent molar volumes of NaCl, Na_2SO_4, $MgCl_2$, and $MgSO_4$ solutions from $0.1\,\text{mol}\,\text{kg}^{-1}$ to saturation and from 273.15 to 323.15 K. J Chem Thermodynam 14:649–662

Lunine JI (1999) Earth: evolution of a habitable world. Cambridge University Press, Cambridge, UK

Lyons WB, Welch KA, Snyder G, Olesik J, Graham EY, Marion GM, Poreda RJ (2005) Halogen geochemistry of the McMurdo Dry Valleys, Antarctica: Clues to the origin of solutes and lake evolution. Geochim Cosmochim Acta 69:305–323

Madigan MT, Marrs BL (1997) Extremophiles. Sci Am, April 1997, pp 82–87

Madigan MT, Oren A (1999) Thermophilic and halophilic extremophiles. Curr Opin Microbiol 2:265–269

Magot M, Ollivier B, Patel BKC (2000) Microbiology of petroleum reservoirs. Antonie van Leeuwenhoek 77:103–116

Marion GM (1997) A theoretical evaluation of mineral stability in Don Juan Pond, Wright Valley, Victoria Land. Antarctic Sci 9:92–99

Marion GM (2001) Carbonate mineral solubility at low temperatures in the Na-K-Mg-Ca-H-Cl-SO$_4$-OH-HCO$_3$-CO$_3$-CO$_2$-H$_2$O system. Geochim Cosmochim Acta 65:1883–1896

Marion GM (2002) A molal-based model for strong acid chemistry at low temperatures (<200 to 298 K). Geochim Cosmochim Acta 66:2499–2516

Marion GM (2007) Adapting molar data (without density) for molal models. Computers & Geosciences 33:829–834.

Marion GM, Catling DC, Kargel JS (2003a) Modeling aqueous ferrous iron chemistry at low temperatures with application to Mars. Geochim Cosmochim Acta 67:4251–4266

Marion GM, Catling DC, Kargel JS (2006) Modeling gas hydrate equilibria in electrolyte solutions. CALPHAD 30:248–259

Marion GM, Farren RE (1999) Mineral solubilities in the Na-K-Mg-Ca-Cl-SO$_4$-H$_2$O system: a re-evaluation of the sulfate chemistry in the Spencer-Møller-Weare model. Geochim Cosmochim Acta 63:1305–1318

Marion GM, Farren RE, Komrowski AJ (1999) Alternative pathways for seawater freezing. Cold Regions Sci Tech 29:259–266

Marion GM, Fritsen CH, Eicken H, Payne MC (2003b) The search for life on Europa: Limiting environmental factors, potential habitats, and Earth analogues. Astrobiology 3:785–811

Marion GM, Grant SA (1994) FREZCHEM: A chemical–thermodynamic model for aqueous solutions at subzero temperatures. CRREL Spec Rept 94-18. USACRREL, Hanover, NH

Marion GM, Grant SA (1997) Physical chemistry of geochemical solutions at subzero temperatures. In: Iskandar, IK, Wright EA, Radke JK, Sharratt BS, Groenevelt PH, Hinzman LD (eds) International symposium on physics, chemistry, and ecology of seasonally frozen soils, Fairbanks, AK, June 1997. CRREL Special Report 97-10, Hanover, NH, pp 349–356

Marion GM, Jakubowski SD (2004) The compressibility of ice to 2.0 kbars. Cold Regions Sci Tech 38:211–218

Marion GM, Kargel JS, Catling DC, Jakubowski SD (2005) Effects of pressure on aqueous chemical equilibria at subzero temperatures with applications to Europa. Geochim Cosmochim Acta 69:259–274

Marion GM, Schulze-Makuch D (2007) Astrobiology and the search for life in the Universe. In: Gerday C, Glansdorff N (eds) Physiology and biochemistry of extremophiles, ASM, Wake Forest, NC, pp 351–358

Max MD, Clifford SM (2000) The state, potential distribution, and biological implications of methane in the Martian crust. J Geophys Res 105:4165–4171

Mazur P (1980) Limits to life at low temperatures and at reduced water contents and water activities. Orig Life 10:137–159

McCaffrey MA, Lazar B, Holland HD (1987) The evaporation path of seawater and the coprecipitation of Br$^-$ and K$^+$ with halite. J Sed Petrol 57:928–937

McCollom TM (1999) Methanogenesis as a potential source of chemical energy for primary biomass production by autotrophic organisms in hydrothermal systems on Europa. J Geophys Res 104:30,729–30,742

McCord TB, Hansen GB, Fanale FP, Carlson RW, Matson DL, Johnson TV, Smythe WD, Crowley JK, Martin PD, Ocampo A, Hibbitts CA, Granahan JC, NIMS Team (1998) Salts on Europa's surface detected by Galileo's near infrared mapping spectrometer. Science 280:1242–1245

McCord TB, Hansen GB, Matson DL, Johnson TV, Crowley JK, Fanale FP, Carlson RW, Smythe WD, Martin PD, Hibbitts CA, Granahan JC, Ocampo A (1999) Hydrated salt minerals on Europa's surface from the Galileo near-infrared mapping spectrometer (NIMS) investigation. J Geophys Res 104:11,827–11,851

McDonald JE (1965) Saturation vapor pressures over supercooled water. J Geophys Res 70:1553–1554

McKay CP, Mancinelli RL, Stoker CR, Wharton RA Jr (1992) The possibility of life on Mars during a water-rich past. In: Kieffer HH, Snyder CW, Matthews MS (eds) Mars. The University of Arizona Press, Tucson, pp 1234–1245

McKay CP, Stoker CR (1989) The early environment and its evolution on Mars: implications for life. Rev Geophys 27:189–214

McKay DS, Gibson EK Jr, Thomas-Keprta KL, Vali H, Romanek CS, Clemett SJ, Chillier XDF, Maechling CR, Zare RN (1996) Search for past life on Mars: possible relic biogenic activity in Martian meteorite ALH84001. Science 273:924–930

McKinnon WB (1997) Sighting the seas of Europa. Nature 386:765–767

McSween HY Jr (1994) What we have learned about Mars from SNC meteorites. Meteoritics 29:757–779

Meyer GH, Morrow MB, Wyss O, Berg TE, Littlepage JL (1962) Antarctica: the microbiology of an unfrozen saline pond. Science 138:1103–1104

Millero FJ (1983) Influence of pressure on chemical processes in the sea. Chem Oceanogr 8:1–88

Millero FJ (2001) Physical chemistry of natural waters. Wiley-Interscience, New York

Millero FJ, Sohn ML (1992) Chemical oceanography. CRC Press, Boca Raton, FL

Mironenko MV, Grant SA, Marion GM, Farren RE (1997) FREZCHEM2: A chemical thermodynamic model for electrolyte solutions at subzero temperatures. CRREL Spec Rep 97-5. USACRREL, Hanover, NH

Mojzsis SJ, Arrhenius G, McKeegan KD, Harrison TM, Nutman AP, Friend CRL (1996) Evidence for life on Earth by 3800 million years ago. Nature 384:55–59

Møller N (1988) The prediction of mineral solubilities in natural waters: A chemical model for the Na-Ca-Cl-SO_4-H_2O system, to high temperatures and concentrations. Geochim Cosmochim Acta 52:821–837

Monnin C (1989) An ion interaction model for the volumetric properties of natural waters: density of the solution and partial molal volumes of electrolytes to high concentrations at 25 °C. Geochim Cosmochim Acta 53:1177–1188

Monnin C (1990) The influence of pressure on the activity coefficients of the solutes and on the solubility of minerals in the system Na-Ca-Cl-SO_4-H_2O to 200 °C and 1 kbar, and to high NaCl concentration. Geochim Cosmochim Acta 54:3265–3282

Morse JW, Mackenzie FT (1990) Geochemistry of sedimentary carbonates. Elsevier, Amsterdam

Morse JW, Marion GM (1999) The role of carbonates in the evolution of early Martian oceans. Am J Sci 299:738–761

Nagornov OV, Chizhov VE (1990) Thermodynamic properties of ice, water, and their mixtures at high pressures (in Russian). Zhurnal Prikladnoi Mekhaniki i Tekhnicheskoi Fiziki 31:41–48

Napier WM (2004) A mechanism for interstellar panspermia. Mon Not R Astron Soc 348:46–51

Navarro-Gonzalez R, Montoya L, Davis W, McKay C (2002) Laboratory support for a methanogensis driven biosphere in Europa. In: Greeley R (ed) Europa focus group workshop 3, p 35.
Available at: http://astrobiology.asu.edu/focus/europa/intro.html

Nealson KH (1997) The limits of life on Earth and searching for life on Mars. J Geophys Res 102:23,675–26,686

Nealson KH, Conrad PG (1999) Life: past, present, and future. Philos Trans R Soc Lond Biol 354:1923–1939

Neilands JB (1957) Some aspects of microbial iron metabolism. Bacteriol Rev 21:101–111

Nelson KH, Thompson TG (1954) Deposition of salts from sea water by frigid concentration. J Marine Res 13:166–182

Nordstrom DK, Alpers CN, Placek CJ, Blowes DW (2000) Negative pH and extremely acidic mine waters from Iron Mountain, California. Environ Sci Technol 34:254–258

Nordstrom DK, Munoz JL (1994) Geochemical Thermodynamics, 2nd edn. Blackwell, Oxford

O'Brien DP, Geissler P, Greenberg R (2002) A melt-through model for chaos formation on Europa. Icarus 156:152–161

Oren A (1988) Anaerobic degradation of organic compounds at high salt concentrations. Antonie van Leewenhoek 54:267–277

Pappalardo RT, Belton MJS, Breneman HH, Carr MH, Chapman CR, Collins GC, Denk T, Fagents S, Geissler PE, Giese B, Greeley R, Greenberg R, Head JW, Helfenstein P, Hoppa G, Kadel SD, Klaasen KP, Klemaszewski JE, Magee K, McEwen AS, Moore JM, Moore WB, Neukum G, Phillips CB, Prockter LM, Shubert G, Senske DA, Sullivan RJ, Tufts BR, Turtle

EP, Wagner R, Williams KK (1999) Does Europa have a subsurface ocean? Evaluation of the geological evidence. J Geophy Res 104:24,015–24,055

Pavlov AA, Kasting JF, Brown LL (2000) Greenhouse warming by CH_4 in the atmosphere of early Earth. J Geophys Res 105:11,981–11,990

Pedersen K (1993) The deep subterranean biosphere. Earth Sci Rev 34:243–260

Petrenko VF, Whitworth RW (1999) Physics of ice. Oxford University Press, Oxford

Petsch ST (2004) The global oxygen cycle. In: Schlesinger WH (ed) Treatise on Geochemistry, Vol. 8, Biogeochemistry. Elsevier, Amsterdam, pp 515–555

Phoenix VR, Konhauser KO, Adams DG, Bottrell SH (2001) Role of biomineralization as an ultraviolet shield: implications for Archean life. Geology 29:823–826

Pierazzo E, Chyba CF (2002) Cometary delivery of biogenic elements to Europa. Icarus 157:120–127

Pierrot D, Millero FJ (2000) The apparent molal volume and compressibility of seawater fit to the Pitzer equations. J Soln Chem 29:719–742

Pitzer KS (1991) Ion interaction approach: theory and data correlation. In: Pitzer KS (ed) Activity coefficients in electrolyte solutions, 2nd edn. CRC Press, Boca Raton, FL, pp 75–153

Pitzer KS (1995) Thermodynamics, 3rd edn. McGraw-Hill, New York

Pledger RJ, Baross JA (1991) Preliminary description and nutritional characterization of a chemoorganotrophic archaeobacterium growing at temperatures of up to 110 °C isolated from a submarine hydrothermal vent environment. J Gen Microbiol 137:203–211

Pledger RJ, Crump BC, Baross JA (1994) A barophilic response by two hyperthermophilic, hydrothermal vent Archaea: An upward shift in the optimal temperature and acceleration of growth rate at supra-optimal temperatures by elevated pressure. FEMS Microbiol Ecol 14:233–242

Plummer LN, Busenberg E (1982) The solubility of calcite, aragonite, and vaterite in CO_2–water solutions between 0–90 °C and an evaluation of the aqueous model for the system CO_2-H_2O-$CaCO_3$. Geochim Cosmochim Acta 46:1011–1040

Plummer LN, Parkhurst DL, Fleming GW, Dunkle SA (1988) A computer program incorporating Pitzer's equations for calculation of geochemical reactions in brines. US Geol Survey Water-Resources Investigations Report, 88–4153

Pollack JB, Roush T, Witteborn F, Bregman J, Wooden D, Stoker C, Toon OB, Rank D, Dalton B, Freedman R (1990) Thermal emission spectra of Mars (5.4–10.5 µm): evidence for sulfates, carbonates, and hydrates. J Geophys Res 95:14,595–14,627

Porco CC et al. (24) (2006) Cassini observes the active South Pole of Enceladus. Science 311:1393–1401

Porter SM, Knoll AH, Affaton P (2004) Chemostratigraphy of Neoproterozoic cap carbonates from the Volta Basin, West Africa. Precambrian Res 130:99–112

Powers DW, Vreeland RH, Rosenzweig WD (2001) Reply to Hazen RM, Roedder E. Nature 411:155

Press WH, Teukolsky SA, Vetterling WT, Flannery BP (1992) Numerical recipes in FORTRAN: the art of scientific computing, 2nd edn. Cambridge University Press, Cambridge, UK

Price PB (2000) A habitat for psychrophiles in deep Antarctic ice. Proc Natl Acad Sci USA 97:1247–1251

Price PB, Sowers T (2004) Temperature dependence of metabolic rates for microbial growth, maintenance, and survival. Proc Natl Acad Sci 101:4631–4636

Prieto-Ballesteros O, Kargel JS (2005) Thermal state and complex geology of a heterogeneous salty crust of Jupiter's satellite, Europa. Icarus 173:212–221

Prieto-Ballesteros O, Kargel JS, Fairen AG, Fernandez-Remolar DC, Dohm JM, Amils R (2006) Interglacial clathrate destabilization on Mars: Possible contributing source of its atmospheric methane. Geology 34:149–152

Priscu JC et al. (11) (1999) Geomicrobiology of subglacial ice above Lake Vostok, Antarctica. Science 286:2141–2144

Priscu JC, Fritsen CH, Adams EE, Giovannoni SJ, Paerl HW, McKay CP, Doran PT, Gordon DA, Lanoil BD, Pinckney JL (1998) Perennial Antarctic lake ice: An oasis for life in a polar desert. Science 280:2095–2098

Psenner R, Sattler B (1998) Life at the freezing point. Science 280:2073–2074

Ptacek CJ (1992) Experimental determination of siderite solubility in high ionic-strength aqueous solutions. PhD dissertation, University of Waterloo, Ontario, Canada

Ramstein G, Donnadieu Y, Godderis Y (2004) Les glaciations du Proterozoique. CR Geoscience 336:639–646

Reardon EJ, Beckie RD (1987) Modelling chemical equilibria of acid minedrainage: The $FeSO_4$-H_2SO_4-H_2O system. Geochim Cosmochim Acta 51:2355–2368

Reed MH (1982) Calculation of multicomponent chemical equilibria and reaction processes in systems involving minerals, gases and an aqueous phase. Geochim Cosmochim Acta 46:513–528

Reed MH, Spycher N (1984) Calculation of pH and mineral equilbria in hydrothermal waters with application to geothermometry and studies of boiling and dilution. Geochim Cosmochim Acta 48:1479–1492

Reynolds RT, Squyres SW, Colburn DS, McKay CP (1983) On the habitability of Europa. Icarus 56:246–254

Richards TW, Speyers CL (1914) The compressibility of ice. J Am Chem Soc 36:491–494

Richardson C (1976) Phase relationships in sea ice as a function of temperature. J Glaciol 17:507–519

Rieder R, Economou T, Wänke H, Turkevich A, Crisp J, Brückner J, Dreibus G, McSween HY Jr (1997) The chemical composition of Martian soil and rocks returned by the Mobile Alpha Proton X-ray Spectrometer: Preliminary results from the X-ray mode. Science 278:1771–1774

Ringer WE (1906) Über die Veränderungen in der Zusammensetzung des Meereswassersalzes beim Ausfrieren. Verh Rijksinst Onderz Zee 3:1–55

Rivkina EM, Friedmann EI, McKay CP, Gilichinsky DA (2000) Metabolic activity of permafrost bacteria below the freezing point. Appl Environ Microbiol 66:3230–3233

Robbins EI, Rodgers TM, Alpers CN, Nordstrom DK (2000) Ecogeochemistry of the subsurface food web at pH 0–2.5 in Iron Mountain, California, U.S.A. Hydrobiologia 433:15–23

Robinson RA, Stokes RH (1970) Electrolyte solutions, 2nd edn (revised). Butterworths, London

Rogers PSZ, Pitzer KS (1982) Volumetric properties of aqueous sodium chloride solutions. J Phys Chem Ref Data 11:15–81

Ross RG, Kargel JS (1998) Thermal conductivity of solar system ices, with special reference to Martian polar caps. In: de Bergh C, Festou M, Schmitt B (eds) Solar system Ices, Kluwer, Dordrecht, pp 33–62

Rothschild LJ, Mancinelli RL (2001) Life in extreme environments. Nature 409:1092–1101

Sanchez-Roman M, McKenzie JA, Vasconcelos C, Rivadenyra M (2005) Bacterially induced dolomite formation in the presence of sulfate ions under aerobic conditions. AGU Fall Meeting, Abstract B13A-1041

Sankaran AV (2003) Neoproterozoic 'snowball earth' and the 'cap' carbonate controversy. Curr Sci 84:871–873

Sassen R, Joye S, Sweet ST, DeFreitas DA, Milkov AV, MacDonald IR (1999) Thermogenic gas hydrates and hydrocarbon gases in complex chemosynthetic communities, Gulf of Mexico continental slope. Org Geochem 30:485–497

Sattler B, Puxbaum H, Psenner R (2001) Bacterial growth in supercooled cloud droplets. Geophys Res Lett 28:239–242

Schaefer MW (1990) Geochemical evolution of the Northern Plains of Mars: Early hydrosphere, carbonate development, and present morphology. J Geophys Res 95:14,291–14,300

Schaefer MW (1993) Aqueous geochemistry of early Mars. Geochim Cosmochim Acta 57:4619–4625

Schidlowski M (2002) Search for morphological and biogeochemical vestiges of fossil life in extraterrestrial settings: Utility of terrestrial evidence, In: Horneck G, Baumstark-Khan C (eds.), Astrobiology: the quest for the conditions of life. Springer, Berlin Heidelberg New York, pp 373–386

Schilpp PA (ed) (1949) Albert Einstein: philosopher-scientist. The Library of Living Philosophers, Evanston, IL

Schleper C, Pühler G, Kühlmorgen B, Zillig W (1995) Life at extremely low pH. Nature 375:741–742

Schlesinger WH (1997) Biogeochemistry: an Analysis of Global Change, 2nd edn. Academic, San Diego

Schopf JW, Packer BM (1987) Early Archean (3.3-billion to 3.5-billion-year-old) microorganisms from the Warrawoona Group, Australia. Science 237:70–73

Schrenk MO, Edwards KJ, Goodman RM, Hamers RJ, Banfield JF (1998) Distribution of *Thiobacillus ferrooxidans* and *Leptospirillum ferrooxidans*: implications for generation of acid mine drainage. Science 279:1519–1522

Schroeter B, Scheidegger C (1995) Water relations in lichens at subzero temperatures: Structural changes and carbon dioxide exchange in the Lichen *Umbilicaria aprina* from continental Antarctica. New Phytol 131:273–285

Schulze-Makuch D, Grinspoon DH, Abbas O, Irwin LN, Bullock MA (2004) A sulfur-based survival strategy for putative phototrophic life in the Venusian atmosphere. Astrobiology 4:11–18

Schulze-Makuch D, Irwin LN (2002) Energy cycling and hypothetical organisms in Europa's ocean. Astrobiology 2:105–121

Schulze-Makuch D, Irwin LN (2004) Life in the Universe. Springer, Berlin Heidelberg New York

Seewald JR (1994) Evidence for metastable equilibrium between hydrocarbons under hydrothermal conditions. Nature 370:285–287

Segerer AH, Burggraf S, Fiala G, Huber G, Huber R, Pley U, Stetter KO (1993) Life in hot springs and hydrothermal vents. Orig Life Evol Biosph 23:77–90

Settle M (1979) Formation and deposition of volcanic sulfate aerosols. J Geophys Res 84:8343–8354

Sharma A, Scott JH, Cody GD, Fogel ML, Hazen RM, Hemley RJ, Huntress WT (2002) Microbial activity at gigapascal pressures. Science 295:1514–1516

Shock EL (1997) High-temperature life without photosynthesis as a model for Mars. J Geophys Res 102:23,687–23,694

Sloan ED Jr (1998) Clathrate hydrates of natural gases, 2nd edn. Marcel Dekker, New York

Smith AG, Pickering KT (2003) Oceanic gateways as a critical factor to initiate icehouse Earth. J Geol Soc Lond 160:337–340

Smith DW (1982) Extreme natural environments. In: Burns RG, Slater JH (eds) Experimental Microbial Ecology. Blackwell, Oxford, pp 555–574

Socki RA, Romanek CS, Gibson EK Jr, Golden DC (2001) Terrestrial aufeis formation as a Martian analog: Clues from laboratory-produced C-13 enriched cryogenic carbonate. In: Lunar and Planetary Science Conference XXXII. Houston, TX. Abstract #2032

Socki RA, Gibson EK Jr, Lauriol B, Clark ID, Romanek CS, Golden DC (2002) Stable isotope enriched carbonates from the karst permafrost region of Northern Yukon, Canada: a Mars analog. In: Lunar and Planetary Science Conference XXXIII. Houston, TX. Abstract #1801

Soina VS, Vorobiova EA, Zvyagintsev DG, Gilichinsky DA (1995) Preservation of cell structures in permafrost: a model for exobiology. Adv Space Res 15:237–242

Speedy RJ (1987) Thermodynamic properties of supercooled water at 1 atm. J Phys Chem 91:3354–3358

Spencer JR, Tamppari LK, Martin TZ, Travis LD (1999) Temperatures on Europa from Galileo photopolarimeter-radiometer: nighttime thermal anomalies. Science 284:1514–1516

Spencer RJ, Møller N, Weare JH (1990) The prediction of mineral solubilities in natural waters: A chemical equilbrium model for the Na-K-Ca-Mg-Cl-SO_4-H_2O system at temperatures below 25 °C. Geochim Cosmochim Acta 54:575–590

Squyres SW, Athena Science Team (2004) Initial results from the MER Athena science investigation at Gusev Crater and Meridiani Planum. Lunar and Planetary Science Conference XXXV. Houston, TX. Abstract #2187

Stainforth D et al. (15) (2005) Uncertainties in predictions of climate response to rising levels of greenhouse gases. Nature 433:403–406

Stark SC, O'Grady BV, Burton HR, Carpenter PD (2003) Frigidly concentrated seawater and the evolution of Antarctic saline lakes. Aust J Chem 56:181–186

Stetter KO (1996) Hyperthermophiles in the history of life. In: Bock GR, Goode JA (eds) Ciba Foundation Symposium 202: Evolution of Hydrothermal Ecosystems on Earth (and Mars?). Wiley, Chichester, UK, pp 1–18

Stetter KO (1999) Extremophiles and their adaptation to hot environments. FEBS Lett 452:22–25

Stetter KO (2002) Hyperthermophilic microorganisms. In: Horneck G, Baumstark-Khan C (eds) Astrobiology: the Quest for the Conditions of Life. Springer, Berlin Heidelberg New York, pp 169–184

Stevenson D (2000) Europa's ocean – the case strengthens. Science 289:1305–1307

Stolz BF, Basu P, Santini JM, Oremland RS (2006) Arsenic and selenium in microbial metabolism. Ann Rev Microbiol 60:107–130

Strom RG (2007) Hot house: global climate change and the human condition. Springer, Berlin Heidelberg New York

Stone R (1999) Permafrost comes alive for Siberian researchers. Science 286:36–37

Stumm W, Morgan JJ (1970) Aquatic chemistry: an introduction emphasizing chemical equilibria in natural waters. Wiley-Interscience, New York

Summit M, Baross JA (1998) Thermophilic subseafloor microorganisms from the 1996 North Gorda Ridge eruption. Deep-Sea Res II 45:2751–2766

Ter Minassian L, Pruzan P, Soulard A (1981) Thermodynamic properties of water under pressure up to 5 kbar and between 28 and 120 °C. Estimations in the supercooled region down to −40 °C. J Chem Phys 75:3064–3072

Thomas DN, Dieckmann GS (2002) Antarctic sea ice—a habitat for extremophiles. Science 295:641–644

Thomson RE, Delaney JR (2001) Evidence for a weakly stratified Europan oceans sustained by seafloor heat flux. J Geophys Res 106:12,355–12,365

Tor JM, Lovley DR (2001) Anaerobic degradation of aromatic compounds coupled to Fe(III) reduction by *Ferroglobus placidus*. Environ Microbiol 3:281–287

Toulmin P III, Baird AK, Clark BC, Keil K, Rose HJ Jr, Christian RP, Evans PH, Kelliher WC (1977) Geochemical and mineralogical interpretation of the Viking inorganic chemical results. J Geophys Res 82:4625–4634

Usiglio MJ (1849) Etudes sur la composition de l'eau de la Mediterranee et sur l'exploitation des sels qu'elle contient. Ann Chim Phys 27:172–191

Van Lith Y, Warthmann R, Vasconcelos C, McKenzie JA (2003) Sulphate-reducing bacteria induce low-temperature Ca-dolomite and high Mg-calcite formations. Geobiology 1:71–79

Ventosa A, Arahal DR, Volcani BE (1999) Studies on the microbiota of the Dead Sea—50 years later. In: Oren A (ed) Microbiology and Biogeochemistry of Hypersaline Environments. CRC Press, Boca Raton, FL, pp 139–147

Vlahakis JG, Chen H-S, Suwandi MS, Barduhn AJ (1972) The growth rate of ice crystals: Properties of carbon dioxide hydrate, a review of properties of 51 gas hydrates. Syracuse University Research and Development Report 830

Vogel G (1999) Expanding the habitable zone. Science 286:7071

Vreeland RH, Rosenzweig WD, Powers DW (2000) Isolation of a 250 million-year-old halotolerant bacterium from a primary salt crystal. Nature 407:897–900

Wadham JL, Bottrell S, Tranter M, Raiswell R (2004) Stable isotope evidence for microbial sulphate reduction at the bed of a polythermal high Arctic glacier. Earth Planet Sci Lett 219:341–355

Wagner W, Saul A, Pruss A (1994) International equations for the pressure along the melting and along the sublimation curve of ordinary water substance. J Phys Chem Ref Data 23:515–525

Wallerstein W et al. (14) (1999) Synthesis of the elements in stars: Forty years of progress. Rev Mod Phys 69:995–1084

Wallis MK, Wickramasinghe NC (2004) Interstellar transfer of planetary microbiota. Mon Not R Astron Soc 348:52–61

Ward PD, Brownlee D (2000) Rare Earth: why complex life is uncommon in the Universe. Copernicus, New York

Webb S (2002) Where Is Everybody? Praxis, New York

Weeks WF, Ackley SF (1982) The growth, structure, and properties of sea ice. CRREL Monograph 82-1. USACRREL, Hanover, NH

Wharton DA (2002) Life at the Limits: Organisms in Extreme Environments. Cambridge University Press, Cambridge, UK

Wilson EO (2002) The Future of Life. Vintage, New York

Wise ME, Brooks SD, Garland RM, Cziczo DJ, Martin ST, Tolbert MA (2003) Solubility and freezing effects of Fe^{2+} and Mg^{2+} in H_2SO_4 solutions representative of upper tropospheric and lower stratospheric sulfate particles. J Geophys Res 108(D14):4434. doi:10.1029/2003JD003420, 2003

Yayanos AA (1995) Microbiology to 10,500 meters in the deep sea. Annu Rev Microbiol 49:777–805

Young GM (2002) Stratigraphic and tectonic settings of Proterozoic glaciogenic rocks and banded iron-formations: relevance to the snowball Earth debate. J African Earth Sciences 35:451–466

Zhilina TN, Zavarzin GA (1994) Alkaliphilic anaerobic community at pH 10. Curr Microbiol 29:109–112

Zolotov MY, Shock EL (2001) Composition and stability of salts on the surface of Europa and their oceanic origin. J Geophys Res 106:32,815–32,827

Zolotov MY, Shock EL (2004) Brine pockets in the icy shell on Europa: Distribution, chemistry, and habitability. Workshop on Europa's Icy Shell: Past, Present, and Future. Houston, TX. Abstract 7028

Index

acidity 2, 40, 76, 79, 84, 88, 123, 132, 156, 175, 178, 180
  acidification 126, 130, 132–136
  hydrochloric acid (HCl) 2, 22, 24, 32, 37–40, 42, 73, 75, 88, 130, 165, 180
  MacInnis convention 88
  nitric acid (HNO_3) 2, 22, 32, 37, 38, 40, 42, 60, 75, 122, 130, 180
  pH 39, 40, 55, 61, 62, 71, 76, 82, 88, 110, 111, 115, 118, 119, 121–123, 127–129, 132, 145, 146, 150, 156, 175, 178–180
  strong acids 2, 19, 24, 40, 88, 122, 175, 178, 180, 184
  sulfuric acid (H_2SO_4) 2, 25, 41, 42, 122, 123, 130, 132, 134, 146, 148, 167
aerosols 4, 81, 101, 112, 121–123
Albert Einstein 1, 159, 160
alkalinity 41, 57, 58, 61, 62, 71, 88, 101, 118–120, 123, 124, 126–129, 132, 133, 135, 137, 138, 145, 175, 176, 179–181
  alkali systems 88, 110, 129, 144
  alkaline systems 33, 41, 69, 71, 76, 82, 101, 126, 129, 135, 141, 142, 144, 145, 175
Antarctica 89, 110, 112, 124
  Don Juan Pond 101, 110, 112, 113
  Dry Valleys 89, 112
astrobiology 2, 79, 157
Atacama Desert 89, 90
atmosphere 40, 67, 85, 89, 91, 94, 108, 113–116, 118, 121–123, 126, 128, 132–137, 139, 156–158, 161, 162, 164–168, 180, 181

banded iron formations 114, 115, 120
brines 41, 74, 85, 86, 91, 92, 98, 106, 107, 110, 120, 126, 129, 132–135, 137, 138, 142, 144–150, 152, 165, 167

chemical equilibrium 10, 16, 21, 33, 34, 49, 50, 52, 54, 62, 121, 122, 146, 170
  disequilibrium 83, 150, 152, 153, 156, 158, 159, 161, 167
  metastable 83, 106, 150–152
  nonequilibrium 150
  saturated 29, 104, 112, 114, 119, 177, 179
  supercooled 25–28, 49, 86, 122, 123
  supersaturated 21, 22, 33, 49, 52, 56, 58, 59, 61, 104, 108, 114, 119, 124, 143, 145, 147
  undersaturated 21, 40, 104, 119, 124
  unstable 4, 83, 96, 118, 128, 150
clouds 2, 81, 97, 112, 120, 123
convection 94, 121, 122, 139, 141, 149, 150
cryovolcanism 102, 107, 139, 149
crystallization
  equilibrium crystallization 22, 23, 76, 77, 107, 178, 180
  fractional crystallization 22, 23, 77, 105–107, 149, 178, 180

Dead Sea 62, 110
deep earth 91, 92
deep ice 124, 158
dehydration 89, 93, 140, 165
desiccation 2, 84, 89, 90, 156
dissolution 21, 90, 93, 103–106, 109, 110, 114, 115, 118, 140, 148

Earth   2, 33, 79–81, 83–85, 88–90, 92, 93, 96–99, 101–103, 107, 109, 110, 112–120, 123, 124, 129, 133, 135, 137, 140, 141, 152, 155–158, 161–166, 168–173, 178, 179
  hothouse   2, 113–120, 163
  snowball   2, 101, 103, 107, 113–120, 178, 179, 186, 188
Enceladus   83, 102, 157, 158, 166–169
equilibrium constants   1, 7–10, 16, 21, 24, 26, 34, 40, 46, 48, 67, 69, 72–74, 178
  dissociation constants   39, 40
  ion associations   39
  solubility products   19, 21, 22, 39–45, 67, 68, 72, 74, 75, 130, 151
Europa   2, 79, 83, 84, 90, 93, 99, 101, 102, 107, 112, 139, 141–150, 157, 158, 161, 162, 166–171, 178
eutectics   2, 23, 24, 33, 34, 41, 56, 57, 62, 64, 65, 68, 75, 76, 86, 103, 104, 106–109, 112, 122, 123, 129–131, 133, 135, 138, 142–151, 176
evaporation   22, 23, 76, 102, 108–110, 126, 129–131, 133, 137, 138, 151, 163, 164, 177, 178, 180, 181
  evaporites   76, 92–94, 102, 103, 110, 139, 140, 151, 152, 165, 171

freezing   1, 5, 15, 23, 24, 29, 57–59, 76, 85, 86, 91, 95–97, 102–110, 112, 114, 116, 117, 120, 121, 123, 124, 126, 129–131, 133, 135–139, 142, 146, 148, 149, 151, 152, 160, 162, 164, 168, 175, 177, 179, 180, 182
FREZCHEM   1–4, 7, 10, 11, 19–27, 30, 34, 35, 37–44, 49–52, 57, 60–64, 67–76, 84, 87, 101–103, 108, 110, 112, 113, 115–117, 120, 121, 123, 125, 130, 132, 137, 143, 144, 150–153, 164, 165, 168, 171, 175, 176, 193

gases
  carbon dioxide (CO_2)   2, 15, 16, 22, 23, 37–49, 55–58, 61, 65–67, 69–71, 108, 113, 115–119, 123, 126, 129, 130, 132–139, 145, 165, 167, 169, 175–181
  clathrates   94, 139–141, 167
  gas hydrates   2, 7, 21–23, 37, 42–49, 52, 56, 57, 67, 70, 71, 85, 93, 94, 114, 116–118, 120, 139, 140, 167, 175–179, 181, 193
  greenhouse   114, 116, 120, 126, 165, 168
  methane (CH_4)   15, 22, 23, 37–39, 42, 43, 45–49, 55, 69, 70, 92–94, 116, 117, 156, 158, 167–169, 175–177, 179–181
  oxygen (O_2)   15, 22, 37–39, 42, 115, 116, 119, 120, 128, 132, 156, 158, 181
geochemical evolution   125, 126, 129, 132, 133, 144–146, 170

heavy metals   86, 88, 164
hydration   83, 109, 130, 140
hydrothermal   79, 81, 84, 85, 91, 92, 96, 140–142, 146, 151, 152, 163, 165, 166

ice   4, 5, 15, 19, 21, 23–30, 34, 35, 49, 52, 55–57, 62, 68, 73, 76, 79, 81, 83, 85, 86, 90, 94, 96–98, 101–108, 110, 113, 115–118, 120–122, 124, 125, 137–139, 141–143, 145–151, 158, 162–171, 176–179, 193
  sea ice   79, 85, 103, 107, 110, 113, 116, 124
Issac Asimov   155

lakes   61, 82, 85, 101, 103, 107, 110, 123, 126, 129, 133, 139, 145, 166
life
  biological activity   81, 82, 84, 86, 87, 89, 92, 93, 98, 110, 112, 122, 143, 151, 152, 165, 167
  dormant   81, 97, 98
  evolution   85, 96–98, 155
  extraterrestrial   79, 97, 155, 156, 158
  habitats   79, 81, 83, 84, 94, 97, 120, 123, 149, 150, 157, 158
  halophiles   110, 143, 145, 146, 149, 152, 163, 164
  hyperthermophiles   84, 92, 93, 96, 163–165
  intelligent   156, 158, 170, 172

limits 79, 81, 82, 84, 87, 89, 92–94, 96, 98, 110, 112, 120–122, 141, 143, 145, 149, 155, 156, 162, 172
microbes 82, 84–86, 89–93, 96, 98, 112, 120, 121, 123, 143, 151, 152, 156, 163, 164, 168, 170, 172
plausibility 155, 157, 158
psychrophiles 86, 163–165
search strategy 2, 155
SETI 156, 158
thermophiles 92
tolerances 82, 84, 85, 87, 110, 141, 156, 163–165
viability 94, 96, 97

Mars 2, 68, 79–81, 83, 84, 93, 99, 101, 102, 123, 125–130, 132–141, 157, 158, 161, 162, 165, 166, 168–172, 179, 190
model
 algorithms 19, 49, 50, 52, 54, 55, 74, 76, 121, 175, 180
 convergence 49, 50, 53–55, 75, 76, 103, 178, 180
 equilibration 22, 40, 49, 55, 56, 62, 118, 132, 137, 138, 179, 180
 extrapolations 25, 27–29, 34–37, 41–43, 49, 68–71, 85
 FORTRAN 2, 75, 77, 175
 limitations 2, 18, 19, 23, 41, 49, 65, 67, 68, 71, 75, 76, 101, 150, 153, 175
 parameterization 2, 13, 15, 16, 18, 19, 21, 24, 28, 29, 33, 35, 36, 42, 43, 57, 60, 62, 65, 67–69, 71, 72, 74, 75, 101, 107, 124, 125, 150, 177, 193
 Pitzer approach 1, 10, 11, 15–17, 19, 39, 61, 71
 pressure dependence 2, 7, 9, 15, 17, 18, 24, 28, 34, 36, 62, 65, 72–74, 135, 148, 149, 193
 surrogates 75, 116, 120, 179
 temperature dependence 8, 9, 15, 16, 24, 36, 193
 validation 2, 29, 33, 56, 57, 60–62, 67, 69, 103, 107, 123
model phases
 gas 4, 40, 42, 48, 116, 176, 177

solid 2, 4, 21–23, 30–37, 39, 42, 45–47, 49, 51, 52, 56, 61, 72, 73, 93, 103, 122, 130, 142, 148, 149, 176–179, 193
solid solution 45, 46
solution 37, 38, 40, 52, 53, 55, 56, 61, 62, 105, 106, 177, 180

oceans 79, 83, 85, 91, 93, 94, 101, 102, 107, 110, 112–120, 129, 130, 133, 135, 136, 141–143, 145–149, 158, 161, 164–168, 178
oxidation 92, 98, 126, 129, 132, 133, 141, 152
oxidation/reduction 83, 141, 165

panspermia 169
Panspermia Hypothesis 98, 99, 172
peritectics 34, 56
permafrost 79, 85, 90, 93, 98, 118, 139–141, 163, 164
precipitation 23, 29, 49, 56, 57, 62, 76, 77, 85, 91, 93, 96, 97, 103–110, 114–116, 118–122, 124, 126, 128–131, 133–138, 143–147, 149–152, 164, 165, 168, 176–181
precipitation-dissolution 2

radiation 2, 79, 82, 84, 85, 89, 90, 97, 98, 123, 141, 156, 157
redox 152, 165
reduction 98, 110, 141, 151
regolith 101, 138–140, 179, 190

salinity 2, 57, 58, 67, 79, 82, 84–87, 110–112, 143, 146, 156, 165
salts
 carbonates 2, 19, 31–33, 39–41, 52–55, 57–59, 61, 62, 71, 75, 92, 96, 103, 110, 114–116, 118–120, 124, 126, 128–130, 132, 133, 135–139, 141, 142, 145, 150, 152, 164, 176, 179, 180
 chlorides 2, 19, 29, 30, 33–35, 39, 40, 49, 52, 53, 56, 57, 60–62, 65–68, 72–76, 86–88, 91, 92, 101, 103, 104, 108–110, 112, 113, 119, 121, 122, 126, 129, 130, 132, 133, 135,

137–139, 143–145, 147–149, 156, 165, 176, 180
  nitrates   2, 19, 31, 33, 52, 60, 73, 75, 180
  sulfates   2, 19, 21, 22, 29–33, 36, 37, 39, 41, 42, 52, 60, 62, 64, 65, 68, 73, 75–77, 86, 90, 91, 103–106, 108–110, 115, 119, 130, 132–135, 137–139, 141–149, 151, 152, 165, 167, 178, 180
seas   116, 118, 165, 166
seawater   22, 24, 57–59, 62, 64, 71, 74, 87, 91, 101–110, 114–118, 120, 123, 124, 129, 132, 133, 150–152, 164, 175, 177, 179, 182
  Gitterman pathway   57–60, 103–108
  Ringer–Nelson–Thompson pathway   57, 58, 103, 104, 106–108, 123
soils   91, 97, 110, 139
Solar System   2, 79, 83, 99, 101, 102, 141, 155–158, 160–162, 165, 168–170, 172
Star Trek   172
Sun   79, 126, 157, 161, 162, 164, 165, 168–170, 173
  red giant   161, 168, 169, 173
  white dwarf   168, 169
systems
  closed   3, 22, 23, 140, 159, 176, 179, 180
  heterogeneous   3, 4, 139, 148, 161, 171
  homogenous   3, 4
  isolated   3, 159
  open   3, 22, 23, 140, 159, 162, 169, 176, 180

thawing   86, 97, 112, 160, 165
thermodynamic laws
  Amagat's law   45
  first law   1, 4–6, 158
  Henry's law   22, 23, 37, 39, 40, 43, 116
  second law   1, 4–6, 158
  van't Hoff equation   9, 34
thermodynamic properties
  activity   1, 7–9, 18, 21, 22, 37, 42, 45, 46, 88, 165, 177, 178

activity coefficient   1, 7–10, 12, 15, 17, 18, 21, 22, 36, 37, 39, 40, 46–48, 57, 60, 68, 71, 73–75, 77, 88, 177
chemical potentials   1, 7, 24, 51
compressibility   9, 18, 26–29, 34–36, 42, 43, 69, 72–74, 193
concentrations   4, 10, 11, 15, 18, 21–23, 29–33, 36, 37, 39, 40, 42, 49, 52–57, 61, 65–69, 73, 75–77, 86, 88, 103–106, 108–110, 112, 114–116, 118, 120–122, 126, 128, 129, 132, 135, 137, 138, 144–146, 149, 152, 156, 175–177, 179–181
Debye–Hückel parameter   1, 10, 11, 16, 19, 68, 71, 72, 193, 201
density   4, 10, 17, 18, 21, 26, 27, 35–37, 62, 64, 75, 87, 91, 96, 110, 116–119, 124, 125, 137, 147, 149, 161, 177
enthalpy (H)   1, 4, 6, 9, 15, 16, 25, 34
entropy (S)   1, 4, 5, 150, 158–161
fugacity   7, 22, 25, 37, 45, 46, 48, 193
fugacity coefficients   37, 43, 45, 55, 69, 70, 76
Gibbs energy (G)   1, 4, 6, 8, 10, 11, 16, 24, 50–52, 62, 106, 153
heat capacities   9, 16, 25
Helmholtz energy (A)   6
internal energy (U)   1, 4–6, 159
molar volumes   9, 16–18, 26–29, 34–36, 42, 69, 71–73, 90, 193
osmotic coefficient   10–12, 15, 17, 21, 24, 67, 177
partial pressures   7, 22, 23, 25, 26, 37, 40, 43, 44, 54, 55, 66, 67, 115, 116, 137, 180
specific volume   4, 16
time   2, 79, 83–85, 90, 96–99, 104–108, 110, 114, 116, 121, 123, 133, 135, 137, 138, 159–161, 163, 164, 166, 169–171
Titan   157, 158, 167, 168

Universe   2–5, 84, 102, 150, 155, 156, 158–162, 169–172

Venus   79, 80, 102, 123, 157, 165

water 2, 4, 5, 7, 10, 11, 16–18, 21–29, 36, 39, 40, 43, 49–51, 55–59, 61, 66, 67, 69, 72, 74, 76, 79, 80, 83–87, 89–98, 101, 102, 106–108, 110, 112, 113, 116–118, 120, 122, 123, 125, 126, 128, 129, 135, 137, 139, 141–143, 148, 149, 151, 152, 155–157, 160–162, 165–168, 170, 171, 175–178, 180, 181, 193

activity (a_w) 7, 10, 11, 17, 18, 21, 22, 24–26, 34, 39, 40, 42–44, 49, 54–56, 66, 67, 74, 82, 83, 86, 87, 89, 110–112, 127, 133, 143, 145, 146, 149, 156, 178

Printing: Krips bv, Meppel, The Netherlands
Binding: Stürtz, Würzburg, Germany